666·767 : 539·374

No..

D1756489

R.P.E. WESTCOTT

B6560

ROCKET PROPULSION ESTABLISHMENT LIBRARY

10/10

Please return this publication, or request a renewal, by the date stamped below.

Name	Date

(4/64) L23964 442077 Wt29280 D7061 10/64 10M T&Co **G871**.   R.P.E. Form 243

DEFORMATION OF CERAMIC MATERIALS

DEFORMATION OF CERAMIC MATERIALS

Edited by

R. C. Bradt and R. E. Tressler
The Pennsylvania State University
University Park, Pennsylvania

PLENUM PRESS • NEW YORK AND LONDON

Library of Congress Cataloging in Publication Data

Main entry under title:

Deformation of ceramic materials.

   Includes bibliographies and index.
   1. Ceramic materials—Congresses. 2. Deformations. (Mechanics)—Congresses.
I. Bradt, Richard Carl, 1938-   II. Tressler, Richard E.
TA430.D4   620.1'4   75-4945
ISBN 0-306-30839-8

Proceedings of a symposium on Plastic Deformation of Ceramic Materials
held at The Pennsylvania State University, July 17-19, 1974

© 1975 Plenum Press, New York
A Division of Plenum Publishing Corporation
227 West 17th Street, New York, N.Y. 10011

United Kingdom edition published by Plenum Press, London
A Division of Plenum Publishing Company, Ltd.
4a Lower John Street, London W1R 3PD, England

All rights reserved

The United States Government has a royalty-free, nonexclusive and
irrevocable license throughout the world for Government purposes to
publish, translate, reproduce, deliver, perform, dispose of, and to
authorize others to do so, all or any portion of this work

Printed in the United States of America

Preface

This volume constitutes the Proceedings of a Symposium on the Plastic Deformation of Ceramic Materials, held at The Pennsylvania State University, University Park, Pennsylvania, July 17, 18, and 19, 1974.

The theme of this conference focused on single crystal and polycrystalline deformation processes in ceramic materials. The 31 contributed papers by 52 authors, present a current understanding of the theory and application of deformation processes to the study and utilization of ceramic materials.

The program chairmen gratefully acknowledge the financial assistance for the Symposium provided by the United States Atomic Energy Commission, The National Science Foundation, and The College of Earth and Mineral Sciences of The Pennsylvania State University. Special acknowledgment is extended to Drs. Louis C. Ianniello and Paul K. Predecki of the AEC and NSF, respectively. Of course, the proceedings would not have been possible without the excellent cooperation of the authors in preparing their manuscripts.

Special appreciation is extended to the professional organization services provided by the J. Orvis Keller Conference Center of The Pennsylvania State University. In particular, Mrs. Patricia Ewing should be acknowledged for her excellent program organization and planning.

Finally, we also wish to thank our secretaries for the patience and help in bringing these Proceedings to press.

University Park   R. C. Bradt

Pennsylvania   R. E. Tressler

July, 1974

Contents

DEFORMATION MECHANISM MAPS - A REVIEW WITH APPLICATIONS

Michael R. Notis

Materials Research Center and the Dept. of Metallurgy &

Materials Science, Lehigh University, Bethlehem,Pa.18015

ABSTRACT

Deformation mechanism mapping has recently evolved as a method
for quantitatively expressing complex deformation equations while
at the same time allowing visual insight into our understanding of
these deformation processes. Techniques for the generation of deformation mechanism maps are described, specific applications in
the literature to date are reviewed, and some applications not
previously considered are presented. These include the examination
of sequential dependent deformation processes, grain boundary sliding, and the use of deformation mechanism maps to interpret pressure-sintering kinetics.

INTRODUCTION

It is always interesting to be able to observe how concepts
develop from very unsophisticated qualitative descriptions to more
precise quantitative approaches. However, in many areas, as our
quantitative ability to handle complex problems increases, our
basic physical understanding of these processes is sometimes easily
lost. I think that the reason there has been so much interest in
deformation mapping over the last year or so is that it overcomes
this problem in a very picturesque way. Deformation mapping is an
approach that allows us to quantitatively express complex deformation equations and at the same time allows us to visualize our
physical concepts about the mechanisms involved.

My purpose today is to describe what deformation mapping is, to review what has been done in the area to date, to describe some applications not previously considered, and to summarize what appears to be the value of these deformation maps for future problems.

In 1968, at an ASM Symposium on High Temperature Creep, Weertman [1] used a figure called a "creep diagram" to demonstrate in a qualitative manner the regions of stress and temperature in which different creep mechanisms could be expected to dominate. Recently, Ashby and coworkers [2,3,4], in a more quantitative approach, have shown that all of the deformation mechanisms may be combined graphically and their relative contributions predicted from a "deformation map." At any given stress and temperature, one mechanism for creep deformation is dominant. This means that on a graphical plot with stress as one axis and temperature as the other (Figure 1), fields may be determined which show the range of stress and temperature over which a particular mechanism predominates. In addition, contours of constant strain rate may be shown (Figure 1). This means that if any pair of the three variables, stress temperature and strain rate is known, the third may then be predicted.

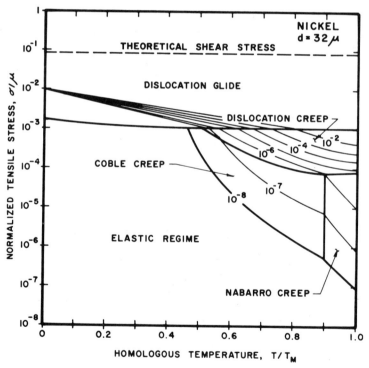

Figure 1 - Deformation Map for Nickel (d = 32μ). Elastic boundaries arbitrarily defined as $\dot{\epsilon} = 10^{-8}$ sec^{-1} (after Ashby [2]).

The boundaries of the fields are obtained by equating consecutive deformation rate equations (each mechanism thus contributes 50% of the deformation rate at a boundary) and solving for stress as a function of temperature. Contours of constant strain rate can be found by summing all of the strain rate contributions and solving for the stress that results in some constant strain rate value as temperature is varied. These calculations can be easily made and the deformation maps may be plotted by a computer if the necessary material constants are known. Maps may be made, for example, for a wide range of stress, temperature, and structural parameters such as grain size, and the results can be examined visually. Conversely, overlays of experimental data onto the deformation map quickly enable the calculation of the relative contribution of each deformation mechanism to the total deformation rate. Similarly, experimental conditions necessary for the observation of a particular deformation mechanism may be easily predicted.

CONSTITUTIVE EQUATIONS FOR DEFORMATION

In order to obtain rate equations for the deformation of materials, Ashby [2] initially considered the following five distinguishable and _independent_ ways that a polycrystalline body may deform. First, a stress which exceeds the theoretical shear strength will cause deformation even in a defect-free crystal. Second, plastic deformation may be achieved through twinning; in early deformation maps [2] twinning was not included because of its limited contribution to deformation but later maps [3] included a contribution that appeared only at high stresses and low temperatures. Third, the glide motion of dislocations can lead to extensive plastic flow. Fourth, at higher temperatures dislocation creep is made possible as the dislocations increase their ability to climb as well as glide. Fifth, at high temperatures, low stresses and small grain sizes, the creep of polycrystalline materials may be controlled by point defects where deformation is accomplished by the diffusional flow of matter from areas near grain boundaries under compressive stresses to those which are under tensile stresses. This diffusional flow may take place through the bulk lattice [5,6] (Nabarro-Herring Creep) or a similar diffusive flow may occur by point defects moving along the grain boundaries [7] (Coble Creep). Although most of these mechanisms will be described in detail during the course of the next few days, I would like to present a short description for each of them in order to develop the rate equations necessary to produce the deformation maps.

The steady state creep (strain) rate due to Nabarro-Herring creep [5,6] is given by:

$$\dot{\epsilon}_{NH} = \frac{13.3 \; D_V \; \Omega \; \sigma}{kT \; d^2} \tag{1}$$

where σ is the stress, D_v is the lattice diffusion coefficient, Ω is the molecular volume, d is the grain size, and kT has its usual meaning. For Coble creep [7] the strain rate is given by:

$$\dot{\epsilon}_C = \frac{47.5\ \Omega\ (D_b\ \delta)\ \sigma}{kT\ d^3} \qquad (2)$$

where $(D_b\ \delta)$ is the product of the grain boundary diffusion co-efficient and the effective grain boundary width.

Raj and Ashby [8] have developed a more general description of deformation by diffusional flow, including both bulk and lattice diffusion such that

$$\dot{\epsilon}_{DIFF} = 14\ \frac{\sigma\Omega}{kT}\ \frac{1}{d^2}\ D_v \left[1 + \frac{\pi\delta\ D_b}{d\ D_v} \right] \qquad (3)$$

Note that if the second term within the brackets is $\ll 1$ this equation reduces to an expression similar to one obtained from the Nabarro-Herring stress-directed diffusional model. Alternatively, if the second term within the brackets is $\gg 1$ this equation yields results very similar to those obtained by Coble for grain boundary diffusion controlled creep.

At high temperatures (above about 0.5 T_M) and at relatively high stresses creep deformation is apparently diffusion controlled but the strain rate is a non-linear function of stress. This has generally been attributed to the ability of dislocations both to climb and glide in this stress-temperature regime. It has been shown [8] that in this regime the creep behavior of many materials can be well described by the semiempirical equation

$$\dot{\epsilon}_{DISL} = A\ \frac{D_v\ \mu\ b}{kT}\ \left(\frac{\sigma}{\mu}\right)^n \qquad (4)$$

where μ is the shear modulus, b is the Burger's vector. The constants A and n are commonly referred to as the Dorn parameters. The exponential constant, n, has been related to the nature of the specific dislocation mechanism proposed to be operative during creep [9]. Recent work by Stocker and Ashby [10] has shown that n and A do not vary independently, and they derive a relationship that can be used to find A when n is known.

At lower temperatures but at approximately the same stress levels applicable for equation (4), it becomes necessary to include

a term that describes atom transport by dislocation core diffusion
[3]. The volume diffusion term (D_v) in equation (4) may be replaced
by an effective diffusion coefficient:

$$D_{eff} \cong D_v \, [1 + \frac{3}{b^2} \, (\frac{\sigma}{\mu})^2 \, \frac{a_c \, D_c}{D_v}] \tag{5}$$

where D_c is the dislocation core diffusion coefficient and a_c is
the cross-sectional area of the dislocation core. Thus, at high
temperatures and at lower stresses lattice diffusion is dominant;
Frost and Ashby [3] have termed this "High Temperature Creep." At
lower temperatures and at higher stresses, dislocation core diffusion is dominant, the strain rate varies as σ^{n+2} rather than σ^n;
this type of dislocation creep has been termed "Low Temperature
Creep." While the more recent deformation maps [3] include a regime
for low temperature dislocation creep, the early maps [2,11,12] ig-
nore this mechanism. Since most of the applications that will be
described here, and that are of interest insofar as steady state
deformation in ceramic materials, occur at higher temperature, I
have chosen to simplify the maps by eliminating low temperature
dislocation creep effects.

At even higher stress levels, the rate of dislocation glide
rather than climb usually controls the deformation rate and this is
usually obstacle limited. However, materials with strong covalent
bonding and many bcc metals demonstrate glide controlled by lattice
resistance (Peierls barrier). Above the cut-off stress for glide
control, Ashby [2] has adopted a rate equation for obstacle limited
flow such that

$$\dot{\varepsilon}_{OBST} = \dot{\varepsilon}_o \, \exp \, [-\frac{b^2 \ell}{kT} \, (S-\sigma)] \tag{6}$$

where ℓ = obstacle spacing

$S = \frac{\mu b}{\ell}$ = flow stress at absolute zero

$\dot{\varepsilon}_o$ = strain rate when $\sigma = S$ (assumed $\approx 10^{+6}$/sec.)

Later work by Frost and Ashby [3,4] gives essentially the same
equation but in slightly different form. As pointed out by Roberts
and Voglewede [13], the upper stress limit for deformation in many
ceramic materials may be determined by the brittle-fracture stress,
σ_F, rather than by the presence of pure dislocation glide. Therefore, on many of the maps presented here, I have included an approximate value of σ_F, rather than including obstacle controlled
glide.

Finally, at stress levels above some fraction of the theoreti-

cal strength, the material will fail even in the absence of crystal
defects. For failure by this mode, Ashby [2] assumes that

$$\left\{\begin{array}{ll} \dot{\epsilon}_{TH} = \infty & \text{above} \\[2em] \dot{\epsilon}_{TH} = 0 & \text{below} \end{array}\right\} \quad \tau = 0.039 \ \mu \quad\quad\quad (7)$$

Another complication that arises in the case of ceramic materials is the dependence of the diffusion constants upon the complex
nature of the diffusing species. Not only must the path for diffusion be considered, as in metals, but creep will depend on the
nature of the rate controlling species (i.e., anion control versus
cation control) and the ambient conditions and purity that delimit
both intrinsic and extrinsic regions for each ionic species. An
initial attempt to include these factors has been made by Stocker
and Ashby [14] for MgO. Recently maps have also been generated for
NiO [12] (Figure 2), CoO [11,17], UO_2 [2,13], olivine [14], NaCℓ
[15], and $MgAl_2O_4$ [16,17]; the interest generated by these maps has
been indicated by their rapid inclusion in at least two review
papers on deformation in ceramic materials [18,19].

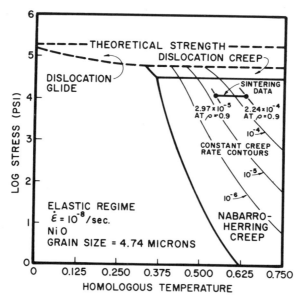

Figure 2 - Deformation Map for NiO (d = 4.74μ). Experimental
pressure sintering data [12] has been superimposed.

APPLICATIONS OF DEFORMATION MAPPING

To a certain extent the principal advantage of deformation mapping is purely pedagogic. As one example, it allows for the easy visual comparison of the properties of different groups of materials; the ability of these maps to contrast the mechanical behavior of the FCC, HCP, BCC and diamond cubic elements has clearly been demonstrated [2]. Similarly, Frost and Ashby [3] have compared the normalized constant strain rate contours calculated for a number of FCC metals and have related the difference in behavior of each metal to its stacking fault energy.

In addition to its use as a pedagogic tool, since deformation maps clearly and quantitatively indicate strain rate contours and the regions in which specific deformation mechanisms are dominant, they also provide a very strong predictive tool. In this context, Frost and Ashby [20,21] have investigated their use as a means of solving a serious creep problem for tungsten filaments in light bulbs. The maps that they generated indicate that for fine grained pure tungsten, under the stress and temperature conditions imposed during operation, the filaments would be in a diffusional creep field with a creep rate approximately six orders of magnitude higher than the acceptable limit. Although the maps indicated that changing to a coarser grained material would shift the controlling creep mechanism to high temperature dislocation creep, the creep rate was still too high. As pointed out by these investigators, a successful solution to the problem could only be found by limiting creep by both mechanisms. A proper solution was found in commercial practice where tungsten is doped with impurities that leave a fine dispersion of voids after processing. This fine dispersion stabilizes an elongated grain structure that limits diffusional creep and inhibits dislocation creep in a manner similar to dispersion hardening. It is interesting to note that an alternate commercial solution to the problem has been to use single crystal tungsten filaments [22]; diffusional creep is decreased by the absence of grain boundaries but the approach does not represent as viable an engineering solution as that indicated by deformation mapping.

Thus deformation mapping provides for the prediction of material behavior for which little or no experimental data exists and, since the dominant regimes for each mechanism can be observed, it prevents incorrect extrapolations from control by one mechanism to another. Because the maps increase our physical understanding of observed relations, they therefore increase the confidence with which data may be extrapolated beyond the range of experiments. It is this predictive-extrapolative feature of deformation mechanism maps that makes them extremely interesting, for example, insofar as applications to nuclear materials. This has already been demonstrated by the generation of maps for UO_2 and UO_2-PuO_2 [13] that are based on experimental data (including in-pile creep effects) rather than theoretically derived rate equations. Combinations of the information output of deformation maps and the reactor-modeling programs already developed [23] should provide a much more satisfying prediction of mechanical properties of nuclear materials than currently exists.

I would like to turn to three areas of application of deformation mapping in which we have been involved. Hopefully, they will help elucidate the variety of applications for which mapping may be used. These are: the inclusion of interdependent deformation mechanisms in the maps, the inclusion of missing mechanisms that have been indicated to be present through microstructural evidence, and finally the use of maps to jointly model deformation and densification during pressure sintering.

INTERDEPENDENT CONSTITUITIVE EQUATIONS

Each of the mechanisms for deformation considered so far have been characterized as independent processes. Namely, the total strain rate due to the combination of these mechanisms is the sum of each individual contribution:

$$\dot{\varepsilon}_T = \dot{\varepsilon}_1 + \dot{\varepsilon}_2 + \ldots \ . \tag{8}$$

These types of mechanisms have been called SIMULTANEOUS-INDEPENDENT [24] or PARALLEL-CONCURRENT [25] modes of deformation. On the other hand, it is possible for two processes to be interdependent in such a way that the one process cannot proceed until the other has taken place to some degree. These types of mechanisms have been called SEQUENTIAL-DEPENDENT [24] or SERIES-SEQUENTIAL DEPENDENT [25] modes of deformation. When additive (independent) processes are involved the fastest is rate controlling; if sequential (dependent) processes are involved the slowest is rate controlling [26,27].

The plastic flow behavior of most materials studied to date indicate agreement with the types of deformation maps previously described. However a few, i.e., α-zirconium and titanium appear to behave in an anomalous manner. For example, the stress dependence of α-zirconium is such that the stress exponent, n, appears to decrease with an increase in stress. Ardell and Sherby [28] have attributed this behavior to the high climb rate that results from the high diffusivity and low elastic modulus of α-zirconium. Dislocation glide could then be assumed as the rate controlling mechanism at lower stresses. Ardell and Sherby define a glide controlled creep rate such that:

$$\dot{\varepsilon}_{GLIDE} = C \exp (\beta\sigma) \exp \{-(Q_k + \eta/\sigma)/RT\} \qquad (9)$$

where C, β and η are constants and Q_k is the kink formation activation energy.

We have considered the problem of the sequential dependent interaction between glide and climb for both α-zirconium and Zircaloy-2 and have incorporated our results into deformation mechanism maps [29]. Equation (4) for dislocation climb and equation (9) for glide were incorporated into the mapping program such that if $\dot{\varepsilon}_{CLIMB}$ was greater than $\dot{\varepsilon}_{GLIDE}$, the deformation rate was set equal to $\dot{\varepsilon}_{GLIDE}$; if $\dot{\varepsilon}_{GLIDE}$ was greater than $\dot{\varepsilon}_{CLIMB}$ the deformation rate was set equal to $\dot{\varepsilon}_{CLIMB}$. The total strain rate was found by adding the smaller of these quantities to the diffusional contribution. The maximum temperature considered for the maps was 862°C, the α(HCP) $\rightarrow$ β(BCC) phase transition temperature, and the temperature scales for the maps were normalized with respect to this temperature.

A deformation map for α-zirconium with a grain size of 60 μ is shown in Figure 3. The values for the constants C, β, η and Q_k were taken from Ardell and Sherby [28] whose experiments were made with 300 μ material; superimposed on the map are experimental values of strain rates (the numbers near the points are log $\dot{\varepsilon}$ values) reported by Bernstein [30]. The agreement between the contours predicted by the map and the experimental values is excellent. An especially significant point is that the constants for the equations were obtained by Ardell and Sherby at quite high temperatures, yet the data points are in agreement with the deformation map at lower temperatures.

Gilbert, Duran and Bement [31] have indicated that conceptually the same deformation mechanisms are applicable in Zircaloy-2 (1.5% Sn, 0.14% Fe, 0.10% Cr, 0.05% Ni, balance Zr) as in α-zirconium. We have therefore included a deformation map of Zircaloy-2 having a grain size of 10 μ in Figure 4; new values for the constants in equations (9), (4), (2) and (1) appropriate to Zircaloy-2 have been used. Again, data from Bernstein [30] are in excellent agreement with the deformation map predictions.

MISSING MECHANISMS

One difficulty with the Nabarro-Herring and the Coble creep theories is that they imply that grain elongation should occur during deformation when in fact this is almost never observed to occur. The models also assume the relaxation of shear stresses at the grain boundary by a viscous relaxation phenomenon but the

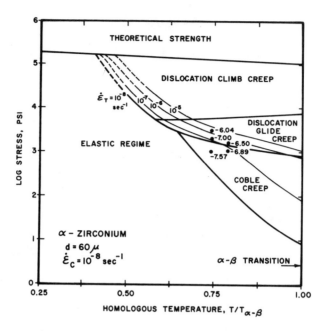

Figure 3 - Deformation Map for α-Zr (d = 60μ). Experimental data from Bernstein [30].

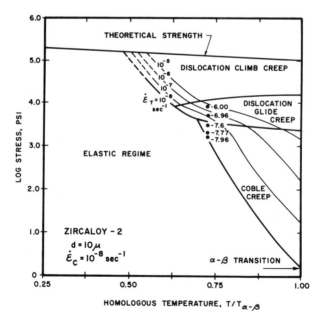

Figure 4 - Deformation Map for Zircaloy-2 (10μ). Experimental data from Bernstein [30].

contribution of this relaxation mechanism is assumed to be unimportant in the deformation process. Creep studies at low stress levels have also revealed that scratch offsetting, grain rotation, and granular surface roughening can occur while an equiaxed grain structure is maintained [32]. These observations are consistent with the description of creep as a grain boundary sliding process. However, in order for sliding to continue it is necessary that an accommodation process occurs. Generally it is the accommodation process which controls the extent and rate of sliding. Raj and Ashby [8] have studied the process of sliding with diffusional accommodation. Their model combined both the grain boundary sliding and the accommodation into a single model where both processes occur simultaneously. Raj and Ashby thus show that diffusional flow and grain boundary sliding are not independent. They are coupled, and the resulting deformation is described either as diffusional creep or as grain boundary sliding with diffusional accommodation [33].

At higher stresses, grain boundary sliding may be accommodated by plastic flow. Actually, two types of grain boundary sliding may be defined. Sliding that takes place on interior grain boundary surfaces must be accompanied by accommodation deformation in adjacent grains (Figure 5a); this type of sliding can be termed accommodated grain boundary sliding. On the other hand grain boundaries that intersect the sample surface (Figure 5b) should slide freely at a rate determined predominantly by the nature of the boundary; this type of sliding can be termed free grain boundary sliding. Raj and Ashby [34] have recently shown that at high temperatures the free sliding rate varies with temperature as D_v while at sufficiently low temperatures the free sliding rate is controlled by boundary diffusion. Matlock and Nix [35,36] have shown that geometrically accommodated grain boundary sliding and free gain boundary sliding can be directly related. Thus, except for the inclusion of a constant term near unity, equations for both accommodated and free grain boundary sliding should be expected to have the same form. In addition, in a porous sample (Figure 6), the accommodational requirements necessary for sliding are considerably relieved.

In metals, grain boundary sliding associated with plastic flow appears to be associated with strain rates that have fairly high stress exponents. Matlock and Nix [36] have correlated the stress exponents associated with grain boundary sliding and stress exponents associated with grain deformation; they indicate that these may be related such that $n_{gbs} \cong n_g - 1$. For oxides such as MgO [37], UO_2 [38], $BaO.6 Fe_2O_3$ [39] and ZrO_2-30 w/o UO_2 [40] there is considerable evidence that plastically accommodated grain boundary sliding is associated with a stress exponent in the range of 1.5 to 2.5.

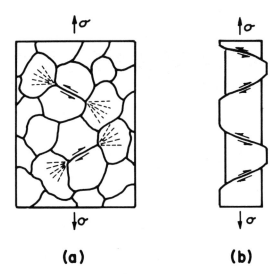

(a)   **(b)**

Figure 5 - Accommodated Grain Boundary Sliding (5a) and Free Grain Boundary Sliding (5b), after Matlock and Nix [36].

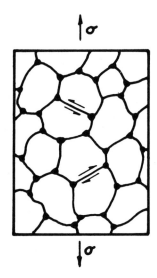

Figure 6 - Sliding in a Porous Compact. Requirements for Accommodation are Considerably Relaxed by Internal Pore Surfaces.

Wire and coworkers [41] have indicated that the region labelled dislocation creep in the deformation mechanism maps should really be subdivided into two regions; one region governed by grain matrix processes and the other governed by the grain boundary sliding contribution through plastic accommodation. Langdon [24] has developed a model for grain boundary sliding by assuming that all the deformation mechanisms, including sliding, operate independently. In this model, sliding occurs by the movement of dislocations along, or adjacent to, the boundary by a combination of climb and glide. The strain rate associated with grain boundary sliding is given as:

$$\dot{\varepsilon}_{gbs} = \frac{\beta \, b^2 \, \sigma^2 \, D_v}{d \, \mu \, k \, T} \qquad (10)$$

where β is a constant approximately equal to one.

We have included grain boundary sliding, as a plastic flow accommodated process, in the deformation mechanism maps by incorporating equation 10. This relation has been chosen because of its availability as a quantitative expression, because of the same stress-exponent (n = 2) and grain size dependence (d^{-1}) as found in at least two other similar models [42,43], and because of the general agreement with the experimentally observed stress dependence in ceramic materials. In order to emphasize the stress and grain size dependence of this mechanism we have chosen to plot a different form of deformation map in Figure 7. This figure shows a "deformation map" for $MgAl_2O_4$ at a constant temperature of 1773°K; rather than a plot in σ-T space, this map is in σ-d space and highlights the microstructural variations more effectively. Constant strain rate contours can still be shown; in addition, because of the log-log nature of the plot and the form of the equations, the boundaries all appear as straight lines rather than curves. For the particular example of $MgAl_2O_4$, shown in Figure 7, grain boundary sliding does not appear to be a major contributor to creep for a grain size below about one micron or below about 500-1000 psi. The constant strain rate contours on this type of plot also highlight the extreme sensitivity of diffusional creep and the insensitivity of dislocation creep to grain size effects.

APPLICATION TO HOT-PRESSING AND SINTERING

As a final application of deformation mapping, we might look at the compaction of a powder under the application of pressure and/or temperature. In a pressure sintering die, the total mass of the powder and the cross-sectional area are constant while the density and total sample height are variable. If the density is given as

$$\rho = \frac{m}{A\ell}$$

where m is the mass of the sample, A is the cross-sectional area of the die and ℓ is the sample height, then

$$\frac{1}{\rho}\frac{d\rho}{dt} = -\frac{1}{\ell}\frac{d\ell}{dt} \tag{11}$$

As shown by Coble [44,45], because the right-hand side of this expression is a linear strain (linear shrinkage) rate which may be obtained from measurement of the ram travel in a pressure-sintering die, the densification rate may be related quantitatively to a kinetic expression of strain rate during creep as a function of stress and temperature if appropriate values for the effective stress in a porous sample can be determined. In addition, Coble has shown [44] that for final stage densification during pressure sintering the driving force should be $[\frac{2\gamma}{r} + \frac{\sigma_a}{\rho}]$, (where γ is surface energy, r is the pore radius, σ_a is the applied stress, and ρ is the relative density) rather than just the surface energy contribution of the pore. In fact, the influence of pressure on changing densification rate can be more significant during most of the closed pore stage than is the surface energy effect.

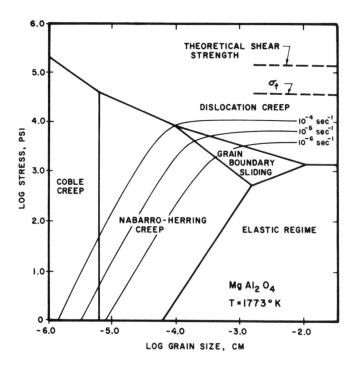

Figure 7 - Deformation Map for MgAl$_2$O$_4$ (T = 1773°K) in σ-d Space Showing Field for Grain Boundary Sliding.

If the stress axis of a deformation map is converted to the effective stress for densification $(\frac{\sigma_a}{\rho} + \frac{2\gamma}{r})$ then it could be used to compare pressure sintering data obtained at constant density levels to the deformation behavior of a material. We have recently used such "densification maps" [17] to jointly model pressure sintering and creep deformation. Superimposed on the map for NiO shown in Figure 2 is the experimental range of stress-temperature-strain rate conditions found during a study of pressure sintering kinetics. It may be seen that the densification rates observed agree very well with those predicted from the map. The pressure sintering data falls within the Nabarro-Herring creep region, and calculations from the experimental data, using the above information, give excellent agreement between effective diffusion constants and <u>cation</u> tracer values with respect to both magnitude and activation energy [12]. Similar results have been obtained with CoO [11]. Markworth [46] has used this same approach to demonstrate the effect of pore radius, grain size, and diffusivity on pressure sintering of UO_2.

It sometimes becomes difficult to use the simple atomistic models available for pressure sintering [44,45], because of either major microstructural changes that occur during processing, or because of a non-linear stress-strain rate dependence over the range of experimental conditions. Our studies, now in progress, for $MgAl_2O_4$ indicate that such non-viscous behavior is indeed observed [16]. Analysis of the pressure sintering data for $MgAl_2O_4$ according to the phenomenological approach of Rossi and Fulrath [47] indicated that a stress exponent of n = 1 was not observed; a hot pressing model developed by Wolfe [40,48] for a solid with non-linear stress-strain rate response was therefore used. According to Wolfe's analysis, for constant applied stress, the change in porosity as a function of time may be given as:

$$\frac{dP}{dt} = A \left[\frac{P(1-P)}{(1-P)^n \ (1-P^{1/n}) \ n} \right] \tag{12}$$

where A is a constant that depends upon the stress level, the stress exponent, and the strain rate. Coble's porosity correction $(\sigma_e - \frac{\sigma_a}{\rho})$ has been incorporated into equation 12. A plot of dP/dt versus the bracketed term in equation 12 should be a linear function for the effective value of n operative over the range of investigation. Our data for $MgAl_2O_4$ hot pressed at 1500°C and 8000 psi is presented on such a plot in Figure 8; the data produces a straight line fit for n = 1.4.

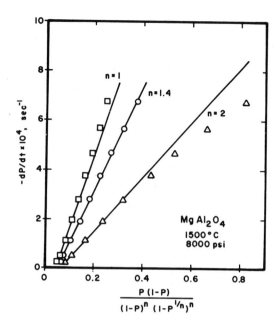

Figure 8 - Wolfe's Non-Viscous Phenomenological Model [48] for MgAl₂O₄ at 1500°C and 8000 psi.

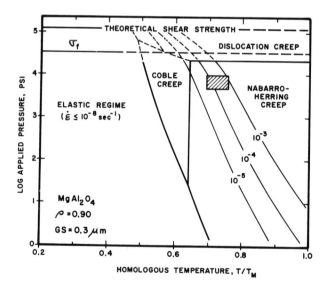

Figure 9 - Deformation Map for MgAl₂O₄ (d = 0.3μ). Experimental Pressure Sintering Data has been superimposed.

The range of stress and temperature investigated during this study has been superimposed on the deformation map for MgAl$_2$O$_4$ shown in Figure 9. This shaded region lies within the Nabarro-Herring creep regime but near the boundaries for both dislocation creep and Coble creep and the experimentally observed creep rates are in good agreement with those predicted from the map. Since tentative results show the Coble contribution not to have a major effect, the total strain rate may be expressed as:

$$\dot{\epsilon}_T = D_{eff} [K_{NH} \sigma + K_{gbs} \sigma^2 + K_{DISL} \sigma^{4.5}] \qquad (13)$$

where D_{eff} is the effective diffusion coefficient and where K_{NH}, K_{DISL} and K_{gbs} are the constants found in equations 1, 4, and 10 respectively. At each temperature, a plot of the total experimental strain rate versus the bracketed term in equation 13 should produce a straight line whose slope is D_{eff}; a plot of log D_{eff} versus inverse temperature may then be used to find the activation energy for the process. Analysis in this manner gave good agreement with anion lattice diffusion control both as to magnitude and activation energy.

The variation in total strain rate (the sum of the terms in equation 13) may be related to the stress according to the relation

$$\dot{\epsilon}_T = K_T \sigma^{"n"} \qquad (14)$$

where "n" is a macroscopic effective stress exponent. This effective stress exponent may be evaluated by first calculating the total strain rates (equation 13) at a number of stress levels and then using equation 14 as the basis for a plot of log $\dot{\epsilon}_T$ versus log σ; the slope of this plot is the effective stress exponent for the multimechanism deformation process. The data for MgAl$_2$O$_4$ at 1500°C with a density of $\rho = 0.87$ was evaluated in this manner over the stress range 5000-10,000 psi; the resultant value of "n," approximately 1.3, is in good agreement with the value obtained from Wolfe's analysis (n = 1.4). A consistent interpretation of the non-viscous behavior observed during pressure sintering is thus made possible through the mapping technique. However, an important point should be made. Although the experimental data (shaded area) in Figure 9 lies entirely within the dominant Nabarro-Herring field, the strain rate contours do not indicate completely viscous response; a mechanism can contribute significantly to the total deformation process as the boundaries of a field are approached, and may even contribute significantly if no regime corresponding to the mechanism appears on the map. This is because the mechanism must contribute at least 50% of the deformation in order to appear. Thus

the grain boundary sliding regime that appears in Figure 7 but not
in Figure 8 can contribute to the total strain rate contours that
do appear in Figure 8.

Two other recent applications of deformation mapping relevant
to sintering phenomena might be mentioned at this point. Stocker
and Ashby [14], in order to explain the deformation that occurs in
the earth's mantle, have included a strain rate contribution due to
fluid phase transport under applied stress for some of their maps.
An extension of this idea could lead to densification maps for
sintering of a material in the presence of a small volume of a
liquid phase (such as MgO-Li$_2$O or mullite). Lastly, Ashby [49] has
reported on the generation of sintering diagrams for initial, intermediate and final stage densification for a number of metallic
systems and for UO$_2$. These maps are in the form of plots of normalized neck radius versus temperature and include contours of
either constant normalized sintering rate or constant time. Because
of the simple way that they demonstrate regions in which rapid
densification would or would not occur, these displays may be used
to design commercial sintering schedules. Specimens held under
conditions that enhance volume and boundary diffusion will sinter
rapidly and could be used where high densities are important;
specimens held under conditions that enhance surface diffusion do
not densify and these conditions could be used, for example, to
produce porous filter materials.

SUMMARY

Deformation mechanism maps provide strong pedagogic, predictive and extrapolative tools for the investigation of the mechanical deformation of materials. They provide a means for quantitatively expressing the behavior of the material while at the same
time increasing our physical understanding of the mechanisms involved.

Applications for deformation mechanism maps include the
generation of information on mechanical properties under conditions
where direct experiment might be extremely difficult; the mapping
technique has already been used to try to understand the behavior
of in-service nuclear materials. Because of the visual insight that
they allow, maps may be helpful in identifying missing or dependent
(sequential) mechanisms that are not easily defined, and they can be
used to understand complex deformation-related processes such as
pressure-sintering.

Deformation maps have been described as "mechanical phase
diagrams." In this light, a map for nickel (after Frost and Ashby
[3]) is shown in Figure 10. On it are shown the areas that have

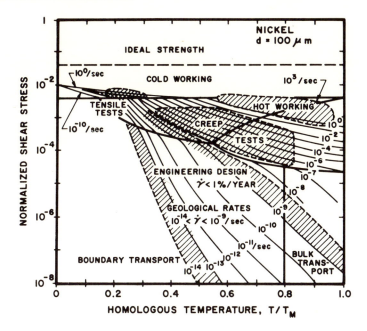

Figure 10 - Deformation Map for Nickel (d = 100μ) (after Frost and Ashby [3]. Areas of commercial interest and laboratory experiments are shaded.

been investigated and the areas that are of interest for commercial use; the areas that are of interest for engineering applications, such as forming and structural design for maximum creep life are the least studied in the laboratory. It is for this reason that the maps are of interest and are useful for understanding the behavior of materials.

ACKNOWLEDGEMENT

The author gratefully acknowledges the Atomic Energy Commission for the support of this work under Contract No. AT (11-1) 2408.

REFERENCES

[1] J. Weertman, "Dislocation Climb Theory of Steady-State
     Creep," Trans. ASM, 61, 681-695 (1968).

[2] M. F. Ashby, "A First Report on Deformation Mechanism Maps,"
     Acta. Met. 20, 887-896 (1972).

[3] H. J. Frost and M. F. Ashby, "A Second Report on Deformation
     Mechanism Maps," ONR Final Report, Harvard University,
     August 1973.

[4] H. J. Frost, "Deformation Mechanism Maps," Ph.D. Thesis,
     Harvard University, January 1974.

[5] F. R. N. Nabarro, "Report of a Conference on the Strength of
     Solids," Physical Society, London, 75-90 (1948).

[6] C. Herring, "Diffusional Viscosity of a Polycrystalline
     Solid," J. Appl. Phys. 21, 437-445 (1950).

[7] R. L. Coble, "A Model for Boundary Diffusion Controlled
     Creep in Polycrystalline Materials," J. Appl. Phys. 34,
     1679-1682 (1963).

[8] R. Raj and M. F. Ashby, "On Grain Boundary Sliding and
     Diffusional Creep," Met. Trans. 2, 1113-1127 (1971).

[9] J. E. Bird, A. K. Mukherjee and J. F.Dorn, "Correlations
     Between High Temperature Creep Behavior and Structure,"
     pp. 255-342 in Quantitative Relation Between Properties and
     Microstructure, D. G. Brandon and A. Rosen, eds., Israel
     Univ. Press (1969).

[10] R. L. Stocker and M. F. Ashby, "On the Empirical Constants
     in the Dorn Equation," Scripta Met. 7, 115-120 (1973).

[11] P. A. Urick and M. R. Notis, "Final Stage Densification
     During Pressure Sintering of CoO," J. Am. Cer. Soc. 56,
     570-574 (1973).

[12] M. R. Notis, "Interpretation of Kinetics of Final Stage
     Pressure Sintering of NiO by Deformation-Mechanism Maps,"
     J. Am. Cer. Soc., 57, 271 (1974).

[13] J. T. A. Roberts and J. C. Voglewede, "Application of
     Deformation Maps to the Study of In-Reactor Behavior of
     Oxide Fuels," J. Am. Cer. Soc. 56, 470-475 (1973).

[14] R. L.Stocker and M. F. Ashby, "On the Rheology of the Upper
      Mantle," Rev. Geophys. and Space Phys. 11, 391-426 (1973).

[15] R. L. Stocker, unpublished work.

[16] R. H. Smoak and M. R. Notis, "Deformation Behavior of
      $MgAl_2O_4$ During Pressure Sintering," Paper 58-B-74, Am. Cer.
      Soc. Mtg., Chicago (1974), Bull. Am. Cer. Soc. 53, 319 (1974).

[17] M. R. Notis, "Pressure Sintering and Creep Deformation--A
      Joint Modeling Approach," Powder Met. Int'l., 6, 82-84 (1974).

[18] G. R. Terwilliger and K. C. Radford, "High Temperature
      Deformation of Ceramics: I, Background," Bull. Am. Cer.
      Soc. 53, 172-179 (1974).

[19] J. Weertman, "Creep Theories for Ceramics," Paper 81-B-74
      Am. Cer. Soc. Mtg., Chicago (1974), Bull. Am. Cer. Soc.,
      53, 320 (1974).

[20] H. J. Frost and M. F. Ashby, "Deformation-Mechanism Maps
      and the Creep of Tungsten Lamp Filaments," ONR Tech. Rept.
      No. 7, Harvard Univ., Oct. 1972.

[21] H. J. Frost and M. Ashby, "Deformation-Mechanism Maps for
      Steady State Flow," J. E. Dorn Memorial Symp. on Rate
      Processes in Plastic Deformation, Metals Engineering Congress,
      ASM (1972); to be published in the proceedings of the
      symposium by ASM.

[22] O. L. Wyatt and D. Dew-Hughes, Metals, Ceramics and Polymers,
      p. 170 Cambridge Univ. Press (1974).

[23] V. J. Jankus and R. W.Weeks, "Life-II - A Computer Analysis
      of Fast Reactor Fuel-Element Behavior as a Function of Re-
      actor Operating History," Nuclear Eng. and Design 18, 83-96
      (1972).

[24] T. G. Langdon, "Grain Boundary Sliding as a Deformation
      Mechanism During Creep," Phil. Mag. 21, 689-700 (1971).

[25] R. C. Gifkins, "Transitions in Creep Behavior," J. Mat'l.
      Sci. 5, 156-165 (1970).

[26] O. D. Sherby and P. M. Burke, "Mechanical Behavior of
      Crystalline Solids at Elevated Temperature," Prog. Mat'l.
      Sci. 13, 325-390 (1968).

[27] F. A. Nichols, "Theory of the Creep of Zircaloy During
      Neutron Irradiation," J. Nucl. Mat'ls. 30, 249-270 (1969).

[28] A. J. Ardell and O. D. Sherby, "The Steady State Creep of
      Polycrystalline Alpha Zirconium at Elevated Temperatures,"
      Trans. AIME, 239, 1547-1556 (1967).

[29] D. B. Knorr and M. R. Notis, "Deformation Mapping of Alpha-
      Zirconium and Zircaloy-2," Senior Project Report, Lehigh
      University, June 1974.

[30] I. M. Bernstein, "Diffusional Creep in Zirconium and Certain
      Zirconium Alloys," Trans. AIME, 239, 1518 1522 (1967).

[31] E. R. Gilbert, S. A. Duran and A. L. Bement, "Creep of
      Zirconium from 50 to 850°C," ASTM STP 458, 210-225 (1969).

[32] C. N. Alquist and R. A. Menezes, "Grain Boundary Sliding
      and the Observation of Superplasticity in Pure Metals,"
      Mater. Sci. Eng. 7, 223-224 (1971).

[33] M. F. Ashby, "Boundary Defects and Atomistic Aspects of
      Boundary Sliding and Diffusional Creep," Surface Sci. 31,
      498-542 (1972).

[34] R. Raj and M. F. Ashby, "Grain Boundary Sliding and the
      Effects of Particles on Its Rate," Met. Trans. 3, 1937-1942
      (1972).

[35] D. K. Matlock and W. D. Nix, "On Interpretation of the
      Effect of Grain Size on High Temperature Strength," Met.
      Trans. 5, 961-963 (1974).

[36] D. K. Matlock and W. D. Nix, "The Effect of Sample Size
      on the Steady State Creep Characteristics of Ni-6 pct W,"
      Met. Trans. 5, 1401-1412 (1974).

[37] J. H. Hensler and G. V. Cullen, "Stress, Temperature, and
      Strain Rate in Creep of Magnesium Oxide," J. Am. Cer. Soc.
      51, 557-559 (1968).

[38] J. T. A. Roberts, "Mechanical Equation of State and High
      Temperature Deformation of Uranium Dioxide," to be published
      in Acta. Met.

[39] M. H. Hodge, W. R. Bitler and R. C. Bradt, "Deformation
      Texture and Magnetic Properties of Ba0.6 Fe$_2$O$_3$," J. Am. Cer.
      Soc. 56, 497-501 (1973).

[40] R. A. Wolfe and S. F. Kaufman, "Mechanical Properties of Oxide Fuels," Westinghouse Report WAPD-TM-587, October 1967 NTIS order No. N6817241 (DA 753707), p. 82.

[41] G. L.Wire, H. Yamada and C. Li, "Mechanical Equation of State and Grain Boundary Sliding in Lead in Monotonic Loading," Acta Met. 22, 505-512 (1974).

[42] M. F. Ashby and R. A. Verall, "Diffusion-Accommodated Flow and Superplasticity," Acta Met. 21, 149-163 (1973).

[43] A. K. Mukherjee, "The Rate Controlling Mechanism in Super-plasticity," Mat'l. Sci. and Eng. 8, 83-89 (1971).

[44] R. L. Coble, "Diffusion Models for Hot Pressing with Surface Energy and Pressure Effects as Driving Forces," J. Appl. Phys. 41, 4798-4807 (1970).

[45] R. L. Coble, "Mechanisms of Densification During Hot Pressing," pp. 329-350 in Sintering and Related Phenomena, G. C. Kuczynski, et al., eds. Gordon and Breach (1967).

[46] A. J. Markworth, "Quantitative Description of Hot Pressing Kinetics," Paper 59-B-74, Am. Cer. Soc. Mtg., Chicago (1974); Bull. Am. Cer. Soc. 53, 319 (1974).

[47] R. C. Rossi and R. M. Fulrath, "Final Stage Densification in Vacuum Hot Pressing of Alumina," J. Am. Cer. Soc. 48, (11), 558-564 (1965).

[48] R. A. Wolfe, "Theory of Hot Pressing for a Solid with Non-linear Viscosity," Paper No 15-J2-67 presented at the annual meeting of the Am. Cer. Soc., May 1-3, 1967; Bull. Am. Cer. Soc. 46, 469 (1967).

[49] M. F. Ashby, "A First Report on Sintering Diagrams," Acta Met. 22, 275-289 (1974).

SLIP AND TWINNING SYSTEMS IN CERAMIC CRYSTALS

G. Y. Chin

Bell Laboratories

Murray Hill, New Jersey 07974

ABSTRACT

The intrinsic plastic resistance to dislocation motion of essentially pure crystals, as pointed out by Gilman, is primarily governed by the nature of bonding and the crystal structure. In this paper these two parameters are used as a basis for reviewing the slip systems in ceramic crystals - the choice of slip systems and the influence of temperature and stoichiometry on the plastic resistance of competing systems. Defects introduced by nonstoichiometry and solute additions exert an additional influence on the plastic behavior. The implications of slip and twinning to polycrystalline behavior - ductility, texture formation and texture strengthening - are discussed.

INTRODUCTION

Crystallographic slip and, in some instances, mechanical twinning are the most important mechanisms of deformation in crystals at temperatures up to about half the melting point. Both processes are characterized by the motion of dislocations[1] and their interaction with defects. As pointed out by Gilman[1], the intrinsic plastic resistance to dislocation motion of essentially pure crystals is primarily governed by the nature of bonding and somewhat less by the crystal structure. The temperature dependence of the plastic resistance is then related to the

25

changes in the nature of bonding with temperature. In this paper, Gilman's idea is amplified in discussing slip and twinning systems in ceramic crystals. First, an up-to-date comparison of metallic, ionic and covalent crystals is made with respect to the dependence of hardness on elastic modulus and on temperature. This is followed by a summary discussion of slip systems in selected ceramic systems. Data from the literature will be reviewed as to the effects of temperature, nonstoichiometry and strain rate on the operation of competing slip systems. Defects introduced by nonstoichiometry and solute additions exert an additional influence on the plastic behavior. Not much is known about mechanical twinning. The primary interest here is twinning in Al_2O_3. In the final sections, a discussion will be made of the implications of slip and twinning systems on polycrystalline plasticity. Ideas such as independence of and competition among slip systems, ductility, texture formation and texture strengthening will be discussed. Some of the broader aspects of deformation of ceramics have been reviewed recently by Terwilliger and Radford.[2]

INTRINSIC PLASTIC RESISTANCE OF CRYSTALS

As indicated by Gilman[1], the elastic modulus is the single most important mechanical characteristic of a crystal. He showed that the hardness of metals is a very small fraction ($\sim 10^{-3}$) of the value of the elastic modulus, while that of the covalent crystals exhibit a large fraction ($\sim 10^{-1}$). The data for ionic crystals, however, did not behave in a simple manner. The present author believes that all three sets of data behave in a simple manner if the temperature dependence of the hardness is taken into account. Gilman's data did not differentiate materials tested at different homologous temperatures. Figure 1 is a plot of hardness as a function of homologous temperature for Si (a covalent material), MgO (ionic) and Cu (metallic). Also included in Fig. 1 are data for TiC, which is thought to be essentially covalent-bonded at low temperature. All data were taken from the paper by Atkins and Tabor.[3] It is noted that up to $T/Tm = 0.5$, the hardness of Si and Cu are mildly dependent on temperature, MgO more so, and TiC the most. If one plots the ratio of the hardness to the shear modulus C_{44} as a function of temperature up to $T/Tm = 0.5$, Fig. 2 is obtained. Values of C_{44} and its temperature dependence were taken from tabulations by Hearmon[4], except that the temperature dependence of C_{44} of TiC is taken to be that of

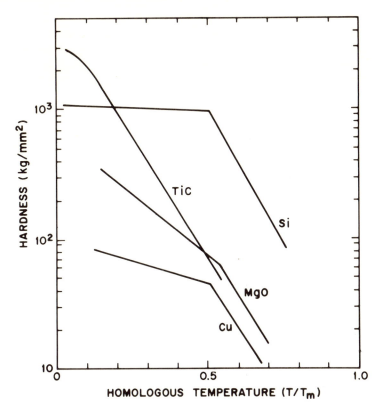

Fig. 1   Hardness as a function of homologous temperature for
         several solids.   From Ref. 3.

the Young's modulus for $TiC_{0.84}$.[5]   The significance of Fig. 2
is as follows:

1.   The value of H/C_{44} is about 0.13 for Si and 0.006 for
     Cu, and is independent of temperature.

2.   From $T/Tm = 0.1$ to 0.5, H/C_{44} of MgO drops from about
     0.025 to 0.006, and that of TiC from about 0.12 to 0.005.

     Thus TiC behaves like covalent Si at low temperature and
changes toward metallic behavior at elevated temperatures.   For
ionic MgO, H/C_{44} is intermediate between covalent and metallic
crystals, but also moves toward the metallic value at $T/Tm = 0.5$.

According to Petty and O'Neill[6], the temperature dependence of
the hardness of cubic metals is essentially the same as that of
the modulus.   Hence the value for Cu in Fig. 2 is expected to
hold for all face-centered cubic metals.   Data for covalent
crystals is limited, but Ge behaves in the same way as Si.[3]
The author is not aware of comparable data to TiC in other
transition metal carbides, nor to MgO in other ionic crystals
with the NaCl structure.   However, the room temperature hardness
of AgCl (NaCl structure) has been measured to be about 5 kg/mm^2.[7]
For AgCl, C_{44} = 637 kg/mm^2[4]; hence H/C_{44} $\sim$0.008. Since room
temperature corresponds to T/Tm = 0.41 for AgCl, a value of H/C_{44} =
0.009 can be read from the MgO curve of Fig. 2, in good agreement
with the measured value.   In addition, as will be shown in Fig. 3,
the H/C_{44} ratio is nearly 0.013 for a number of alkali halides at
room temperature, which corresponds to a value of T/Tm $\sim$ 0.3.
This value compares very well with that of 0.0125 as read from
the MgO curve.   Thus it is believed that the MgO curve is representative of ionic crystals with the rock salt structure.

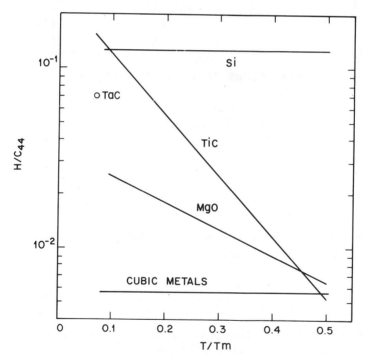

Fig. 2   Ratio of hardness to shear modulus C_{44} as a function
         of homologous temperature for data of Fig. 1. TaC point
         from Ref. 23.

With Fig. 2 in mind, Fig. 3 shows a plot of room temperature
hardness as a function of C_{44} for a variety of covalent crystals,
ionic crystals of NaCl structure, and f.c.c. metals, in a manner
similar to that of Gilman. The hardness of metals is taken from
Petty and O'Neill[6] and refers to Vickers hardness on polycrystalline material. The hardness of ionic crystals refers to
Knoop hardness of single crystals with indenter aligned in <001>
direction; data from Chin et. al.[8] except LiF and MgO data from
Brookes et. al.[9] and NaF data from Barker.[10] The hardness of
covalent crystals generally refers to Knoop hardness of single
crystals; data from Ivan'ko.[11] Values of C_{44} are taken from
Hearmon[4], except InAs, InP and GaP data from Drabble.[12] For
the covalent crystals and for the metals, since H/C_{44} is independent of temperature (Fig. 2), materials with different melting
points can be plotted into a single curve. For the ionic crystals,
$T/Tm \sim 0.3$ as stated above. Note the value for MgO uncorrected
for temperature lies considerably higher from the curve. In
Gilman's plot, he included the uncorrected room temperature values
of MgO as well as those of TiC, NbC, ZrC and VC. The carbides,
as Fig. 2 shows, behave differently from the ionic crystals. With
the temperature dependence of hardness corrected and with the
carbides removed from the curve, the ionic crystals thus also
show a linear dependence of hardness with elastic modulus as
indicated in Fig. 3.

The nature of bonding in transition metal carbides such as
TiC is still somewhat controversial. Results of detailed measurements by Ramquist[13], based on electron spectroscopy chemical
analysis (ESCA) and on x-ray emission and absorption spectra,
indicate essentially covalent bonding between the Ti 3d states and
the C 2p states, with partial election transfer from the 3d to the
2p levels. Earlier energy band calculations[14,15] had predicted
the sign of this electron transfer, although the model by Lye[16]
indicates the opposite. This question was recently discussed by
Novotny et. al.[17]

SLIP SYSTEMS IN CERAMIC CRYSTALS

General Considerations

Since bonding and crystal structure govern the level of the
intrinsic plastic resistance they obviously also determine to a
major degree the choice of slip systems. In metals, where slip

systems have been studied most extensively, the slip direction is
usually the closest-packed direction. The slip plane is generally
the closest-packed plane. Because dislocations tend to lower
their energy by dissociating into partial dislocations separated
by stacking faults, the generalization cannot be carried too far.
For ceramic crystals, strong directional bonding as in the case of
covalent crystals, or electrostatic interactions as in the case of
ionic crystals, places severe restrictions on the configuration
of ions at the dislocation core and further limits the choice of
slip systems. A reasonably comprehensive list of slip systems has

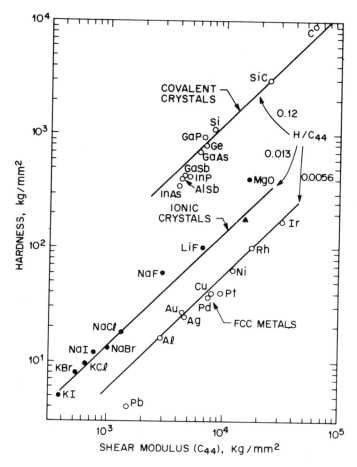

Fig. 3   Hardness as a function of C_{44} for three classes of solids
measured at room temperature. The point (▲) is taken from
MgO curve of Fig. 2 corresponding to $T/T_m = 0.3$ applicable
to the other ionic crystals shown.

been compiled and shown in Table 1, together with a more limited list of twinning systems. Although twinning is a less frequently observed event, it has not received much attention either, except in the cases of Al_2O_3 and crystals of the diamond structure. Hence it is likely that in the future additional twinning systems will be discovered in those materials listed in Table 1.

Slip systems are generally identified by metallographic analysis of slip traces and dislocation etch pits as well as by transmission electron microscopy analysis of the dislocations. For the more brittle materials, hardness indentation is often used to generate plastic flow which would otherwise be suppressed by cleavage. In this way, slip and twinning systems which are found active only at high temperatures can be forced to operate at lower temperatures. Salient examples in this regard are findings of dislocation generation and motion at room temperature in Si[46,47] and Al_2O_3.[48] For a comprehensive review of the use of this technique for studying single crystal plasticity, the reader is referred to the paper by Hockey.[49]

Diamond Cubic and Sphalerite Structures

Deformation of crystals of these two structures has been extensively reviewed by Alexander and Haasen.[49] Slip occurs on closest-packed {111} planes and in closest-packed <110> directions. Since the diamond cubic structure consists of two f.c.c. lattices shifted by an (a/4) <111> vector, however, the dislocations are more complicated than those of f.c.c. metals. In the sphalerite structure, the two f.c.c. sublattices are occupied by two different kinds of atoms, such as In and Sb of the III-V semiconductors. This situation leads to a polarity of stacking of {111} planes with consequent differences in mechanical and electrical behavior.

Transition Metal Carbides with NaCl Structure

The transition metal carbides having the NaCl structure are often called interstitial compounds because the small carbon atoms practically fit into the octahedral interstices of the much larger transition metal ions. As pointed out previously, they are essentially covalent solids. This is manifested by a high H/C_{44}

TABLE 1

Slip and Twinning Systems of Some Ceramic Crystals

(Listed in order of increasing CRSS; difference in CRSS generally decreases at high temperature)

a. Slip Systems

Material	Structure	Slip System	Remarks	Ref.
C(diamond),Si,Ge	diamond cubic	{111} $\langle\bar{1}10\rangle$		18
GaAs,InSb	sphalerite	{111} $\langle\bar{1}10\rangle$		18
MgAl$_2$O$_4$, (Mn,Zn)Fe$_2$O$_4$	spinel	{111} $\langle\bar{1}10\rangle$, {110} $\langle\bar{1}10\rangle$	{110} predominate in MgO·2 Al$_2$O$_3$(20)	19-21
TiC,ZrC,HfC, VC,NbC,TaC,UC	rock salt	{110} $\langle\bar{1}10\rangle$, {111} $\langle\bar{1}10\rangle$	UC,Ref. 18; at RT {111} predominant in TaC and quite active in NbC	22-26
NaCl,NaBr,KCl,KBr AgCl,LiF,NaF,MgO CaO,MnO,MnS,CoO,FeO	rock salt	{110} $\langle\bar{1}10\rangle$, {001} $\langle\bar{1}10\rangle$, {111} $\langle\bar{1}10\rangle$		18
PbS,PbTe,PbSe	rock salt	{001} $\langle\bar{1}10\rangle$, {110} $\langle\bar{1}10\rangle$	slip in $\langle100\rangle$ appears unlikely or minor	27-28
CsBr,Tl(Br,I)	cesium chloride	{110} $\langle001\rangle$		18

a. Slip Systems (con't)

Material	Structure	Slip Systems	Remarks	Ref.
CaF_2, BaF_2, SrF_2 UO_2, ThO_2	fluorite	{001}<$\bar{1}$10> {110}<$\bar{1}$10> {111}<$\bar{1}$10>	{111} appears to be cross-slip plane only.(30)	18,29-32
TiH_2, ZrH_2	fluorite	{111}<$\bar{1}$10>		33,34
TiO_2	rutile	{101}<10$\bar{1}$> {110}<001>		18
Al_2O_3	hexagonal	(0001)<11$\bar{2}$0> {11$\bar{2}$0}<1$\bar{1}$00> {$\bar{1}$012}<10$\bar{1}$1> {$\bar{1}$101}<10$\bar{1}$1> (structural cell)	most prominent systems only. See Ref. (35).	35
C(graphite)	hexagonal	(0001)<11$\bar{2}$0>		18
TiB_2, ZrB_2	hexagonal	{10$\bar{1}$0}<11$\bar{2}$0> (0001)<11$\bar{2}$0>		36,37
WC	hexagonal	{10$\bar{1}$0}[0001] {10$\bar{1}$0}<11$\bar{2}$0>		38,89
βSi_3N_4	hexagonal	{10$\bar{1}$0}[0001]		40,41

b. Twinning Systems

Material	Twinning Systems	Ref.
Si,Ge,InSb,GaSb,ZnS	$\{111\}\langle11\bar{2}\rangle$ $\{123\}\langle41\bar{2}\rangle$	42
PbS	$\{441\}\langle\bar{1}\bar{1}8\rangle$	28
Al_2O_3	$(0001)\langle1\bar{1}00\rangle$ $\{10\bar{1}1\}\langle10\bar{1}\bar{2}\rangle$ (morphological cell)	43 44
C(graphite)	$\{11\bar{2}1\}$	45

ratio at low temperature, see Fig. 2. At high temperature, the
strength decreases rapidly so that they behave like metals. The
slip systems are generally $\{111\}<\bar{1}10>$ at high temperature, as
in f.c.c. metals. Near room temperature, slip also occurs in
$\{110\}<\bar{1}10>$ systems. The latter appear to be dominant with the
exception of NbC[25] and TaC.[22,23] Evidence for a change in
slip mode has come from etched dislocation patterns around
hardness indentations, and from the study of Knoop hardness
anisotropy. Figure 4 shows the Knoop hardness anisotropy of TiC
at three different temperatures.[25] (The $TiC_{0.8}$ curve[24] will
be discussed later.) The Knoop indenter was aligned at different
directions for an impression on the (001) plane. At room
temperature, the hardness is largest in the <110> direction and
smallest in <100> while at 610°C the situation is reversed. The
high temperature hardness anisotropy is typical of slip in
$\{111\}<\bar{1}10>$ systems and the low temperature anisotropy is typical
of slip in $\{110\}<\bar{1}10>$ systems. At 250°C, the hardness anisotropy
takes on a mixed appearance, suggesting activity on both systems.

It should be pointed out that the use of hardness anisotropy
as an indicator of slip systems must be done with caution, as
different slip modes could give rise to the same anisotropy. It
has been shown, for example, that BaF_2, SrF_2, Al, AgCl and NaCl
all show similar hardness anisotropies despite different primary
slip systems.[7] Sodium chloride and other alkali halides, in
fact, exhibit a higher Knoop hardness in the <100> direction[50],
in contrast to other ionic crystals such as MgO and LiF, even
though the primary slip system is $\{110\}<\bar{1}10>$ for all. The deformation caused by the hardness indentation is generally complex,
involving slip and/or twinning on several systems to accommodate
the imposed strain. Thus the hardness anisotropy is by no means
obviously related to the easy slip systems[9], but more likely
reflects the hard systems if such systems are needed for the
imposed deformation.

Theoretically, Van der Walt and Sole[51] used a hard sphere
model to consider the expected slip modes of covalent crystals
with NaCl structure. They concluded that in the case of strong
bonds between unlike atoms, the expected slip planes are in
the direction $\{110\} \mapsto \{111\} \mapsto \{100\}$ as the ratio of the radius of
the small atom to that of the larger atom is increased. The slip
direction is $<\bar{1}10>$. This conclusion appears to hold in the
transition metal carbides as the radius ratio becomes higher
toward the NbC and TaC end.[24] (Although VC is expected to have
a slightly larger radius ratio than that of NbC and TaC, stoichiometry cannot be obtained.[24]) Related to the trend in radius
ratio, but perhaps on a more basic level, is the decreased
covalent bonding for NbC and TaC as compared to the others. At
room temperature, TaC has a H/C_{44} ratio below that of TiC, Fig. 2.

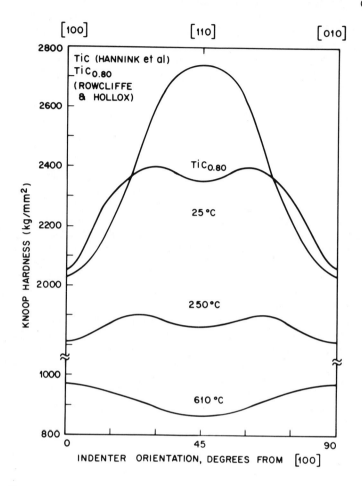

Fig. 4 Knoop microhardness as a function of orientation of the
long axis of indentor for indentation on (001) plane of
TiC at three different temperatures. From Ref. 25 except
$TiC_{0.8}$ data from Ref. 24.

The effect of nonstoichiometry on slip behavior will be discussed
in a later section.

Ionic Crystals with the NaCl Structure

Although the expected slip system is {001}<$\bar{1}$10> in terms of
close packing this system is observed only in the highly polarized
(or less ionic) lead compounds (PbS, PbTe, and PbSe). For the
highly ionic (or less polarized) compounds such as LiF and MgO,

electrostatic interactions favor {110}<$\bar{1}$10> as observed. This
point has been discussed in detail[27,52,53] and will not be
elaborated here. At elevated temperature other slip systems,
{100}<$\bar{1}$10> and {111}<$\bar{1}$10>, also become activated. This point will
be taken up later in connection with temperature effects.

Cesium Chloride Structure

For ionic crystals such as CsBr, the slip system is {110}<001>.
Rachinger and Cottrell[54] have considered slip in the CsCl
structure and concluded that the expected slip system changes from
{110}<001> to {110}<111> as the bonding changes from ionic to
metallic, in general agreement with observations. The {110}
plane is the closest-packed plane; the closest-packed direction
is <001> for CsCl, but <111> if the structure becomes disordered
(b.c.c.).

Fluorite Structure

Although {110} is the densest-packed plane and <$\bar{1}$10> is the
densest-packed direction, slip generally occurs in {001} planes
again as a result of electrostatic considerations. The {110}
plane becomes operative at elevated temperatures. An interesting
anomaly appears to exist in ZrH_2[33] and TiH_2[34], which slip on
{111}<$\bar{1}$10> systems at about 100^{6}C and above. This observation is
puzzling, although Irving and Beevers[34] presented some fairly
convincing arguments for diffusion of the small hydrogen ions in
the wake of dislocation motion on {111} planes.

Saphire

The mechanical behavior of this material has been studied
extensively. Snow and Heuer[35] have reviewed the observed slip
systems. Those listed in Table 1 are considered the most
prominent ones. As in h.c.p. metals where the c/a ratio is large,
basal slip is generally the easiest. Prism slip, {11$\bar{2}$0}<1$\bar{1}$00>
(all slip systems based on structural unit cell), operates when
the applied stress axis is perpendicular to the c-axis.[55]
Pyramidal slip, either {$\bar{1}$012}<10$\bar{1}$1>[56] or {$\bar{1}$101}<10$\bar{1}$1>[57,58],
is favored for stress applied parallel to the c-axis, where the
resolved shear stress for basal or prism slip is zero.

Kronberg[43] has made detailed analysis of the basal
dislocations in Al_2O_3. He concluded that the 1/3 <11$\bar{2}$0> slip
vector is energetically favored to split into four partials of
the 1/6 <1$\bar{1}$00> type. Thus far transmission electron microscopy
operating at 1 MeV has failed to reveal the expected splittings.[59]

Other Hexagonal Structures, Spinels, Rutile

Graphite is characterized by strong covalent bonding in the
basal plane and weak Van der Waals bonding between basal planes,
with c/a ∿2.73.[60] Consequently, slip occurs easily on (0001) in
the <$11\bar{2}0$> direction. Tungsten carbide has a small c/a ratio,
0.972.[39] The observed slip systems are {$10\bar{1}0$}[0001] and
{$11\bar{2}0$}[0001][38,39], the reverse of the usual case of large c/a
material. Diborides of titanium, zirconium, and hafnium also have
small c/a ratios (∿1.1).[37,61] The primary slip system is
{$10\bar{1}0$}<$11\bar{2}0$>, while the secondary slip system is (0001)<$11\bar{2}0$>.[36,37]
Not much is known about the slip behavior of the potentially
important βSi_3N_4. The {$10\bar{1}0$}[0001] slip system is based on dislocation observations of Butler[40] and of Evans and Sharp.[41] βSi_3N_4
has a c/a ratio of ∿0.38.[62]

The remaining list of Table 1 consists of spinels with
{111}<$\bar{1}10$> as the primary slip system and {110}<$\bar{1}10$> as the
secondary system.[19-21] Hornstra[63] has analyzed the dislocations. Rutile (TiO$_2$) is a tetragonal crystal with slip systems
{101}<$10\bar{1}$> and {110}<001>.[64]

Twinning Systems

As mentioned previously, twinning has not been studied
extensively. A rather detailed examination on crystals of the
diamond and sphalerite structures was made by Churchman et. al.[42]
They found that surrounding hardness indentations are twins of the
{111}<$11\bar{2}$> and {123}<$41\bar{2}$> types. These twins were observed in
crystals deformed at intermediate temperatures, and reasoned that
slip takes over exclusively at high temperatures in general trend
with other materials. The theory of {123} twinning was considered
by Bullough.[65]

Twinning has also been studied in some detail in Al$_2$O$_3$. The
most important system is rhombohedral twinning {$10\bar{1}1$}<$10\bar{1}2$>
(morphological notation)[44] together with basal twinning.[43]
Conrad et. al.[66] studied the temperature and strain rate dependence of these two twinning systems by compressing crystals with
the c-axis oriented at 60 deg to the compression axis. The results
are shown in Fig. 5. Twinning is favored at low temperatures and
high strain rates, with rhombohedral twinning predominating.
Becher and Palmour[67] have also observed transition from
rhombohedral twinning to "rhombohedral" slip during compression
parallel to the c-axis, at temperatures above 1550°C.

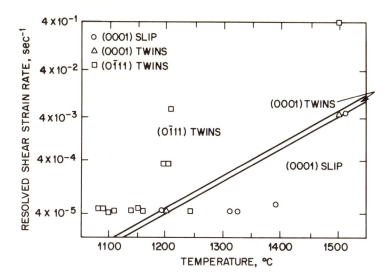

Fig. 5 The effect of temperature and strain rate on the activity
        of slip and twinning systems in Al_2O_3. From Ref. 66.

Influence of Temperature and Strain Rate on Slip

     As the testing temperature is increased, the degree of ionic
or covalent bonding is decreased, as can be seen in the hardness
behavior of Fig. 2. This means that those slip systems listed in
Table 1 which are less favorable at room temperature as a result
of bonding restrictions are expected to become active at high
temperature. Examples for the temperature dependence of the
critical resolved shear stress for slip (CRSS) for $\{110\}<\bar{1}10>$ and
$\{001\}<\bar{1}10>$ systems are shown in Fig. 6 for MgO[68] and in Fig. 7
for CaF_2.[69] The ratio of CRSS for $\{001\}$ to $\{110\}$ slip in CaF_2
decreases from about 6 at 250°C to about 3 at 400°C. These ratios
appear from Fig. 7 to be independent of strain rate for the two
rates used, although the value of CRSS itself is changed substantially. The effect of strain rate on the CRSS of different slip
systems needs further study.

Influence of Non-Stoichiometry and Solute Additions

     In ceramic crystals, stoichiometry plays a major role in alter-
ing the behavior of slip systems. Substantial nonstoichiometry
exists in some oxides and carbides, particularly at high temperature. The oxides TiO_2, VO_2 and FeO have been reviewed recently by
Nadeau.[70] Brittain discussed the defects in TiO_2 in detail.[71]

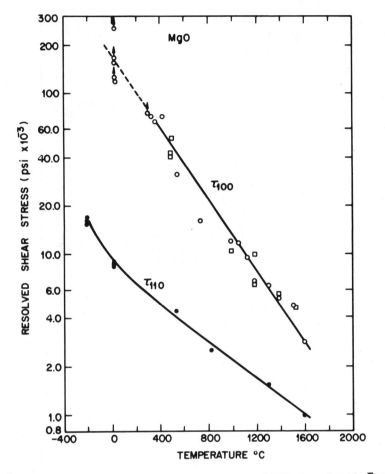

Fig. 6   The temperature dependence of the CRSS for {110}<$\bar{1}$10>
         and {001}<$\bar{1}$10> slip modes in MgO.   From Ref. 68.

The general conclusion is that lattice defects resulting from
nonstoichiometry consist of complex combinations of vacancies and
interstitials whose configurations generally change with temperature. Considerable short range ordering among the defects often
exists. In the case of transition metal carbides such as NbC[26]
and VC[72], long range ordering also takes place. Thus depending
on the nature of the defect structure the slip behavior can be
altered.

Ashbee and Smallman[64] have found that oxygen deficiency in
TiO$_2$ hardens the {101}<10$\bar{1}$> system at all temperatures tested.
The {101}<10$\bar{1}$> dislocations are dissociated, and it is thought that

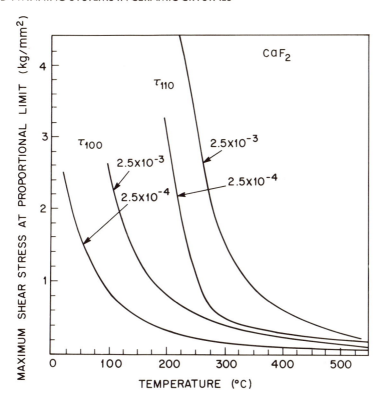

Fig. 7   The temperature and strain rate dependence of the CRSS for
{110}<$\bar{1}$10> and {001}<$\bar{1}$10> slip modes in CaF$_2$.   From
Ref. 69.

oxygen vacancies precipitate onto the stacking faults, thereby
pinning them.[73] Hirthe and Brittain[74], on the other hand,
have observed softening on the {1$\bar{1}$0}<001> slip system in oxygen
deficient TiO$_2$. The {110}<001> dislocations are not dissociated,
and Brittain has determined from internal friction measurements
that defects caused by nonstoichiometry interact more strongly
with dissociated dislocations than with nondissociated ones.[71]

     In UO$_2$, Nadeau observed that excess oxygen decreases the CRSS
for the {1$\bar{1}$0}<$\bar{1}$10> system while the {001}<$\bar{1}$10> system is much less
affected.[29]   The temperature dependence of the CRSS for the two
slip modes is shown in Fig. 8, for crystals of UO$_{2.001}$ and UO$_{2.10}$
stoichiometry. The slip behavior of UO$_2$ was also studied by
Sawbridge and Sykes;[30] their values of CRSS agree quite well
with those of Nadeau. The effect of stoichiometry was further

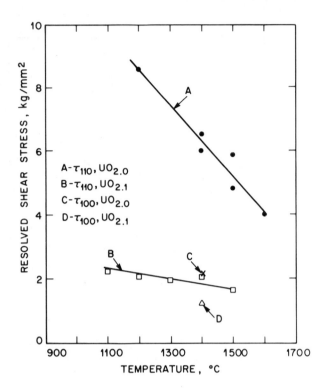

Fig. 8 Effect of stoichiometry on the temperature dependence
        of the CRSS for {110}<$\bar{1}$10> and {001}<$\bar{1}$10> slip modes
        in UO$_2$. From Ref. 29.

studied by Yust and McHargue[31,75], who proposed a model by which
oxygen clusters in UO$_{2.06}$ promote cross-slip from {001} onto
{111} planes. Sawbridge and Sykes[30] had evidence earlier that
both {001} and {110} systems could cross-slip onto {111} planes.
The cross-slip mechanism could then be responsible for reducing
the CRSS of {110} slip drastically as observed by Nadeau.

Newey and Radford[20] have found in the spinel system
MgO·Al$_2$O$_3$, alumina-rich materials become softer as the CRSS for
the primary {111}<$\bar{1}$10> systems is reduced. In the MgO·2 Al$_2$O$_3$
stoichiometry, the secondary {110}<$\bar{1}$10> system actually became
dominant. They commented that the lack of crystal perfection in
this material may have caused the anomaly.

In the transition metal carbides, room temperature hardness
indentations indicate a strength decrease in carbon deficient
samples except NbC and TaC, where a large increase is observed.[76]
Knoop hardness anisotropy studies indicate a trend for increased
{111}<110> slip activity (at the expense of {110}<$\bar{1}$10>) in
TiC[24], as may be seen from the TiC$_{0.8}$ curve in Fig. 4. For TaC,
similar study shows increased {110}<1$\bar{1}$0> at the expense of
{111}<$\bar{1}$10>.[23] These changes appear to come from variations in
bonding with carbon content. Ramquist[76] have found that the
heat of formation, which is related to bonding, is decreased in
TiC, ZrC, HfC, and VC but increased in NbC and TaC as the carbon
content is decreased, consistent with the hardness trends.

Besides possible changes in dislocation mechanisms and
bonding behavior, defect-enhanced diffusion at elevated temperatures is expected to cause softening.[70] Thus the effects are
complicated.

It is generally known that solute additions tend to increase
the CRSS, particularly those that result in defect clusters
because of the necessity for charge compensation. Alkali halides,
for example, are hardened by divalent ions much more so than by
monovalent ions.[77-79] While monovalent ions enter as substitution for the alkali ions, the divalent ion generally associates
itself with an alkali ion vacancy (for charge balance) in a
neighboring site, forming a dipole.[80] Fleisher[81] has proposed
a theory of hardening based on the interactions of dislocations
with the tetragonal strain field of the dipole. A more recent
theory due to Gilman[82], based on the results of Chin et.
al.[8,83] and based on the change in electrostatic energy on
shearing the dipole, accounts for the hardening quantitatively
without the use of disposable parameters. Not much is known about
the effect of solutes on the behavior of competing slip systems.
In the alkali halide study cited above[8,83], the yield stress is
increased dramatically but the hardness only modestly. Since the
yield stress refers to the softer {110}<$\bar{1}$10> system and the hardness is governed by the harder {001}<$\bar{1}$10> system, the results
imply that divalent solutes raise the CRSS of {001} slip to a
less extent than {110}.

IMPLICATIONS OF SLIP SYSTEMS TO POLYCRYSTALLINE PLASTICITY

Much has already been discussed regarding polycrystalline
plasticity[84-86] that only a short summary need be made.
Briefly, because of constraints from its neighbors, a grain
embedded inside a polycrystalline aggregate must be able to undergo

any arbitrary shape change if fracture is to be avoided. An
arbitrary shape change is specified by six independent strain
components, which are reduced to five in the case of no volume
change, such condition being satisfied by simple shear. Hence a
crystal must have five independent slip or twinning systems.
Embodied in the idea of five independent slip systems are such
notions as the availability of mobile dislocations and the inter-
penetrability of the slip systems.

Those crystals lacking five independent slip systems are
almost assuredly brittle, while those possessing five independent
systems can be ductile if cleavage or grain boundary cracking occurs
at stresses considerably higher than the yield stress. Actually
there are probably few crystals truly lacking five independent
slip systems. As stresses are built up, harder systems generally
become operative if the stress level for cracking is not reached
first. For this reason, ductility is extended in a hardness
indentation where hydrostatic compression suppresses cracking.
Similarly, crack propagation is delayed in fine-grained samples.
Thus rather than saying that LiF is brittle at room temperature
because it lacks five independent slip systems, it is preferable to
say that LiF is often brittle because cleavage generally occurs
at stress levels below those necessary for operating five independent slip systems. This idea is not new, but is worth repeating.[84]

If several sets of slip systems are available to a crystal,
and if all these sets are needed to provide five independent slip
systems, the yield stress of the polycrystal is generally governed
by the harder set. This is quite clearly illustrated in Fig. 9
for data on MgO as provided by Langdon and Pask.[87] The two
dashed curves correspond to two extremes in single crystal behavior.
The higher curve is for operation of the harder $\{001\}<110>$ systems
when a single crystal is compressed in the $<111>$ orientation. The
lower curve, for the softer $\{110\}<\bar{1}10>$ system, is obtained by $<001>$
compression. Several types of polycrystalline samples have yield
strength close to the $<111>$ curve, that is, the flow is governed
by the harder $\{001\}<\bar{1}10>$ system. Similar results were pointed out
by Nadeau for UO_2[29], where the yield strength of the polycrystalline material is closer to the harder $\{110\}<\bar{1}10>$ system than to
the softer $\{001\}<\bar{1}10>$ system. Since there are two $\{110\}<\bar{1}10>$
independent systems and three $\{001\}<\bar{1}10>$ systems, both are required
for general plasticity.

The requirement for operation of five independent slip systems
in polycrystalline plasticity presents an interesting challenge
with regard to the selection of active systems in a given deformation. The most workable model in this regard is that of Taylor[88]
as generalized by Bishop and Hill[89] and Chin and Mammel.[90] If
slip is governed by a critical resolved shear stress criterion,

then the active slip systems are those which minimize the internal
work done by slip or, equivalently, the external work done by the
applied stresses. The most convenient procedure for carrying out
the minimization has been done using linear programming techniques
in conjunction with an electronic computer.[91] The case for slip
in <$\bar{1}$10> direction with possible slip planes {001}, {111} and
{110} has been analyzed in detail and numerical solutions given
for the cases of deformation by uniaxial flow (free tension or
compression) and by axisymmetric flow (constrained tension or
compression, as in an aggregate).[91-93] The results of the
calculations provide the following information of interest: (1) the
active slip systems, (2) the applied axial stress σ_{xx} necessary
for yielding, expressed as the Taylor factor $M = \sigma_{xx}/\tau$ where τ is
the CRSS for slip, and (3) the lattice rotation. Knowledge of
the orientation dependence of the yield stress is basic to the
control of texture for increased strength. Information on lattice
rotation leads directly to predictions of deformation texture
development. Some of the results are given below.

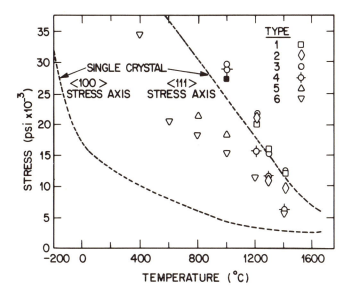

Fig. 9 The yield strength of several types of polycrystalline MgO
        samples as a function of temperature. Dashed curves refer
        to extremes in single crystal behavior. From Ref. 87.

Competition Among {110}, {100} and {111}<110> Slip Modes

Since slip is governed by a CRSS criterion, a given slip
system generally becomes active if the resolved shear stress on
this system is greater than others provided the CRSS is equal
among all systems. On the other hand, if the CRSS is not equal,
as is the general case where more than one slip mode are available,
a given slip system may not be able to operate even under maximum
resolved shear stress if its CRSS is too high compared with others.
When considering slip among {110}<110>, {110}<110> and {111}<110>
systems, it is recognized that the number of independent slip
systems is 2, 3 and 5 respectively. Hence an arbitrary deformation
can be accomplished by one of the following slip mode combinations:
1) {111}, 2) {111} + {110}, 3) {111} + {100}, 4) {100} + {110},
and 5) {111} + {100} + {110}. The conditions for the activation of
these combinations have been obtained[93] and the results shown
graphically in Fig. 10. Figure 10 shows, for example, that {110}
slip cannot occur under any circumstances if $\tau_{111}/\tau_{110} < 2\sqrt{6} =$
0.81. In Sawbridge and Sykes' study on UO_2, the values for {100},
{110} and {111} slips are determined to be 27, 66 and 50 MN/m^2
respectively.[30] According to Fig. 10, only {100} and {111} slip
systems should be active. The observation of {100} slip thus
suggests that the three slip modes do not follow a CRSS criterion.
Sawbridge and Sykes pointed out that {111} is mainly a cross-slip
plane for {100} and {110} systems; hence the situation is not
simple.

Axisymmetric Flow for Mixed Slip on {110}<110> and {100}<110>
Systems

With the values of CRSS confined to specific regions of
Fig. 10, one can calculate the Taylor factor and lattice rotation
for any imposed deformation. All regions of Fig. 10 have been
so treated. As an example of a situation encompassing several
solids over a large temperature range, the results for axisymmetric
flow for mixed slip on {110}<110> and {100}<110> systems will be
summarized. Unlike free tension or compression where only the
axial strain is specified, the incremental strain components for
axisymmetric flow with respect to the specimen axes x, y, z are
specified as:

$$d\varepsilon_{yy} = d\varepsilon_{zz} = -d\varepsilon_{xx}/2, \quad d\varepsilon_{yz} = d\varepsilon_{zx} = d\varepsilon_{xy} = 0$$

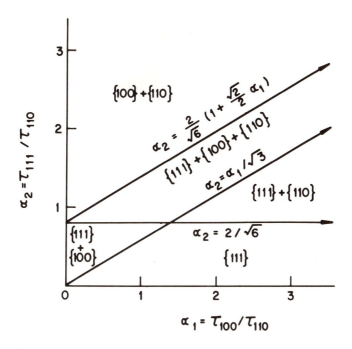

Fig. 10 Deformation mode diagram showing regions of allowed slip
         modes as a function of the CRSS ratios α_1 and α_2. From
         Ref. 93.

with x being the axial direction. For cubic crystals, the axial
direction is represented by a point inside the standard stereographic triangle [100]-[111]-[110]. Figures 11 to 13 show computer
plotted contours of the Taylor factor (here expressed as σ_{xx}/τ_{110},
where τ_{110} is CRSS for $\{110\}$ slip) inside the orientation triangle,
for values of $\alpha_1 = \tau_{100}/\tau_{110} = 0.1, 1.0,$ and 10 respectively. It
may be seen that the triangle is divided into four regions, each
region governed by a different set of slip systems as indicated in
Table 2.   As expected, both $\{100\}$ and $\{110\}$ slip systems are
involved. The number for each region refers to a particular stress
state that activates the requisite slip systems.[94] It turns out
that the analytical solutions for M are quite simple[94]; these
are also listed in Table 2.

     Examination of Figs. 11 to 13 shows that where $\{100\}$ slip is
easier ($\alpha_1 < 1$) such as CaF_2, the strength is highest in the <100>
corner and decreases toward <110> and <111>. The reverse situation
is true where $\{110\}$ slip is easier ($\alpha_1 < 1$) such as MgO, although

TABLE 2

Active Slip Systems and Associated M Values for

the Four Regions of Fig. 14. (from Ref. 94)

Region	Active Slip Systems	$M = \sigma_{xx}/\tau_{110}$
1	1,2,3,4,7,10	$3\cos^2\phi\cos^2\theta - 1 + (3\sqrt{2}\alpha_1/2)[\cos^2\phi\cos\theta\sin\theta + \cos\phi\sin\phi(\cos\theta+\sin\theta)]$
9	1,2,-5,-6,7,10,12	$3\cos^2\phi \quad -2 + (3\sqrt{2}\alpha_1/2)[\cos^2\phi\cos\theta\sin\theta + \cos\phi\sin\phi(\cos\theta+\sin\theta)]$
15	1,2,3,4,7,8,-11,12	$3\cos^2\phi\cos^2\theta - 1 + 3\sqrt{2}\alpha_1\cos^2\phi\cos\theta\sin\theta$
21	1,3,-5,-6,7,8,-11,12	$3\cos^2\phi \quad -2 + 3\sqrt{2}\alpha_1\cos^2\phi\cos\theta\sin\theta$

1-(10ī)[101], 2-(101[10ī], 3-(110)[1ī0], 4-(1ī0)[110], 5-(011)[0ī1], 6-(0ī1)[011], 7-(010[101], 8-(010)[10ī], 9-(001)[1ī0], 10-(001)[11c], 11-(100)[0ī1], 12-(100)[011]. System -5 means (011)[01ī], etc.

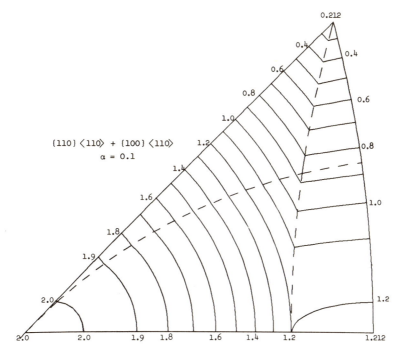

Fig. 11 Computer plotted contours of constant Taylor factor as a
function of axial orientation for axisymmetric flow.
Case of $\alpha_1 = \tau_{100}/\tau_{110} = 0.1$. From Ref. 92.

here the <110> orientation is stronger than <111> (Fig. 13). The
situation at <110> is interesting in that the Taylor factor and
the associated slip systems not obvious on the basis of free tension or compression. For the latter cases, the deformation can be
accomplished by slip on either {100} or {110} planes (critical
value of α_1 is $\sqrt{2}$). For axisymmetric flow, however, both slip
modes are required and the strength is mainly attributed to the
harder mode. In fact, only at <100> can axisymmetric flow be
attained with pure {110} slip and at <111> with pure {100} slip,
as special cases of the stress states listed in Table 2.

The lattice rotations for axisymmetric tension are shown in
Fig. 14. The sense of rotation for compression is merely the
reverse of that for tension. It should be pointed out that for a
given axial orientation, the same slip systems must be activated
regardless of the value of α_1. This result comes from the fact
that neither slip mode by itself possesses five independent slip
systems. Thus, although the value of M is changed with a change

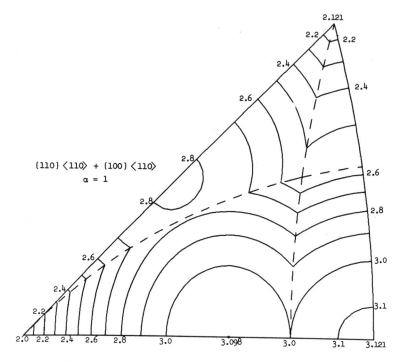

Fig. 12 Same as Fig. 11, case of $\alpha_1 = 1$.

in α_1, the lattice rotations are not. Consequently Fig. 14 applies to all values of α_1.

In Fig. 14 each ray emanating from a point represents the change in axial orientation for an axial strain of 5 pct, for one of the basic combinations of five independent slip systems with minimum M. Since the CRSS is reached in 6, 7 and 8 slip systems, Table 2, more than one basic combination of five systems can have the same minimum. Thus several rays are possible as reflected in Fig. 14. By further combining the basic combinations, rotations in intermediate positions between the rays can be obtained. The overall pattern, however, indicates a rotation toward <111>. In compression, the expected rotation is toward the <100>-<110> zone.

Information on the strength anisotropy and deformation texture development in polycrystalline ionic materials is scanty. Observations of a <100> fiber texture in hot extruded MgO and CaO appeared related to a recrystallization texture as the grains were found to be equiaxed.[95] Hunt and Lowenstein[96], however, had reported a strong <111> fiber texture in extruded UO_2 rods, which is in accord with the prediction of Fig. 10. The grains were found to be elongated. Recently there is much interest in

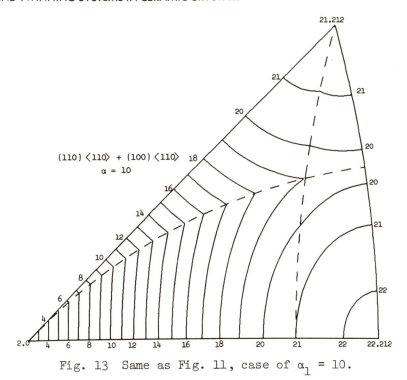

Fig. 13   Same as Fig. 11, case of $\alpha_1 = 10$.

exploring polycrystalline alkali halides for use as laser window
materials.[97]   Several laboratories have investigated the use of
hot forging on single crystals to convert to polycrystalline
materials of fine grain size with clean grain boundaries. Besides
increased strengthening due to a refinement in grain size, there
appears to be significant texture strengthening as well. The
optical properties, being isotropic, are not affected by the
texture.

The most significant example of texture strengthening among
all materials is probably that of graphite fibers. Here a
judicious selection of thermal and mechanical treatments starting
with appropriate textured fibers has resulted in almost perfect
alignment of the basal planes parallel to the fiber axis.[98]
Values of Young's modulus as high as 60×10^6 psi and tensile
strength as high as 430×10^3 psi have been attained. Recent work
has indicated that fibers processed under clean room conditions
to minimize flaws exhibit consistently high values of tensile
strength in long lengths.[99]

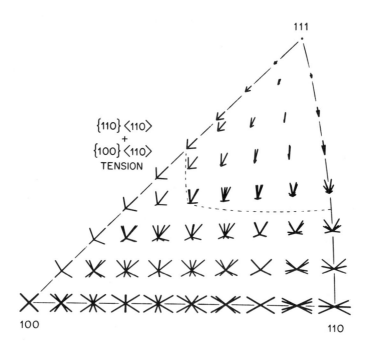

Fig. 14   Lattice rotation in axisymmetric tension for {110}<110>
          and {100}<110> slip; all values of α_1. From Ref. 92.

CONCLUDING REMARKS

     In this review, the slip and twinning systems of the more
prominent ceramic materials have been tabulated. The behavior of
these systems with respect to structural and environmental parameters, particularly stoichiometry and temperature, has been
discussed. An overall picture of deformation of ceramics,
especially in the technically important regime of high temperature
and low strain rate, a knowledge of other deformation mechanisms
such as dislocation climb and vacancy diffusion controlled flow
is required. Also needed is an understanding of the recovery
and recrystallization aspects of these materials. Hot working
of ceramics is a proven technique.[95] One should thus be able
to use deformation for microstructural control more than is done
now. This is especially true in the area of control preferred
orientation, which has exploited but little. Besides improving
on the mechanical strength mentioned previously, a textured
material should prove useful in improving such physical properties
as remanent polarization in piezoelectric transducers, the remanent
magnetization of square-loop magnetic memory cores, and the energy

product of permanent magnets. The recent work of Hodge et. al.[100] shows promise of improving the energy product of hexagonal ferrites through deformation induced texture. Mizushima[101] has found that a Mg-Zn ferrite single crystal recording head has the best combination of wear and signal output when moving on a {110} plane in a <100> direction. Perhaps fine-grained textured polycrystalline material could be fabricated for this purpose. Textured ferrites or garnets are expected to reduce the resonance line widths useful in microwave applications. Thermal shock resistance could be improved by texturing as anisotropic thermal contraction among grains is minimized. Other examples may be named, but these should suffice to point out the expanded possibilities for property improvements brought about by texture control.

REFERENCES

1. J. J. Gilman, Australian J. Phys. 13, 327 (1960).

2. G. R. Terwilliger and K. C. Radford, Ceramic Bulletin 53, 172, 465 (1974).

3. A. G. Atkins and D. Tabor, Proc. Roy. Soc. Lond. A292, 441 (1966).

4. R. F. S. Hearmon, in Landolt - Börnstein New Series, Group III/Vol. 1, K. H. Hellwege and A. M. Hellwege, eds. (Springer-Verlag, Berlin 1966).

5. R. H. J. Hannink and M. J. Murray, J. Mater. Sci. 9, 223 (1974).

6. E. R. Petty and H. O'Neill, Metallurgia 60, 25 (1961).

7. G. Y. Chin, M. L. Green, L. G. Van Uitert and W. A. Hargreaves, J. Mater. Sci. 8, 1421 (1973).

8. G. Y. Chin, L. G. Van Uitert, M. L. Green and G. Zydzik, Scripta Met. 6, 475 (1972).

9. C. A. Brookes, J. B. O'Neill and B. A. W. Redfern, Proc. Roy. Soc. Lond. A322, 73 (1971).

10. J. D. Barker, Electro-Optical Systems Design, Oct., 1972, p.32.

11. A. A. Ivan'ko, Handbook of Hardness Data, Publication TT70-50177, U.S. Department of Commerce, National Technical Information Service, Springfield, Va.

12. J. R. Drabble, in Physics of III-V Compounds, Ch. 5,
    R. K. Willardson and A. C. Beer, eds. (Academic Press, N.Y.,
    1966).

13. L. Ramquist, J. Appl. Phys. 42, 2113 (1971).

14. V. Ern and A. C. Switendick, Phys. Rev. 137, A1927 (1965).

15. J. B. Conklin, Jr. and D. J. Silversmith, Int. J. Quantum
    Chem. 25, 243 (1968).

16. R. G. Lye, in Solid State Chemistry (National Bureau of
    Standards Special Publication 364, 1972), p.567.

17. H. Nowotny, H. Boller and G. Zwilling, as in Ref. 16, p.487.

18. A. Kelly and G. W. Groves, Crystallography and Crystal
    Defects (Addison-Wesley, Reading, Mass. 1970), pp.175-6.

19. M. H. Lewis, Phil. Mag. 17, 481 (1968). [spinel]

20. C. W. A. Newey and K. C. Radford, in Anisotropy in Single
    Crystal Refractory Compounds, Vol. 2, F. W. Vahldiek and
    S. A. Mersol, eds. (Plenum Press, N.Y., 1968), p.321. [spinel]

21. T. Ito, J. Amer. Ceram. Soc. 54, 24 (1971). [$(Mn,Zn)Fe_2O_4$]

22. D. J. Rowcliffe and W. J. Warren, J. Mater. Sci. 5, 345
    (1970). [TaC]

23. D. J. Rowcliffe and G. E. Hollox, J. Mater. Sci. 6, 1261
    (1971). [TaC,HfC]

24. D. J. Rowcliffe and G. E. Hollox, J. Mater. Sci. 6, 1270
    (1971). [TiC,VC]

25. R. H. J. Hannink, D. L. Kohlstedt and M. J. Murray, Proc.
    Roy. Soc. Lond. A326, 409 (1972). [TiC,VC,ZrC,NbC]

26. G. Morgan and M. H. Lewis, J. Mater. Sci. 9, 349 (1974).
    [Nb_6C_5]

27. J. J. Gilman, Acta Met. 7, 608 (1959). [PbTe]

28. K. D. Lyall and M. S. Paterson, Acta Met. 14, 371 (1966).
    [PbS]

29. J. S. Nadeau, J. Amer. Ceram. Soc. 52, 1 (1969). [UO_2]

30. P. T. Sawbridge and E. C. Sykes, Phil. Mag. 24, 33 (1971). [UO_2]

31. C. S. Yust and C. J. McHargue, J. Amer. Ceram. Soc. 54, 628 (1971). [UO_2]

32. J. W. Edington and M. J. Klein, J. Appl. Phys. 37, 3906 (1966). [ThO_2]

33. K. G. Barraclough and C. J. Beevers, J. Mater. Sci. 4, 518, 802 (1969). [ZrH_2]

34. P. E. Irving and C. J. Beevers, J. Mater. Sci. 7, 23 (1972). [TiH_2]

35. J. D. Snow and A. H. Heuer, J. Amer. Ceram. Soc. 56, 153 (1973). [Al_2O_3]

36. J. S. Haggerty and D. W. Lee, J. Amer. Ceram. Soc. 54, 572 (1971). [ZrB_2]

37. S. A. Mersol, C. T. Lynch and F. W. Vahldiek, in Anisotropy in Single Crystal Refractory Compounds, Vol. 2, F. W. Vahldiek and S. A. Mersol, eds., (Plenum Press, N.Y., 1968), p.41. [TiB_2]

38. D. N. French and D. A. Thomas, Trans. TMS-AIME 233, 950 (1965). [WC]

39. L. Pons, in Anisotropy in Single Crystal Refractory Compounds, Vol. 2, F. W. Vahldiek and S. A. Mersol, eds., (Plenum Press, N.Y. 1968), p.393. [WC]

40. A. G. Evans and J. V. Sharp, J. Mater. Sci. 6, 1292 (1971). [Si_3N_4]

41. E. Butler, Phil. Mag. 21, 829 (1971). [Si_3N_4]

42. A. T. Churchman, G. A. Geach and J. Winton, Proc. Roy. Soc. Lond. A238, 194 (1956).

43. M. L. Kronberg, Acta Met. 5, 507 (1957).

44. A. H. Heuer, Phil. Mag. 13, 379 (1966).

45. E. J. Freise and A. Kelly, Proc. Roy. Soc. A264, 269 (1961).

46. V. I. Nikitenko, M. M. Myshlyaev and V. G. Eremenko, Sov. Phys.-Sol. State 9, 2047 (1968).

47. D. J. Rowcliffe, Brown Boveri Research Center, Baden, Switzerland, submitted to J. Mater. Sci., 1974.

48. B. J. Hockey, in The Science of Hardness Testing and Its Research Applications, J. H. Westbrook and H. Conrad, eds. (ASM, Metals Park, Ohio, 1973), p.21.

49. H. Alexander and P. Haasen, in Solid State Phys., Vol. 22, F. Seitz, D. Turnbull and H. Ehrenreich, eds. (Academic Press, N.Y., 1968), p.28.

50. G. Y. Chin, L. G. Van Uitert, M. L. Green and G. Zydzik, Scripta Met. 6, 503 (1972).

51. C. M. Van der Walt and M. J. Sole, Acta Met. 15, 459 (1967).

52. M. J. Buerger, Amer. Mineral. 15, 21, 35 (1930).

53. H. Mueller, Amer. Mineral. 16, 237 (1931).

54. W. A. Rachinger and A. H. Cottrell, Acta Met. 4, 109 (1956).

55. D. J. Gooch and G. W. Groves, Phil. Mag. 28, 623 (1973).

56. P. D. Bayer and R. E. Cooper, J. Mater. Sci. 2, 301 (1967).

57. D. J. Gooch and G. W. Groves, J. Mater. Sci. 8, 1238 (1973).

58. R. E. Tressler and D. J. Barber, J. Amer. Ceram. Soc. 57, 13 (1974).

59. G. V. Berezhkova, Yu. M. Gerasimov, V. G. Govorkov and E. P. Kozlovskaya, Phys. Stat. Soc. (a)21, K61 (1974).

60. For example, C. S. Barrett and T. B. Massalski, Structure of Metals, 3rd Ed. (McGraw-Hill, N.Y., 1966), p.627.

61. L. Bsenko and T. Lundström, J. Less Common Metals 34, 273 (1974).

62. S. Wild, P. Grieveson and K. H. Jack, in Special Ceramics, Vol. 5, P. Popper, ed. (Brit. Ceram. Res. Assoc., 1972), p.385.

63. J. Hornstra, in Materials Science Research, Vol. 1, H. H. Stadelmaire and W. W. Austin, eds. (Plenum Press, N.Y., 1963), p.88.

64. K. H. G. Ashbee and R. E. Smallman, Proc. Roy. Soc. A275, 195 (1963).

65. R. Bullough, Proc. Roy. Soc. A241, 568 (1957).

66. H. Conrad, K. Janowski and E. Stofel, Trans. TMS-AIME 233, 255 (1965).

67. P. Becher and H. Palmour III, J. Amer. Ceram. Soc. 53, 119 (1970).

68. C. O. Hulse, in Anisotropy in Single Crystal Refractory Compounds, Vol. 2, F. W. Vahldiek and S. A. Mersol, eds. (Plenum Press, N.Y., 1968), p.307.

69. P. L. Pratt, C. Roy and A. G. Evans, in Materials Science Research, Vol. 3, W. W. Kriegel and H. Palmour III (Plenum Press, N.Y., 1966), p.225.

70. J. S. Nadeau, in Anisotropy in Single Crystal Refractory Compounds, Vol. 2, F. W. Vahldiek and S. A. Mersol, eds. (Plenum Press, N.Y., 1968), p.361

71. J. O. Brittain, in Anisotropy in Single Crystal Refractory Compounds, Vol. 2, F. W. Vahldiek and S. A. Mersol, eds. (Plenum Press, N.Y., 1968), p.95.

72. R. H. J. Hannink and M. J. Murray, Acta Met. 20, 123 (1972).

73. K. H. G. Ashbee, R. E. Smallman and G. K. Williamson, Proc. Roy. Soc. Lond. A276, 542 (1963); Proc. Brit. Ceram. Soc. No. 1, 201 (1964).

74. W. M. Hirthe and J. O. Brittain, J. Amer. Ceram. Soc. 46, 411 (1963).

75. C..S. Yust and C. J. McHargue, J. Amer. Ceram. Soc. 56, 161 (1973).

76. L. Ramquist, Jernkont. Ann. 153, 159 (1969).

77. A. Edner, Z. Phys. 73, 623 (1932).

78. H. Shoenfeld, Z. Phys. 75, 442 (1932).

79. W. Metag, Z. Phys. 78, 363 (1932).

80. J. S. Cook and J. S. Dryden, Aust. J. Phys. 13, 260 (1960).

81. R. L. Fleischer, Acta Met. 10, 835 (1962); J. Appl. Phys. 33, 3504 (1962).

82. J. J. Gilman, J. Appl. Phys. 45, 508 (1974).

83. G. Y. Chin, L. G. Van Uitert, M. L. Green, G. J. Zydzik and T. Y. Komentani, J. Amer. Ceram. Soc. 56, 369 (1973).

84. G. W. Groves and A. Kelly, Phil. Mag. 8, 877 (1963).

85. S. M. Copley and J. A. Pask, in Materials Science Research, Vol. 3, W. W. Kriegel and H. Palmour, III; eds. (Plenum Press, N.Y., 1966), p.189.

86. R. J. Stokes, in Ceramic Microstructures, R. M. Fulrath and J. A. Pask, eds. (Wiley, N.Y., 1968), p.379.

87. T. G. Langdon and J. A. Pask, J. Amer. Ceram. Soc. 54, 240 (1970).

88. G. I. Taylor, J. Inst. Metals 62, 307 (1938); Stephen Timoshenko 69th Anniversary Volume, (MacMillian Co., N.Y., 1938), p.218.

89. J. F. W. Bishop and R. Hill, Phil. Mag. 42, 414, 1298 (1951).

90. G. Y. Chin and W. L. Mammel, Trans. TMS-AIME 245, 1211 (1969).

91. G. Y. Chin and W. L. Mammel, Trans. TMS-AIME 239, 1400 (1967).

92. G. Y. Chin and W. L. Mammel, Met. Trans. 4, 335 (1973).

93. G. Y. Chin and W. L. Mammel, Met. Trans. 5, 325 (1974).

94. G. Y. Chin, Met. Trans. 4, 329 (1973).

95. R. W. Rice, in High Temperature Oxides, Part III, A. M. Alper, ed. (Academic Press, N.Y., 1970), p.235.

96. J. G. Hunt and P. Lowenstein, Am. Ceram. Soc. Bull. 43, 562 (1964).

97. Proc. Conf. High Power Infrared Laser Window Materials, AFCRL-71-0592 (1971), AFCRL-TR-73-0372 (1973), AFCRL-TR-74-0085 (1974), Air Force Cambridge Research Laboratories, Bedford, Mass. See particularly the papers in the Processing section in AFCRL-TR-74-0085(2).

98. W. Watt, Proc. Roy. Soc. Lond. A319, 5 (1970).

99. R. Moreton, S. Driffill and W. Watt, RAE Tech. Rept. 73048 (1973), Royal Aircraft Establishment, England.

100. M. H. Hodge, W. R. Bitler and R. C. Bradt, J. Amer. Ceram. Soc. 56, 497 (1973).

101. M. Mizushima, IEEE Trans. Mag. MAG-7, 342 (1971).

REVIEW OF DIFFUSIONAL CREEP OF Al_2O_3

R. M. Cannon and R. L. Coble

Massachusetts Institute of Technology

Cambridge, Massachusetts

ABSTRACT

The diffusional creep mechanisms are reviewed including the extensions of the theories for application to ceramic materials. The creep data for Al_2O_3 are critically analysed to show the relative conditions where normal and dislocation controlled diffusional creep are important. The behavior at intermediate stresses, when both slip and diffusional creep occur is also discussed.

INTRODUCTION

Although the basic diffusional creep theories are nearly twenty five years old, until rather recently their greatest applicability has been to sintering and in allowing measurements of surface tension by zero creep method. The diffusional creep mechanisms have been interesting theoretically because they can be derived with a high degree of rigor since arbitrary assumptions regarding the deformation structure are minimal; as a result explicit expressions for the creep rate exist which have no disposable parameters and so are susceptible to qualitative experimental confirmation. More recently, it has been realized that diffusional creep may be the dominant mechanism for many long time applications, and is not simply limited to very low stresses at temperatures near the melting point. Recent studies, as reviewed by Jones (1969), Bird et al (1969), and Burton and Greenwood (1970), have shown quantitative agreement between theory and experimental creep and diffusion data for metals.

In the case of ceramics, it has become widely accepted that
diffusional creep is an important deformation mechanism. This
view has resulted from a realization of the relative difficulty
of slip on secondary systems in many ceramics and from the accumulation of considerable steady state creep data, for many oxides and
other ceramics, which at least qualitatively conform to diffusional
creep kinetics, e.g. Coble (1965), and Passmore and Vasilos (1968).
However, detailed analysis of the data has often indicated considerable disagreement among the results of various investigators and
with the predictions from self diffusion data. These discrepancies
result in part from the possibility of multiple diffusion paths along
boundaries and dislocations, as well as through the lattice, and, in
addition to the extreme sensitivity of the diffusion coefficients to
impurities and stoichiometry variations. These conditions often
make it difficult to be certain whether the cation or anion will be
the slower and so rate controlling. Consequently, plausible explan-
ations are available for many disparities, but the results and
inferences from them have been viewed with considerable skepticism.

This paper will review the basic physical processes involved
in diffusional creep and the resultant kinetic equations. The creep
data for Al_2O_3 will then be analysed in detail to indicate the
range of stress; temperature and grain size over which diffusional
creep is important and, further, to indicate values of the controlling diffusivities. Consideration will also be given to the causes
for failure to satisfy certain of the kinetic predictions, particularly proportionality between stress and strain rate.

THEORY

The initial considerations of diffusional creep were for
single component materials and so the theories were most directly
applicable to pure metals. However, the extension to ceramics (or
alloys) is often straight forward requiring only a determination
of the effective diffusivity, except where there is a shift away
from diffusion as the rate limiting step.

Basic Mechanism

Upon application of a normal stress on a surface of a body,
such as a grain in a polycrystal, for which the surfaces can act as
perfect sources or sinks of point defects, the chemical potential
of the atoms and of the point defects will be altered by the external work which is done when an atom moves from a normal lattice
position to create a vacancy or an interstitial. This will result
in an alteration of point defect concentrations which will depend
upon the sign and magnitude of the normal stress. For a single
component material the vacancy concentration will be raised at a

surface with a normal tensile stress by the amount

$$\Delta C = C_o \frac{\sigma_n \Omega}{kT} \tag{1}$$

for the case where

$$\frac{\sigma_n \Omega}{kT} < 1 \tag{2}$$

where C_o is the stress free equilibrium concentration of vacancies, σ_n is the normal stress, Ω the atomic volume and kT is Boltzmann's constant times the temperature. Non-linear terms will become important if the condition on $\sigma_n \Omega/kT$ is not satisfied. The result, as shown in Figure 1, is that under any non-hydrostatic stress the vacancy concentrations on orthogonal faces will be different. This will eventually cause a steady state concentration gradient between the orthogonal surfaces and a steady state mass flux from surfaces under normal compression to those under normal tension. If the dominant defects are interstitials rather than vacancies the result is similar except that the interstitial concentration gradient is opposite in sign to that for vacancies.

There will be an initial transient of decreasing strain rate during which the concentration gradient drops from a very high value near the surfaces to the steady state value. This period will also be accompanied by a redistribution of the stress which will drop to zero in the region of the grain corners and increase to a maximum at the centers of the grain faces so that the concentration gradients are nearly uniform along each surface. The exact stress distribution will be determined by the compatibility requirement that coherency is maintained across the grain boundaries, unless cracking is occurring. This transient has been analyzed, considering only lattice diffusion, by Lifshitz and Shikin (1965) and shown to be dominated by the diffusional transients if the body is effectively elastically rigid. For this case the longest relaxation time, τ, will be

$$\tau = \frac{d^2}{KD_d} \tag{3}$$

where d is the grain size, and D_d the defect diffusivity, and K is a geometric constant. If elastic strains are important, as indicated by the condition

$$\lambda = \frac{\Omega C_o E}{kT} << 1 \tag{4}$$

where E is the elastic modulus, then the stress redistribution will

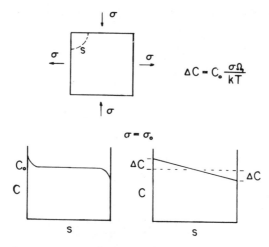

$$\Delta C = C_o \frac{\sigma \Omega}{kT}$$

Figure 1   Defect concentration gradients expected at short times
(left) and at steady state (right) in a body with an applied shear
stress as shown.

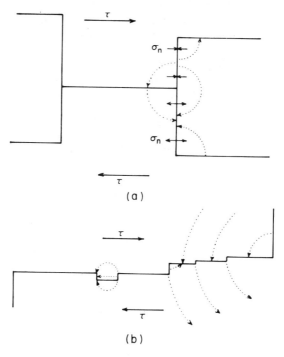

(a)

(b)

Figure 2   Idealization of diffusional creep as a grain boundary
sliding problem showing the mutual requirements for sliding along
boundaries and diffusional transport of mass between orthoganal
boundaries. The dotted lines denote the direction of mass transport.

be much slower with the longest relaxation time given by

$$\tau_E = \frac{\tau}{\lambda} \tag{5}$$

Analysis of the creep transient in MgO by Gordon and Terwilliger (1972) indicates that the elastic strains are important in the relaxation. Even for this slower case the transient strain is only a few times the elastic strain and so is very much smaller than typically found in dislocation creep. Upon removal of the stress much of this transient strain will be recovered with a similar relaxation time, as observed by Beauchamp, et al (1962) and Chang (1966) for Al_2O_3 and BeO. For a constant strain rate application, the transient will be manifested by a gradual increase in stress to the steady state value.

In principal, the mass transport can occur by diffusion through the lattice, controlled by D_ℓ, and also by diffusion along the grain boundary, in a region of width δ where the diffusivity is enhanced to a value of D_b. The diffusion through the grains and along the boundaries can occur independently of each other within the restriction that the local stress and total flux will be such as to preserve coherency at the boundaries. Raj and Ashby (1971) have shown that even in cases where lattice diffusion dominates the steady state creep, boundary diffusion will be important during the transient in relaxing the initially high stresses or fluxes at grain corners or ledges.

The steady state problem of grain elongation by diffusion has been analyzed by Nabarro (1948), Herring (1950), and Lifshitz (1963), and others for the case of lattice diffusion control and by Coble (1963) and others for the case of dominance by grain boundary diffusion. For polycrystalline bodies the results are all found to be of the form

$$\dot{\varepsilon} = \frac{14\sigma\Omega D_\ell}{kTd^2} + \frac{14\pi\sigma\Omega\delta D_b}{kTd^3} \tag{6}$$

except for thin foils, Gibbs (1966), or wires, Herring (1950), where the grain size is larger than the thickness. The numerical constants vary slightly depending upon the assumed grain shape, but for approximately equiaxed grains the results are rather insensitive to the details of the assumption; the values given are for σ and $\dot{\varepsilon}$ as tensile stress and strain rate, respectively.

Herring (1950) recognized that the stress redistribution would be affected if grain boundary sliding can allow relaxation of all shear stresses in the plane of the boundary and that this would result in an increase in the strain rate by a factor of 5/2. However, it was subsequently shown by Lifshitz (1963) and by Gibbs (1965) that grain boundary sliding is actually required to maintain compatability at all the grain boundaries. This comes from the need

for grains to shift their relative positions with respect to each
other as they elongate; it is accomplished by concurrent sliding
along some boundaries while mass is added to or removed from other
boundaries by diffusion.

It has more recently been shown, particularly by Raj and
Ashby (1971) and Stevens (1972), that this dual process can be
alternatively reviewed as grain boundary sliding with diffusional
accomodation. For boundary sliding to occur it must be accomodated
by deformation of the grains, particularly at triple points, ledges,
or other points of constraint. This accomodation can occur by
diffusion instead of by slip as suggested by Gifkins et al (1966,
1968). This can be visualized in Figure 2a where the problem is
idealized to show sliding along the horizontal grain faces which is
accomodated by removal or addition of mass to the vertical boundaries by diffusion. Ashby et al (1970) further refined the analysis
and showed, as suggested in Figure 2b, that the rate of sliding will
be limited by diffusion between the highest steps, which generally
are further apart. For an equiaxed polycrystal the limiting obstacles are the neighboring grains. A calculation, Raj and Ashby
(1971), of the creep rate, based on diffusion limited boundary
sliding for equiaxed grains, gives the same results as calculated
from grain elongation, Eqn. (6). In fact, these are simply alternative descriptions of the same process. Both considerations, however require that the atomic resistance to sliding is relatively
easy so that the diffusive transport is the rate limiting step.

An analysis by Green (1970) shows quantitatively that a
gradual effective hardening may occur as deformation procedes, resulting in elongated grains and an increase in the diffusion distances. This elongation may, of course, not be obtained in practice
since grain boundary migration will tend to maintain an equiaxed
grain shape. It has also been shown by Ashby and Verrall (1973)
and W. R. Cannon and Nix (1973) that if significant rearrangement
of the grains occurs as a result of grain boundary sliding the total
strain will be appreciably higher than that calculated in Eq. (6).
This would result from grains changing their neighbors causing an
increase in the number of grains along the tensile direction in the
body A calculation by Ashby and Verrall (1973), based upon a
particular model of grain rearrangement by sliding, and with the
necessary accomodation by diffusion, gave a strain rate expression
similar to Eqn. (6) but with the coefficient higher by a factor of
about 7. However, if significant grain rearrangement does not occur,
W. R. Cannon (1972) has shown that Eqn. (6) is correct and does not
need to be modified to account for the strain resulting from the required sliding accomodation. At the low strains typically achieved
in the testing of ceramics Eqn. (6) probably provides the best estimate of the creep rate. At high strains modifications from the opposing effects of grain elongation and grain rearrangement may be
important.

Concurrent grain growth under load will give rise to an

effective hardening as the diffusion distances increase. Terwilliger, Bowen, and Gordon (1970) and Tigai and Zisner (1968) have shown that this effect can be very important in pure MgO, and that it can be quantitatively assessed in some cases. Contrary to the results of Terwilliger et al (1971), Cannon and Rhodes (1971) found that the grain growth in Al_2O_3 can be strain enhanced so that the application of the results of stress free grain growth studies may not provide a satisfactory prediction of the effective hardening.

To summarize, diffusional creep will occur by a flux of point defects and the creep rate will be as given by Eqn.(6) provided that:
a) the grain boundaries are the principle sources and sinks for defects, and
b) they are sufficiently good sources that the defect concentrations at the boundaries are at the equilibrium value for the local stress and temperature, and
c) the boundary sliding necessary to satisfy compatibility occurs easily.

If either of the latter two conditions are not satisfied then the creep rate will be slower than predicted by Eqn.(6), a condition often referred to as interface control. Ashby (1969), Burton (1972), and Ashby and Verrall (1973) have derived simple models to account for a significant chemical potential drop at the boundaries to create point defects. These models generally predict non-linear stress-strain rate behavior which in some cases may involve an effective threshold stress, or Bingham type behavior. Qualitatively similar behavior has been reported in ceramics by Heuer et al (1969) and R. M. Cannon et al (1975) for fine grained Al_2O_3, by Crouch (1972) for Fe_2O_3, by Burton and Reynolds (1973) for UO_2, by Seltzer and Talty (1974) for ZrO_2, and by Passmore et al (1966) for fine grained MgO. There is also considerable similar behavior in the superplastic metals and for fine grained metals with fine, refractory particles at the grain boundaries. Particularly for the ceramics the data indicate that at moderate stresses the behavior approaches a linear stress-strain rate relation, if slip does not intervene; but at low stresses the strain rate is lower than predicted causing a range of non-linear stress dependence. There are still sufficient unexplained features that the problem cannot yet be regarded as well understood.

Edge dislocations can also act as sources and sinks for point defects, and so diffusion between orthogonal sets of edge dislocations can lead to climb of the dislocations giving a macroscopic strain. However, for steady state diffusional creep to occur the dislocations can not simply climb out of the crystal, but must be regenerated. Nabarro (1967) analyzed this problem assuming that the dislocations were arrayed in a stable Frank network, with the free segments climbing as Bardeen-Herring (1951) sources to give continuous dislocation multiplication. The resultant strain rate is non-linear in stress since the network size becomes finer with

increasing stress. Nabarro's result for tensile stress and strain
rate is:

$$\dot{\varepsilon} = \frac{\Omega}{2\pi b^2 \ln \frac{4G}{\pi\sigma}} \frac{\sigma^3 D_\ell}{G^2 kT} [1 + \frac{4}{3} \frac{D_c}{D_\ell} (\frac{\sigma}{G})^2] \tag{7}$$

where G is shear modulus, b the Burgers vector and D_c is the core
diffusion coefficient which is assumed to be effective in a radius
b around the dislocation; the term $\ln \frac{4G}{\pi\sigma}$ is about 10 for most cases.
The contribution from enhanced pipe diffusion along dislocations
normally would be more important at lower temperatures. Groves
and Kelly (1969) have shown that deformation of a polycrystal by
this mechanism will require 6 independent slip systems so that
climb will be required by dislocations on secondary as well as
primary systems.

An alternative expression for the climb rate has been derived
by Weertman (1968) and by Nix et al (1971). Using this, Weertman
then obtained the following expression:

$$\dot{\varepsilon} = \frac{\pi\beta^2\Omega}{20b^2} \frac{\sigma^3 D_\ell}{G^2 kT} \tag{8}$$

for control by lattice diffusion, where β is the coefficient in the
Taylor relation

$$\rho = (\frac{\beta\sigma}{bG})^2 \tag{9}$$

which was assumed to describe the stress dependence of the dislocation density, ρ. Weertman assumed the β values typically found
from creep tests of metals deforming by climb and glide were applicable since in both cases the internal stress fields of the dislocations are an appreciable fraction of the applied stress. Bird
et al (1969) show that β is about unity. For many studies Weertman's
equation gives a strain rate an order of magnitude faster than
Nabarro's. If the dislocation density is independent of stress, then
a linear stress dependence might result, but deviations would be
expected at low stresses which are insufficient to activate the
Bardeen-Herring sources.

Because of the high strain rate sensitivity, diffusional creep
will generally only be important at low stresses where slip becomes
relatively slower. However, since it can provide isotropic deformation it can be particularly important in strongly anisotropic
materials such as non-cubic ceramics and metals. In these materials, at stresses between the primary and secondary flow stresses,
diffusion can provide strain components not achievable by slip on
primary slip systems, and so can provide essential satisfaction

of the von Mises condition. For these intermediate stresses, slip
will occur in favorably orientated grains in which the critical
shear stress is reached on the easy slip planes. The creep rate
will still be controlled by the diffusional creep since it will be
the slower process compared to primary slip. Although the strain
rate will be slightly faster than predicted by the diffusional
creep equations, (6) or (7), the functional dependencies will be
approximately as predicted. At coarser grain sizes where the
Nabarro climb mechanism will be the more important of the diffusional modes, climb will be required on these secondary dislocations under conditions for which they do not glide.

Diffusional mass transport can also contribute to crack growth
(or shrinkage, i.e. sintering) under stress. The slow growth of
small pores on grain boundaries under a normal tensile stress has
been well documented for metals as well as ceramics. Small cracks
or pores at grain boundaries which develop from grain boundary
sliding or from failure to satisfy the von Mises criterion, can
grow by diffusion at boundaries where there is a higher vacancy
concentration, Eqn.(1), due to the normal tensile stress.

When the void fraction becomes sufficiently high, the diffusion required to accomodate grain boundary sliding is considerably reduced and much higher strain rates result. Therefore,
highly porous bodies or bodies undergoing significant cracking,
generally show only approximate agreement with the theoretical
constituitive relations and may exhibit much higher creep rates.
Since the early analyses of pore growth by diffusion of Balluffi
and Seigle(1957) and Hull and Rimmer (1959) there have been many
extensions and refinements as discussed recently by Weertman
(1974).

Modifications for Ceramics

For ceramics both the cations and anions must diffuse and at
steady state the total fluxes of the two components must be in
the stoichiometric ratio. Since the diffusivities of the two
components will generally be significantly different, initially
there will be a greater flux of the faster species. This will
cause a slight charge separation which produces an internal
electric field which will enhance the slower species and retard
the faster one. At steady state the flux of defects on each
sublattice will be given by expressions of the form

$$j_i = - \frac{D_d^i}{\Omega_i kT} \; (\Delta c_i + q_i \, c_i \, \nabla \phi) \tag{10}$$

where q_i is the effective charge on the ith defect, $-\nabla \phi$ is the

Nernst field and D_d^i is the defect diffusivity. For transport in the compound $A_\alpha O_\beta$ the steady state requirement is that at the defect sources both species must be accumulated at the stoichiometric ratio. This requires that at boundaries:

$$\frac{(j_\ell^a)_{\overline{b}} \; \delta \nabla \cdot j_b^a}{(j_\ell^o)_{\overline{b}} \; \delta \nabla \cdot j_b^o} = \frac{\alpha}{\beta}$$

(11-a)

where $(j_\ell^i)_b$ is the normal lattice flux at the boundary, and j_b^i is the flux in the boundary; alternatively at dislocations:

$$\frac{\int j_\ell^a \cdot \overline{r}_c d\theta - \pi r_c^2 \nabla \cdot j_c^a}{\int j_\ell^o \cdot \overline{r}_c d\theta - \pi r_c^2 \nabla \cdot j_c^o} = \frac{\alpha}{\beta}$$

(11-b)

where j_c^i is the flux along the dislocation with core radius r_c. These requirements specify $\nabla\phi$. Solutions for several different creep, sintering, or surface diffusion problems have been given by Ruoff (1965), Readey (1966), Lifshitz et al (1965, 1967), Blakely and Li (1966), Reijnen (1970), Gordon (1973) and Fryer (1972). The exact solutions have been for cases assuming the same diffusion path for both species. They indicate that the previously presented expressions for diffusional creep are correct with the following substitutions:

$$D_{eff} = \frac{(\alpha+\beta) D_e^a D_e^o}{\beta D_e^a + \alpha D_e^o} = \frac{D_e^a D_e^o}{X_O D_e^a + X_A D_e^o}$$

(12)

and

$$\Omega = \frac{\Omega_c}{\alpha + \beta}$$

(13)

where Ω_c is the volume of the molecule $A_\alpha O_\beta$. For boundary defect sources;

$$D_e^i = D_\ell^i + \frac{\pi \delta D_b^i}{d}$$

(14)

and for climb controlled diffusional creep:

$$D_e^i = D_\ell^i + \frac{4}{3} \left(\frac{\sigma}{G}\right)^2 D_c^i$$

(15)

When the two ions diffuse by different paths, the solutions have
only been approximate, Gordon (1973), but they appear to be
reasonable. Blakely and Li (1966) have shown, however, that where
volume diffusion is only important near the surfaces or boundaries,
the lattice diffusivities will be significantly changed by the
space charge cloud.

It can be seen that the creep rate will be controlled by the
slower diffusing species along its fastest path.[1] These relations
hold whether the diffusivities are extrinsic or intrinsic as long as
the slow species is properly identified.[2] Consequently, aleovalent
impurities which alter the diffusivity of the slow species will
similarly affect the diffusional creep rate.

Very significant effects on the creep rates can be achieved
by atmosphere changes which alter the stoichiometry of the base
material or alter important impurities. This has been demonstrated
for MgO and Al₂O₃ by Terwilliger et al (1970), and Hollenberg
and Gordon (1973), for transition metal oxides and UO₂ as reviewed
by Clauer et al (1971) and Seltzer et al (1971), and shown by
Crouch (1972) for Fe₂O₃. Processes such as precipitation or defect
association which alter the self diffusion coefficient will also
influence the creep rate. However, uncharged defects or complexes
which contribute to self diffusion, but not to electrical conduction
will likely not contribute to creep.

At steady state there is constant charge separation to provide
the required Nernst field. When diffusion occurs on alternative
paths for the two ions, further charge separation does not develop
since the flux accumulation everywhere on the boundaries will be
such as to satisfy the neutrality (stoichiometry) condition given
by Eqn.(11) as well as the strain compatibility requirements.
Lifshitz et al (1965, 1967) have shown that the required charge
separation results from modifications of the space charge cloud
and surface charge normally at each source; for most low stress
cases this will require only very small changes of the zero stress
values of the surface charge and space charge. In solving a
similar problem Charles (1969) has shown that the steady values of
the charge separation and of the field develop rapidly from

1
    It is seen that for A as the slow species the usual practice of
using $D \rightarrow D_e^a$ and $\Omega \rightarrow \Omega_a = \Omega_c/\alpha$ properly accounts for the ambipolar
coupling.

2
 Pasco and Hay (1973) have erroneously concluded that the extrinsic
diffusivity may not be the proper value; this resulted from neglecting the ambipolar coupling field which acts on all of the extrinsic
defects even though the concentration gradient given by equation (1)
is correct only for C_o taken as the thermal defects in excess of
those required to compensate impurities.

movement of the fast species with a relaxation time given by:

$$\tau_a = \frac{L^2}{D_d^f} \tag{16}$$

where L is the Debye distance, typically of the order of 10^{-6}-10^{-5}
cm for ceramics and D_d^f is the defect diffusivity of the fast species;
a similar result can be inferred for the present problem. Comparison with Eqn. (3) and (5) shows that the ambipolar coupling will
develop rapidly compared to the other transients. Redistribution of
impurities in the electric and stress fields will be much slower,
however.

 Although these derivations have been for ionic conductors,
they are also applicable to ceramics which are electronic conductors, since the same coupling field must be developed to obtain
the required flux equality. Any mobile electronic defects present
will move in response to the field. However, counter gradients in
the concentrations of holes or electrons will develop so that the
net flux in both electronic defects, which are also given by relations similar to Eqn. (10), will go to zero. This will require
larger values of the local nonstoichiometry (charge separation)
necessary to obtain the coupling field. For semiconductors (or
alloys) in which the conductivity is so high that electric fields
are non-existant, the same steady state solutions still apply, as
shown by Herring (1950), but the coupling force becomes the counter
gradient in chemical potential which developes from concentration
gradients. The more nearly the compound or alloy approaches an
ideal solution, the greater will be the local compositional deviations and the greater will be the transients, but the steady state
solutions remain the same.

 For $A\ell_2O_3$ it has been found, Paladino and Coble (1963), that:

$$D_\ell^o \ll D_\ell^a \qquad \text{and} \tag{17}$$

$$\delta D_b^o \gg \delta D_b^a \tag{18}$$

and so for the normal boundary mode:

$$D_{eff} = \frac{5(D_\ell^a + \dfrac{\pi \delta D_b^a}{d})\dfrac{\pi \delta D_b^o}{d}}{3(D_\ell^a + \dfrac{\pi \delta D_b^a}{d}) + 2\dfrac{\pi \delta D_b^o}{d}} \tag{19}$$

The net result is shown by a schematic plot of steady state strain rate versus grain size in Figure 3; here several values of the ratio of the anion and cation boundary diffusivities have been assumed to illustrate the effect on the creep rate kinetics which will be observed. At fine grain sizes creep is controlled by cation boundary diffusion; for coarse grain sizes it is controlled by oxygen boundary diffusion, and at intermediate grain sizes by cation lattice diffusion. The oxygen lattice diffusion is sufficiently slow that it likely will never be limiting except for extremely large grain sizes where dislocation mechanisms would become dominant. It can be seen that a region of clear dominance by cation lattice diffusion will only be found if the oxygen boundary diffusion is sufficiently larger than the cation boundary diffusion. In this case the total range of strain rate necessary to experimentally observe all three ranges would not be achievable at a single temperature.

There is currently little theory or comprehensive data to indicate the effects of impurities on boundary diffusion in Al_2O_3 except when high diffusivity impurity phases are present in the boundaries. Although there is not yet complete agreement regarding the lattice defect chemistry of Al_2O_3, Brook et al (1971) and Hollenberg and Gordon (1973) have shown that the assumption of Frenkel defects on the cation sub-lattice as the predominant defect provides the simplest explanation for impurity effects on ionic conduction and diffusional creep. The effect of impurities on the cation lattice diffusivity can be analyzed from the defect chemistry using this assumption. Neglecting defect complexes, divalent and trivalent cation impurities then lead to the following reactions:

$$3MO + Al_{Al} \rightarrow 3M'_{Al} + Al_i^{\cdot\cdot\cdot} \tag{20}$$

and

$$3NO_2 \rightarrow 3N_{Al}^{\cdot} + 6O_O + V_{Al}''' \tag{21}$$

and for extrinsic conditions charge neutrality requires for the two cases

$$[Al_i^{\cdot\cdot\cdot}] = \frac{1}{3}[M'_{Al}] \tag{22}$$

and

$$[V_{Al}'''] = \frac{1}{3}[N_{Al}^{\cdot}] \tag{23}$$

that is, in both cases the defect concentration is increased proportionally to the impurity additions. Diffusion results from

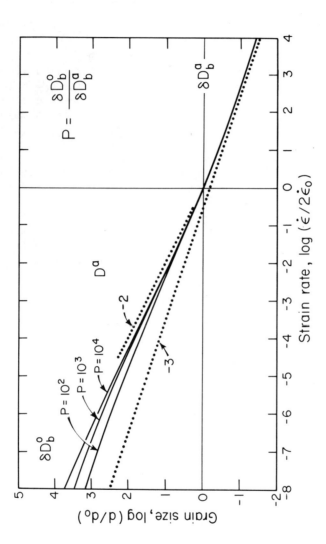

Figure 3   Schematic plot of strain rate versus grain size for Al_2O_3 showing the effective grain size dependence to be expected for various values of P, the ratio of the boundary diffusivities $\delta D_b^o/\delta D_b^a$. The dotted lines show the slopes of -2 and -3 expected for purely lattice or purely boundary control, respectively.

either cation interstitials or vacancies, depending upon which is in excess and faster. For additions of fixed valence, such as MgO, at values below the solubility limit the defect content will be fixed and the activation energy for creep or diffusion will be that for defect mobility.

For additions above the solubility limit the equilibrium relations must also be satisfied which are:

$$\frac{[Al_i^{\cdot\cdot\cdot}][M_{Al}']^3}{[MO]^3} = \frac{\frac{1}{3}[M_{Al}']^4}{[MO]^3} = e^{-\Delta G_1/kT} \tag{24}$$

giving

$$3[Al_i^{\cdot\cdot\cdot}] = [M_{Al}'] = (3)^{1/4}[MO]^{3/4}e^{-\Delta G_1/kT} \tag{25}$$

and for NO$_2$ additions

$$\frac{[V_{Al}''']\,[N_{Al}^{\cdot}]^3}{[NO_2]^3} = \frac{\frac{1}{3}[N_{Al}^{\cdot}]^4}{[NO_2]^3} = e^{-\Delta G_2/kT} \tag{26}$$

giving

$$3[V_{Al}'''] = [N_{Al}^o] \doteq (3)^{1/4}[NO_2]^{3/4}e^{-\Delta G_2/kT} \tag{27}$$

where ΔG_1 and ΔG_2 are free energies of solution. For this case the activation energy for creep or diffusion will include both the energy of solution, ΔH_i, and of defect mobility, ΔH_m, i.e.

$$\Delta H = \Delta H_m + \Delta H_i \tag{28}$$

Similar reasoning indicates that oxygen lattice diffusion by a vacancy mechanism, will be increased by divalent impurities and reduced by tetravalent impurities.

CREEP OF Al$_2$O$_3$

Relation of Polycrystalline and Single Crystal Deformation

Aluminum oxide is extremely anisotropic and it has been amply demonstrated that slip on the basal system is much easier than on

the non-basal system. Figure 4 presents the flow stress versus
temperature for the various slip systems from tests on single
crystals. The stresses plotted are tensile yield tests on single
crystals. The stresses plotted are tensile yield stresses for 60°
crystals for basal slip 1/3 $\leq 11\bar{2}0>$ (0001), for 90° crystals for
prismatic slip, 1/3 <1$\bar{1}$00>{11$\bar{2}$0}, and for 0° crystals for pyramidal
slip, 1/3 <01$\bar{1}$1> on one or both of {1$\bar{1}$02} or {10$\bar{1}$1}; these represent
Schmid factors of 0.4 to 0.5. In all cases lower yield stresses
are plotted except for the higher pyramidal values of Gooch and
Groves (1973) which are flow stresses. Activation analyses of the
single crystal yield behavior have not conclusively identified the
barriers which control slip. However, it has been inferred that
the lower yield stresses may be representative of the Peierls
stress for the non-basal systems, Tresseler and Michael (1974)
and Govorkov et al (1972), and for basal slip at lower temperatures,
Conrad (1965). For temperatures above about 1600°C climb around
obstacles may be limiting for basal slip, Conrad (1965) and
Govorkov et al (1972). Then it is still uncertain whether there
is an appreciable contribution to the apparent basal yield stress
from forest interactions which would not be present in fine grained
materials where annihilation at the boundaries prevents a significant
increase in dislocation density. All of the data plotted are for
nominally pure samples, the significant hardening from small alloying additions reported by Radford and Pratt (1970) is very likely
from precipitates and so would not affect polycrystalline grains
where the precipitates are typically at triple points. The basal
and pyramidal flow stresses have strain rate sensitivities in the
range n = 4-7 and so all the data in Figure 4 were taken for
$\dot{\varepsilon}$ = 4 x 10^{-5}sec^{-1}. The strain rate dependence for the prismatic
yield stresses is quite small, Gooch and Groves (1973).

Steady state flow stresses for 1 and 13μ polycrystalline
material at the same strain rate are also plotted. For the 13μ
material and for 7 to 34μ material, Folweiler (1961) demonstrated
that the kinetics indicated diffusional creep for the entire range
of stress investigated from considerably below to well above the
basal yield stress. The stress exponents were typically n = 1.1-
1.2 which is consistent with an increasing contribution of basal
slip at higher stresses. The stress and grain size dependence
reported by others, particularly W. R. Cannon (1971), confirm
this same behavior for grain sizes up to at least 50μ. The 1μ
material from Heuer et al (1969) and R. M. Cannon et al (1975)
exhibited conformance to diffusional creep kinetics at the high
stresses, with n decreasing to less than 1.1 as the stress increased
up to the highest achievable values of nearly 50 ksi ($\sigma/G \approx 2$ x 10^{-3}).
At the lower stresses the strain rate became slower than for a
linear dependence and stress exponents increased to n $\approx$ 2 at the
lowest strain rates achieved, indicating a shift to control by one
of the interfacial processes. The few data points at stresses
above 100,000 psi for polycrystalline materials were reported by

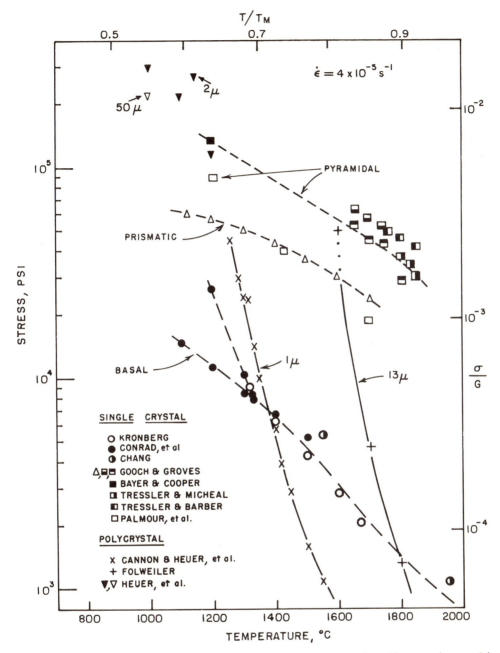

Figure 4 Flow stress versus temperature curves for the various slip systems in sapphire and for fine grained polycrystalline Al_2O_3. The value of G was taken at 1500°C. The polycrystalline samples, ∇, $\blacktriangledown$, tested under hydrostatic constraint were at $\dot{\varepsilon} = 10^{-6} s^{-1}$.

Heuer et al (1970, 1971) and Snow (1972) from tests performed under
very high hydrostatic constraints to eliminate cracking. The
initial yield stresses are plotted and are seen to be in agreement
with the extrapolated curves of pyramidal slip as would be required
for slip on five independent systems.

Examination by transmission microscopy of polycrystalline
Al_2O_3 after creep has been reported by Heuer et al (1969, 1970,
1971), Snow (1972), Becker (1971), and R. M. Cannon et al (1975).
In all cases few dislocations were found within grains for samples
where the flow stress conformed approximately to that predicted
by Eqn.(6), even when the flow stress was above that for basal
yield. There is however, evidence to indicate that slip, particularly on the basal system, does occur at stresses above the basal yield,
but that the boundaries serve as the primary sources and sinks for
dislocations and that there is little increase in the dislocation
density during creep. Appreciable increases in dislocation density
are apparent after several percent creep, in coarser grained materials, ($\sim 100\mu$), W. R. Cannon (1971), or in fine grained materials
at stresses approaching the prismatic flow stress, Becker (1971).
The hydrostatically constrained polycrystalline samples showed
high dislocation densities with typical cold worked structures of
dislocations in tangled arrays; both basal and pyramidal dislocations were present.

Fine Grained, MgO Doped Al_2O_3

Since the original work by Folweiler (1961), which demonstrated
that deformation of fine grain Al_2O_3 was controlled by diffusional
creep over a wide range of temperature and stress, there have been
more than a dozen additional creep studies on Al_2O_3 most of which
tended to confirm the original finding but raised questions as to
the controlling diffusivity and path and the effects of impurities.
All the results must be considered together to obtain a consistent
view of the controlling species.

The spurious effects of unknown impurities can be suppressed
by initially considering only those materials doped with MgO which
will dominate the defect structure. The controlling diffusivity can
be inferred by assuming one species and diffusion path to be dominant and calculating the apparent diffusivity from the stress-strain
rate data using Eqn.(6). This has been done alternately assuming
cation lattice control and then cation boundary control. The
results are plotted in Figures 5 and 6 respectively for all of the
available creep data on MgO doped Al_2O_3 with grain sizes below 70μ
and densities over about 97% of theoretical. For these conditions
the kinetics have been shown to have the nearly linear stress dependence and the strong grain size dependence required by the diffusion creep theories; the studies where this was not directly
demonstrated have been under conditions where it would be expected.

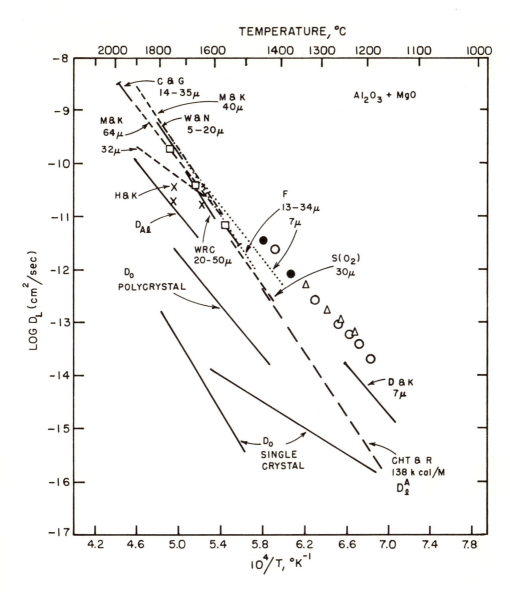

Figure 5 Comparison of self-diffusion measurements and apparent
lattice diffusivities calculated from creep tests of Al₂O₃ doped
with MgO from the results of Folweiler (1961), Warshaw and Norton
(1962), Coble and Guerard (1963), Stuart (1964), Dawihl and
Klingler (1965), Hewson and Kingery (1967), Mocellin and Kingery
(1971), W. R. Cannon (1971) and R. M. Cannon et al (1975), represent
1.2μ - O, Δ, 2.7μ - ●, and 15μ -□ grain sized samples.

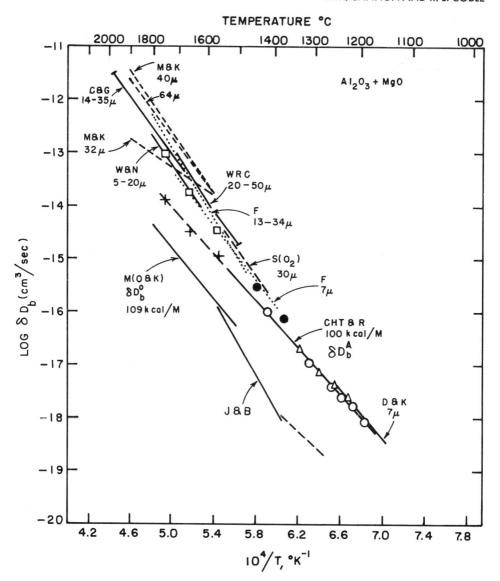

Figure 6   Comparison of apparent cation boundary diffusivities
calculated from creep tests of Al_2O_3 doped with MgO and compared
with boundary diffusivities inferred from sintering by Johnson
and Berrin (1967) and from self-diffusion measurements by Mistler
(1967).   The + points represent the contribution of cation boundary
diffusion to the creep rate of the 15μ samples obtained by sub-
tracing the contribution from lattice diffusion calculated from
the curve marked D_ℓ^A in Figure 5.

The diffusivities for the 1µ material of R. M. Cannon et al (1975)
were calculated from the high stress data where the behavior was
indicated to be in the Newtonian, diffusion limited regime.

R. M. Cannon et al (1975) analysed their data, which covered
1-15µ grain sizes, assuming both cation boundary and lattice diffusion to be important. The results indicated cation boundary control for the 1µ material and cation lattice control for the larger
grained materials, but with a contribution from boundary diffusion
especially at the lower temperatures. The lines marked D_ℓ^a and δD_b^a
on these and the following figures are the resultant values of
aluminum diffusivity indicated by this analysis. But for a few important exceptions, it can be seen that all of the available data
for coarser grained material agree with this inferred value of D_ℓ^a
to within a factor of two.[3]

One major exception is the data of Mocellin and Kingery
(1971). The 32µ samples had an MgO concentration of 0.043% by
weight which is below the solubility limit determined by Roy and
Coble (1968). At the high part of the temperature range investigated the 64µ sample with 0.083% MgO were also below the solubility
limit. The remainder of the data for MgO doped Al₂O₃ in Figures
5 and 6 was above the solubility limit. Although the other data
represent a considerable range of MgO content, there is no apparent
effect when it exceeds the solubility limit. Except when variable
valence impurities were present there is no effect of test atmosphere
apparent. Two of the three data points of Hewson and Kingery (1967)
do not exhibit a good fit; these had been reported to demonstrate
that increasing the MgO content reduced the creep rate. This hypothesis must be rejected on the basis of the considerable other
evidence to the contrary, and further since the MgO concentrations
for these samples were above the solubility limit and so should
have had the same MgO content in the lattice.

The second exception is the 7µ material which at low temperatures shows an increasing deviation due to the contribution of
boundary diffusion which has a lower activation energy. This contribution can be seen in Figure 6 where the data indicate that
below 1300°C the 7µ material, is dominated by boundary diffusion;
the data agree well with the 1µ results. One of the sets of data
for the 1µ material, represented by Δ, are for a high purity,
undoped material. Surprisingly, there seems to be no difference in
the cation boundary diffusivity compared to that for the MgO doped
materials. The δD_b values calculated for the coarser grained data
scatter above the δD_b^a line indicating the decreasing contribution

---

[3]   The scatter of nearly two orders of magnitude previously seen in
such complications for Al₂O₃ was reduced by using a consistent value
of 3/2 L for the grain size, and further rechecking the calculations
to ensure that the same values of the coefficients and of Ω were
used; see the appendix for further details.

of cation boundary diffusion. The data for the coarsest grained
samples do not appear to be converging in a manner which would
conclusively indicate they are significantly limited by boundary
oxygen diffusion and so they only provide an indication of a lower
bound to the boundary oxygen diffusivity.

Thus, for grain sizes below that where oxygen diffusion
becomes limiting, the diffusional creep of Al_2O_3 doped with MgO
(or presumably other divalent impurities) in excess of the solubility
limit is described by a combination of cation lattice and cation
boundary diffusivities with activation energies of 138 and 100
kcal/mole, respectively. As expected, this value of lattice diffusivity is higher than the measured self diffusion coefficient for
oxygen, Oishi and Kingery (1960). The lattice diffusivity and its
activation energy are also higher than the measured aluminum self
diffusion coefficients for undoped Al_2O_3 (114 kcal/mole), Paladino
and Kingery (1962), indicating that the cation lattice diffusivity
is increased by the MgO dopant.

Fine Grained Al_2O_3, Other Dopants

There are several studies of creep in Al_2O_3 with other dopants
which are compared in a similar manner in Figures 7 and 8. Hollenberg and Gordon used both Fe and Ti dopants and systematically varied
the P_{O_2} to change the Fe^{+2}/Fe^{+3} or the Ti^{+4}/Ti^{+3} ratios. Limited
data on different grain sizes indicated that both these materials
were in a range of lattice diffusion control. It can be seen from
Figure 7 that the creep rates increase as the amount of either
divalent or tetravalent impurity is increased. Analysis by these
authors showed that the atmosphere dependence was inconsistent
with lattice oxygen control or with a Schottky defect model for
cation diffusion. The results are satisfactorily explained by a
Frenkel defect on the cation sublattice which results in an increased defect concentration from either aleovalent impurity. Some
of the Fe doped samples tested at low stresses exhibited n values
very close to one as predicted by theory; however, many of the Fe
doped and Ti doped samples had n = 1.1-1.3; it is not clear whether
this represents a contribution from slip or a shift to interface
controlled processes because of the high diffusivities obtained by
doping.

Also shown are data for Ni doped Al_2O_3. The tests of Crosby
and Evans (1973) were in a strongly reducing atmosphere and the
high lattice diffusivities indicate the presence of considerable
Ni^{+2} in solution. The inferred value of D_ℓ is qualitatively consistent with the results for Fe doped Al_2O_3. Sugita and Pask (1970)
tested in air and so most of the Ni in solution may have been trivalent. These data for 3μ samples indicate strain rates higher
that predicted by lattice diffusion, but which agree closely with

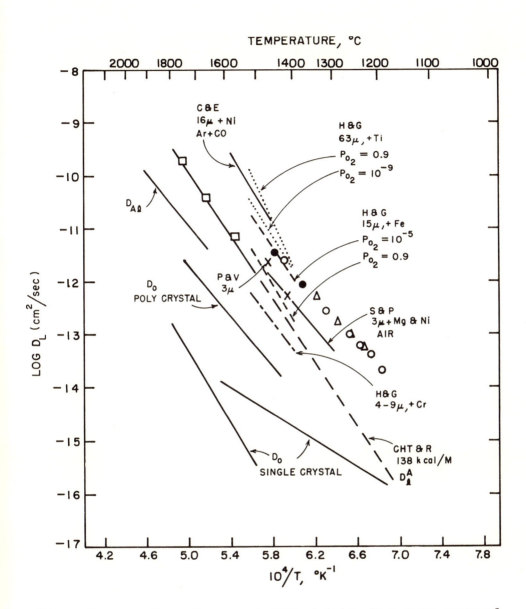

Figure 7 Comparison of apparent diffusivities from creep tests of
Al$_2$O$_3$ with various additives from the results of Passmore and
Vasilos (1966), Sugita and Pask (1970), Hollenberg and Gordon (1973)
and Crosby and Evans (1973).

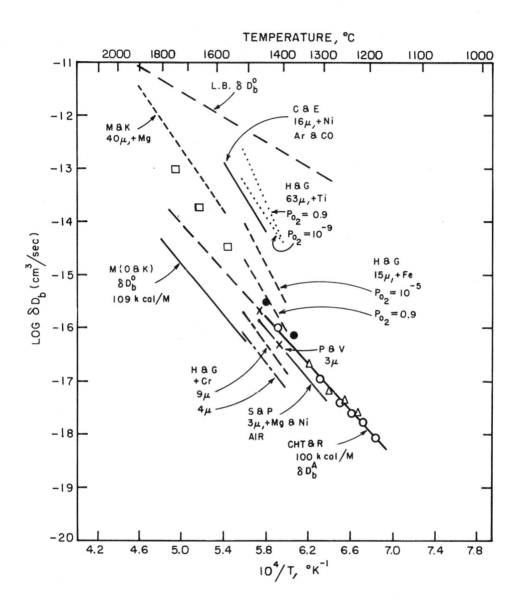

Figure 8   Comparison of apparent boundary diffusivities from creep tests of Al_2O_3 with various additives.

the previous boundary diffusion values. These investigators observed no grain size dependence as a result of grain growth during
testing. This comparison, however, indicates the magnitude of the
creep rates can be satisfactorily explained by diffusional creep;
it is likely that the considerable cavitation which occurred during
testing obscured the effects of grain growth on the creep rate.

It may be expected that undoped material or material doped
with trivalent impurities such as Cr_2O_3 should exhibit lower
strain rates. This has been reported by Hollenberg and Gordon
(1973) as seen in Figure 7. These data indicated a squared grain
size dependence, consistent with lattice diffusion, and a lower
activation energy. The calculated diffusivity agrees quite well
with the extrapolated values of Al self diffusion. It is, however,
highly unlikely that these values represent intrinsic diffusion,
but is more likely that the diffusivity is extrinsic and controlled
by the other impurities which are present. Perhaps some caution
should be taken in this interpretation, however, since the strain
rates are slower than would be predicted for boundary diffusion
as shown in Figure 8. This may indicate that the cation boundary
diffusivity is reduced by the Cr_2O_3; alternatively the fact that
these data exhibited a strain rate sensitivity of n = 1.3 may
suggest that the creep rate has been retarded because of an interface control problem.

In general, interpretation of creep data for undoped materials
is difficult because of the uncertainty about their impurity content
amd its effect on the diffusivity. The data of Crosby and Evans
(1973) for undoped material (not shown) gave lattice diffusivities
higher than expected for MgO doping suggesting the presence of
considerable low valence impurity as a result of the reducing
atmosphere. The undoped Al_2O_3 tested by Engelhardt and Thümmler
(1970) which had 5% porosity and showed a linear stress dependence
at low stresses give a lattice diffusivity (not shown) which is
very close to that for MgO doped Al_2O_3 suggesting, again, the
presence of appreciable divalent impurity. The test results of
Passmore and Vasilos (1966) for undoped 3μ material exhibit good
agreement with the cation boundary diffusivity found in Figure 6
for both MgO doped and undoped Al_2O_3

Coarse Grained Al_2O_3

In an attempt to find evidence for control by oxygen boundary
diffusivity, the inferred diffusion coefficients were calculated
for all the available test data for material of 75μ or larger.
These data are shown in Figure 9. It can be seen that none of the
values fall below the cation lattice diffusivity as would be expected if they were in a diffusional creep regime with limitation by
oxygen diffusion. Instead the results indicate creep rates equal to
or faster than those predicted by diffusional creep. W. R. Cannon

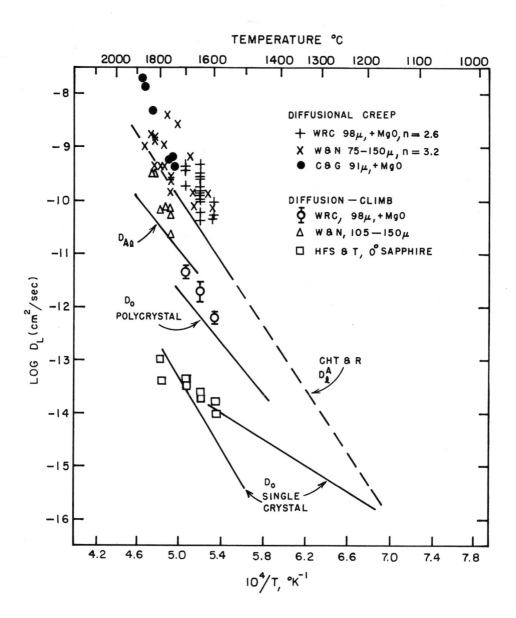

Figure 9   Apparent lattice diffusivities inferred from tests of
course grained Al_2O_3 alternatively assuming the normal diffusional
creep model and the dislocation climb models.

(1971), and Warshaw and Norton (1962) observed strain rate sensitivities of 2.6 and 3-4 indicating a different mechanism than simple diffusional creep. Significant cavitation was observed in the coarse grained specimens of Warshaw and Norton (1962) and Coble and Guerard (1963) which contributed at least in part to the high strain rates. However, the results of W. R. Cannon (1971) cannot be similarly discounted since they were done in compression and indicated very little density change. Further, these specimens provided etch pit evidence of slip and slip bands and microscopic evidence for grain boundary sliding.

W. R. Cannon and Sherby (1973) suggested that the strain rate dependence of n ≃ 3 may indicate a contribution of diffusional creep by the Nabarro climb mechanism. This seems plausible since the stresses used were below those necessary for non-basal slip although they were above the basal flow stresses. The required diffusivity for deformation by the climb mechanism has been calculated using equation (8) with β = 1 for all of these data. Some of the results are shown in Figure 9. The results for the data of W. R. Cannon (1971) fall considerably below the D_ℓ^a line as anticipated since oxygen lattice diffusion should be rate controlling based on the measured stress exponent. These values are, however, considerably above the self diffusion values for oxygen. The calculated values for the other creep data were even higher presumably due to cavitation with many of them above the D_ℓ^a line; only those for the coarser grain samples of Warshaw and Norton are plotted. The calculated diffusivities must also be high because of contributions from grain boundary sliding, from slip, and perhaps from normal diffusional creep. In addition, the MgO addition would be expected to increase the oxygen diffusivity compared to that for the undoped Al_2O_3 used for the self diffusion measurements.

Heuer et al (1971) reported test results for 0° sapphire at stresses below most of the measured values for pyramidal slip. The measured stress exponents were in the range of 3 to 5 which are also below those found by Gooch and Groves (1974), Tressler et al (1973, 1974) and Palmour et al (1968). Heuer et al (1971) suggested that the observed creep rates may involve a significant contribution from the Nabarro climb mechanism. The required diffusivities, calculated using Eqn.(8), and are shown in Figure 9. The agreement with measured oxygen diffusivities of Oishi and Kingery (1960) is remarkable.[4] The low activation energy and the tendency toward high values of the stress exponent at lower temperatures suggest a

---

[4] The fact that the calculated values of D_ℓ are lower than D_ℓ^a from measurements of self diffusion or diffusional creep of medium grained materials provides a further confirmation of the hypothesis that for normal diffusional creep the oxygen diffusivity at the boundaries is faster than the cation diffusion.

contribution of pipe diffusion, particularly at the lower temperatures; self diffusion measurements of Anderson (1967) also indicated pipe diffusion to be important for oxygen, A further uncertainty in the quantitative confirmation of this mechanism, arises
from the microscopic examination of the sapphire samples which showed
evidence of slip on non-basal systems and also indicated that the
density of non-basal dislocations was $\sim 6 \times 10^7 cm^{-2}$ which was
more than an order of magnitude below that predicted by Eqn.(9)
for the stresses used.

DIFFUSIVITIES

Cation Lattice Diffusivity

The creep results indicate that the cation diffusivity is
extrinsic for all the materials tested, and that it is increased
by both divalent and tetravalent impurity additions which is at
least qualitatively consistent with a cation Frenkel defect
model in which cation diffusion occurs by both vacancies or interstitials. Since the MgO solubility in Al_2O_3 has been measured,
Roy and Coble (1968), and shown to have an apparent activation
energy of 61 kcal/mole, the activation energy for cation defect
diffusion can be determined from the creep data for MgO doped
Al_2O_3.

It can be determined separately from the data of Mocellin
and Kingery (1971) for unsaturated MgO additions, and, using
Eqn. (28), from the diffusivity determined for the MgO saturated
material which was

$$D_\ell^a = 1.36 \times 10^5 \exp\left(\frac{-138,000 \text{ cal/mole}}{RT}\right) cm^2/s \qquad (29)$$

These values are compared in Table I with other available determinations of the cation mobility from the literature.

Table I

Cation Mobility Activation Energy

Method	ΔH kcal/mole
Creep - MgO Saturated	77
Creep - MgO Unsaturated	63
Ionic Conductivity, (Kitazawa & Coble (1974)	57
Ionic Conductivity, (Brook et al (1971)	67
Color Boundary Migration, (Jones et al (1969)	80

These values which are all between 57 and 80 kcal/mole, probably represent both interstitial and vacancy activition energies.
These may be similar in Al_2O_3, although somewhat higher values
would be expected for vacancy mobility. The color boundary mobility was determined from an oxidation experiment in which cation
vacancies would seem more likely to be the dominant defect. The
14 kcal/mole difference between the unsaturated and saturated creep
results may be a real effect due to dissociation of defect complexes which would be more important in the saturated samples with
higher concentrations of MgO. The ionic conductivity measurements
were both for oxidizing conditions which would be metal deficient
for pure Al_2O_3; however, the value reported by Brook et al (1971)
includes results from divalent impurity doped crystals in which
interstitials were expected to be dominant.

Reference to Figure 7 shows that for the Fe, Ni and Ti dopants the activation energies for lattice diffusivity were also in
the range of 140-160 kcal. These cases do not represent precipitation, but rather a temperature dependence of the impurity
valence state, which in a similar manner results in a significant
contribution to the activation energy as has been shown in greater
detail by Hollenberg and Gordon (1973). For the Ti doped material
tested in high P_{O_2} precipitation was also occuring resulting in an
apparent activation energy of about 205 kcal/mole. Ti, Bagley
et al (1970), and Fe, Rao and Cutler (1973), have been reported to
increase the rate of initial sintering of Al_2O_3, whereas MgO,
Jorgenson (1965), has been observed to slow the initial rate. The
cause of these differences in behavior compared to creep are uncertain as are some differences in the magnitudes of boundary and
lattice diffusivities deduced from sintering. The different defect
concentrations in the space charge region near the boundaries, may
account for some of the differences.

Boundary Diffusivities

For the MgO doped materials and some undoped Al_2O_3 the cation
boundary diffusivity was found to be:

$$\delta D_b = 8.60 \times 10^{-4} \exp\left(\frac{-100,000 \text{ cal/mole}}{RT}\right) \text{ cm}^3/s \qquad (30)$$

It may be surprising that such close agreement for the boundary diffusivity should be found for Al_2O_3 with and without the MgO doping
and for both sintered and hot pressed materials. The spinel phase
which forms is non-wetting and is found as small grains at the
triple points and so does not form a grain boundary film. The
undoped, 1μ material used by R. M. Cannon et al (1975) had only
100-300 ppm total cation and anion inpurities; if this is assumed
to be uniformly distributed around the grain boundaries it re-

presents less than 1/2 of a monolayer for a 1μ grain size and so cannot account for a second phase film at the boundaries. For the other 1-3μ materials the background impurity content is only 2 to 3 times the above value and so a boundary film would also not be expected. For the sintered materials of 7μ or greater similar grain boundary diffusivities suggest that high diffusivity boundary films are also not important. It may be that most of the SiO_2 impurity is taken into solution due to compensated solubility with the excess MgO; this would reduce the amount of low melting grain boundary phase to a sufficiently low level that it is not important. For undoped materials with higher impurity levels, boundary films could be expected to contribute to high values of cation boundary diffusivity.

Examination of all of the available data for MgO doped Al_2O_3 fails to provide a conclusive indication of a shift to control by oxygen diffusion. The doped materials of Hollenberg and Gordon (1973) similarly did not indicate a shift to control by oxygen diffusion. Crosby and Evans (1973) reported a grain size dependence of 2.7 for the undoped materials which were from 15-45μ, which may indicate a contribution of boundary diffusion, however there were insufficient data to indicate whether it is from enhanced cation boundary diffusion or from oxygen boundary diffusion control. The highest calculated values of δD_b for the coarsest materials shown in Figure 8 gave an indication of the magnitude of δD_b^O which must exist. The line shown as a lower bound to δD_b^O was calculated assuming that at the highest apparent values of δD_b^a the oxygen and aluminum diffusivities were just equal for the data of Mocellin and Kingery (1971) and the Ti doped Al_2O_3 of Hollenberg and Gordon (1973).

It is not obvious, of course, that δD_b^O will be the same for the various dopants and so the indicated value is only an approximate bound. For the material with divalent dopants the amount of silicate phase at the boundaries would likely be reduced due to compensating solubility effects. For the Ti doped material this would not be the case. Recently Gordon (1974) reported $\delta D_b^O = 3.4 \times 10^{-14}$ at 1450°C from coarse grained Fe doped Al_2O_3. This value would suggest higher values of δD_b^O for Ti doped Al_2O_3 than for divalent doped Al_2O_3. These results do indicate that δD_b^O is at least 10^2 times δD_b^a.

There is very little other data on δD_b for Al_2O_3 with which to compare these results. On Figure 6 and 8 are shown a value of δD_b^O with ΔH = 109 kcal/mole which was calculated by Mistler (1967) from the self diffusion data of Oishi and Kingery (1960) using a Harrison volume fraction model. It is apparent that this value is too low indicating the simple model was not applicable and that the effective diffusivities measured were in a range where the boundaries were saturated and lattice diffusion was limiting. Johnson and Berrin (1965) deduced values of δD_b from sintering data and obtained values, Figure 6, which were lower in magnitude and higher in activation energy than the values of δD_b^a found here.

These values may be indicative of cation boundary diffusion, but are likely in error due to neglect of surface transport in the analysis. Kitazawa and Coble (1974) measured values of the defect diffusivity of oxygen in the grain boundaries by a permeativity technique; they found values much higher than those indicated here for δD_b^o and with activation energies of 20-50 kcal/mole. Further, they found a marked dependence of MgO doping on the apparent oxygen defect diffusivity.

There are four different determinations of the surface diffusivity for Al_2O_3. Robertson and Chang (1966), Robertson and Eckstrom (1969) and Shackelford and Scott (1968) determined wD_s from grain boundary grooving measurements. Yen and Coble (1972) determined wD_s from pore spheroidization measurements. These results gave activation energies from 75 to 130 kcal/mole, but were within an order of magnitude of each other in absolute magnitude over most of the temperature range. Most of these results lie in a band which ranges from 2 to 10 times higher than the δD_b^a values shown in Figure 8, and are below the δD_b^o value. It is likely that all of those represent the surface diffusion of Al as the slower species. Chang (1967) and Robertson (1969) have pointed out that other oxides, especially UO_2 for which there is good data, also exhibit a similarity between the values of surface and grain boundary cation diffusivity.

The absolute value indicated for δD_b^o is quite high, i.e. in the range of 10^{-12}cm³/s. A value of δ of 10^{-5}cm which may be possible for the space charge cloud would infer a value of D_b^o of 10^{-7}cm²/s which seems very high for transport in unfaulted lattice. Alternatively, assuming δ of 10^{-7}cm, representative of the structurally disordered region, would infer a value of D_b^o of 10^{-5}cm²/s. This value is typical of that expected for a liquid, and so may be plausible for the distorted, core region of the boundary. It seems likely that there must be a significant contribution to the boundary diffusivity from the core. A similar analysis for the cation boundary diffusivity gives D_b values which are several orders of magnitude lower. These may still suggest a considerable contribution of the distorted region to the boundary diffusivity, because for samples below the isoelectric temperature the cation interstitial concentration is depressed in the space charge cloud around the boundary, Kingery (1974). The activation energy of 100 kcal/mole is appreciably higher than that for cation mobility in the lattice and so implies a significant temperature dependence of the boundary defect concentration or structure.

SUMMARY

Diffusional creep with the grain boundaries as defect sources has been analysed to a high degree of rigor compared to other creep theories; the modifications necessary for application to steady

state creep of ceramics are direct and easy to apply. The major
uncertainty in the theory remains the exact value of the numerical
coefficient due to uncertainty about the extent of grain rearrangement which may occur. For the Nabarro climb model for diffusion
with dislocation sources there is greater uncertainty as to the
detailed accuracy of the theoretical creep rate equations.

Analysis of the creep rate for Al_2O_3 indicates that for low
stresses the creep is dominated by diffusional creep. At intermediate stresses, which are above the basal yield, but below the
much higher non-basal yield stresses, slip will occur on favorably
orientated basal slip systems. However, as earlier suggested by
Groves and Kelly (1969) and Gooch and Groves (1974 b), it appears
likely that much of the strain required to satisfy the von Mises
condition results from diffusional creep. For medium and fine
grained materials this is still the normal grain boundary controlled diffusional creep. In this case the kinetics appear to be
reasonably well described by the theoretical constituitive
equations, although the contribution of basal slip results in a
small increase in the creep rate which causes stress exponents
slightly greater than unity. The regime of interface control,
where diffusion is not the rate limiting step, is only partially
understood, but it can be a problem at very fine grain sizes, or
when diffusion is very rapid, giving rise to stress exponents
appreciably greater than one. For coarser grained materials diffusional creep by the Nabarro climb mechanism becomes faster. A
constituitive relation which properly accounts for the contributions of glide, climb, and grain boundary sliding does not
appear to be available, although Eqns.(7) or (8) provide ap-
proximations. Only at stresses approaching the non-basal yield
stresses which are between σ/G of 10^{-3}–10^{-2} does it appear that
polycrystalline deformation without cracking can occur without an
important contribution from diffusional creep, even at temperatures approaching the melting point.

When analyzed together, the creep results of fine and medium
grained Al_2O_3 are in satisfactory agreement with each other and
with the predictions of the diffusional creep relations. For MgO
doped Al_2O_3 control is by cation diffusion at least up to a 60μ
grain size. The lattice diffusivity appears to be enhanced by MgO
additions, and when the MgO is above the solubility limit the creep
data indicated:

$$D_\ell^a = 1.36 \times 10^5 \exp\left(\frac{-138,000 \text{ cal/mole}}{RT}\right) cm^2/s$$

The boundary diffusivity is insensitive to MgO doping and is:

$$\delta D_h^a = 8.60 \times 10^{-4} \exp\left(\frac{-100,000 \text{ cal/mole}}{RT}\right) cm^3/s$$

The hypothesis, initially advanced by Paladino and Coble (1963), that boundary oxygen diffusion is rapid relative to aluminum diffusion is confirmed by the present analysis.

ACKNOWLEDGEMENTS

The authors acknowledge with pleasure many constructive and pertinent discussions with W. H. Rhodes and A. H. Heuer. Thanks are also extended to K. H. Bowen, M. F. Yan and C. F. Yen for critical comments regarding the manuscript. Our appreciation is also extended to A. Mocellin for helping to correct an error in his reported data. Finally the sponsorship of the AEC, under Contract No. AT(11-1)2390 is gratefully achnowledged.

APPENDIX

There has been a signifiacnt scatter in the values of D_ℓ and δD_b reported by various investigators which has been due in large part to differences in the values of the coefficients and of Ω used for the strain rate equation. For all the curves of D_ℓ and δD_b in Figures (5-9) the coefficients in Eqn.(6) were used with $\Omega_a = 2.12 \times 10^{-23} cm^3$. For the values of D_ℓ calculated for the climb mechanism Eqn.(8) was used with $\Omega_o = 1.52 \times 10^{-23} cm^3$ and $b = 4.93 \times 10^{-8} cm$ which is the average of 1/3 <11$\bar{2}$0> and 1/3 <01$\bar{1}$1>. All grain sizes used in calculations or in the text were converted to $G = 1.5\bar{L}$, if $\bar{L}$, the mean linear intercept, had been used by the original author.

Much of the confusion has persisted because the values used were often not reported or were misrepresented and often no strain rate data were reported which would allow a check. In most cases the diffusivities were recalculated from strain rate data. However, as this data often was in theses which are not generally available, the following summarize the corretions made to the various sets of results.

Cannon, R. M., et al (1975): Used same values

Cannon, W. R., (1971): Grain size measurements from several micrographs indicated the values reported were $\bar{L}$. Changing coefficients and using G gave $D_\ell = 1.61D$ (Reported).

Coble and Guerard, (1963): Re-examination of the original data indicated a factor of 12 error in most of the strain calculations. Recalculation brought the data for 14-35μ samples of MgO doped, MgO - CrO₃ doped, and Lucalox into good agreement with an average value of D_ℓ about 12 times that previously reported.

Crosby and Evans, (1973): Calculations were from reported $\dot{\varepsilon}$ and G data.

Dawihl and Klingler, (1965): Strain rates and D_ℓ were calculated
     from G and deflection curves.
Folweiler, (1961): D_ℓ calculated from $\dot{\epsilon}$ and G data are about 1/3
     those reported. R. F. personally confirmed G = 1.5$\underline{L}$
     has been reported.
Hewson and Kingery, (1967): In the M.S. Thesis of C. W. H. it is seen
     that (G/2) and Ω_a has been used in the calculations. Measure-
     ments from micrographs indicated G was reported. Correcting for
     the proper coefficient gave D_ℓ = 2.86D (Reported).
Hollenberg and Gordon, (1973): Calculations from the $\dot{\epsilon}$ and G
     data in the Ph.D. Thesis of G. W. H. generally gave similar
     results to the reported values of D_ℓ.
Mocellin and Kingery, (1971): A. M. personally confirmed Ω_a and G
     had been used; however, the reported value of δD_b was in error
     by a factor of 10 too low. This is corrected along with small
     changes for the coefficients used in Eqn. (6).
Stuart, (1964): Measurements of micrographs (3rd Quarterly Report
     indicate that $\underline{L}$ was reported. Calculations then confirmed
     D_ℓ = 4.29 D (Reported).
Sugita and Pask, (1970): Measurements of Micrographs indicated G was
     reported; D_ℓ was calculated from $\dot{\epsilon}$ and G data.
Warshaw and Norton, (1962): In the Ph.D. Thesis of S. I. W. it is
     seen that (L/2) was used in the calculations; measured from
     micrographs confirmed that $\underline{L}$ was reported. Calculations from
     $\dot{\epsilon}$ data confirmed D = 4.24 D (Reported).

REFERENCES

Anderson, R. L., (1967), "Measurement of Oxygen Self-Diffusion in
     Single Crystal Aluminum Oxide by Means of Proton Activation,"
     Sc.D. Thesis, M.I.T., Cambridge, Mass.
Ashby, M. F., (1969), "On Interface-Reaction Control of Nabarro-
     Herring Creep and Sintering," Scripta Met., 3, 837.
Ashby, M. F., Raj, R. and Gifkins, R. C., (1970), "Diffusion-
     Controlled Sliding at a Serrated Grain Boundary," Scripta
     Met., 4, 737.
Ashby, M. F. and Verrall, R. A., (1973), "Diffusion-Accomodated
     Flow and Superplasticity," Acta Met., 21, 149.
Bagley, R. D., Cutler, I. B. and Johnson, D. L., (1970), "Effect
     of TiO_2 on Initial Sintering of Al_2O_3," J. Am. Ceram. Soc.
     53, 136.
Balluffi, R. W. and Seigle, L. L., (1957), "Growth of Voids in
     Metals During Diffusion and Creep," Acta Met., 5, 449.
Bardeen, J. and Herring, C., (1951), "Diffusion in Alloys and the
     Kirkendall Effect," in Atom Movements, A.S.M., Cleveland.
     Ohio, p. 87.
Bayer, P. D. and Cooper, R. E., (1967), "A New Slip System in
     Sapphire," J. Mater. Sci., 2, 301.

Beauchamp, E. K., Baker, G. S. and Gibbs, P., (1962), "Impurity Dependence of Creep of Aluminum Oxide," Technical Report ASD TR 61-481, Contract AF 33 (616)-6832.

Becker, P. F., (1971a), "Deformation Substructure in Polycrystalline Alumina," J. Mater. Sci., $\underline{6}$, 275.

Becker, P. F., (1971b), "Deformation Behavior of Alumina at Elevated Temperatures," in Ceramics in Severe Environments, Mater. Sci. Res. V5, ed. W. W. Kriegel and H. Palmour III, Plenum Press, New York, p. 315.

Bird, J. E., Mukherjee, A. K.and Dorn, J. E., (1969), "Correlations Between High-Temperature Creep Behavior and Structure," in Quantitative Relation Between Properties and Microstructure, ed. D. G. Brandon and A. Rosen, Israel U. Press, p. 255.

Blakely, J. M. and Li. C. Y., (1966), "Changes in Morphology of Ionic Crystals Due to Capillarity," Acta. Met., $\underline{14}$, 279.

Brook, R. J., Yee, J. and Kröger, F. A., (1971), "Electrochemical Cells and Electrical Conduction of Pure and Doped Al_2O_3," J. Am. Ceram. Soc., $\underline{54}$, 444.

Burton, B., (1972), "Interface Reaction Controlled Diffusional Creep: A Consideration of Grain Boundary Dislocation Climb Sources," Mater. Sci. Eng., $\underline{10}$, 9.

Burton, B. and Greenwood, G. W., (1970), "The Contribution of Grain-Boundary Diffusion to Creep at Low Stresses," Met. Sci. J., $\underline{4}$, 215.

Burton, B. and Reynolds, G. L., (1973a), "The Diffusional Creep of Uranium Dioxide: Its Limitation by Interfacial Processes: Acta Met., $\underline{21}$, 1073.

Burton, B. and Reynolds, G. L., (1973b), "The Influence of Deviations from Stoichiometric Composition on the Diffusional Creep of Uranium Dioxide," Acta Met., $\underline{21}$, 1641.

Cannon, R. M., Heuer, A. H., Tighe, N. J. and Rhodes, W. H., (1975), "Plastic Deformation in Fine-Grained Alimina," to be published.

Cannon, R. M. and Rhodes, W. H., (1970), "Deformation Processes in Forging Ceramics, Summary Report, Contract NASW-1914.

Cannon, W. R., (1971), "Mechanisms of High Temperature Creep in Polycrystalline Al_2O_3," Ph.D. Disertation, Stanford, University.

Cannon, W. R., (1972), "The Contribution of Grain Boundary Sliding to Axial Strain During Diffusional Creep," Phil. Mag., $\underline{25}$, 1489.

Cannon, W. R. and Nix, W. D., (1973), "Models for Grain Rearrangement Resulting from Grain Boundary Sliding," Phil. Mag. $\underline{27}$, 9.

Cannon, W. R. and Sherby, O. B., (1973), "Third-Power Stress Dependence in Creep of Polycrystalline Nonmetals," J. Am. Ceram. Soc., $\underline{56}$, 157.

Chang, R., (1960), "Creep of Al_2O_3 Single Crystals," J. Appl. Phys., $\underline{31}$, 484.

Chang, R., (1966), "Diffusion-Controlled Deformation and Shape Changes in Nonfissionable Ceramics," in Proc. of Conf. on Nuclear Applications of Nonfissionable Ceramics, eds. A. Boltax and J. H. Handwerk, Am. Nuclear Soc., Hinsdale, Ill., p. 101.

Charles, R. J., (1969), "Stress Induced Binary Diffusion in a Solid,"
    J. Electrochem. Soc. 116, 1514.
Clauer, A. H., Seltzer, M. S. and Wilcox, B. A., (1971). "The
    Effect of Nonstoichiometry on Creep of Oxides," in Ceramics
    in Severe Environments, Mater. Sci. Res. V5, ed. W. W.
    Kriegel and H. Palmour III, Plenum Press, New York, p. 361.
Coble, R. L., (1963), "A Model for Boundary Diffusion Controlled
    Creep in Polycrystalline Materials, J. Appl. Phys., 34, 1979.
Coble, R. L., (1965), "Deformation Behavior of Refractory Com-
    pounds," in High Strength Materials, ed. V. F. Zackay, John
    Wiley, New York, p. 706.
Coble, R. L. and Guerard, Y. H., (1963), "Creep of Polycrystalline
    Aluminum Oxide," J. Am. Ceram. Soc., 46, 353.
Conrad, H., (1965), "Mechanical Behavior of Sapphire," J. Am.
    Ceram. Soc. 48, 195.
Conrad, H., Stone, G. and Janowski, K. (1965), "Yielding and Flow
    of Sapphire in Tension and Compression," TMS-AIME, 233, 889.
Crosby, A. and Evans, P. E., (1973), "Creep in Pure and Two Phase
    Nickel Doped Alumina, J. Mater. Sci., 8, 1573.
Crouch, A. G. , (1972), "High Temperature Deformation of Polycrys-
    talline Fe_2O_3," J. Am. Ceram. Soc. 55, 558.
Dawihl, V. W. and Klingler, E., (1965), "The Influence of Compres-
    sive Stresses on the Creep Behavior of Sintered Alumina
    Bodies in the Temperature Region of 1200°C," Ber. Dt. Keram.
    Ges., 42, 270.
Englehardt, V. G. and Thümmler, F., (1970), "High Temperature Creep
    Tests by 4-Point Loading II: Measurements of Polycrystalline
    Alumina," Ber. Dt. Keram. Ges. 47, 571.
Folweiler, R., (1961), "Creep Behavior of Pore-Free Polycrystalline
    Aluminum Oxide," J. Appl. Phys., 32, 773.
Fryer, G. M., (1972), "Theory of Stress Induced Diffusion in Ionic
    Crystals," Proc. Roy. Soc. Lond., A327, 81.
Gibbs, G. B. (1965), "Fluage par Diffusion dans les Solids
    Policristallins, "Mem. Sci. Rev. Met., 62, 781.
Gibbs, G. B., (1967/1968), "The Role of Grain Boundary Sliding in
    High Temperature Creep," Mater. Sci. Eng., 2, 269.
Gifkins, R. C., (1968), "Diffusional Creep Mechanisms," J. Am.
    Ceram. Soc., 51, 69.
Gifkins, R. C. and Snowden, K. U., (1966), "Mechanism for 'Viscous'
    Grain Boundary Sliding," Nature, 212, 916.
Gooch, D. J. and Groves, G. W., (1972), "Prismatic Slip in Sap-
    phire," J. Am. Ceram. Soc., 55, 105.
Gooch, D. J. and Groves, G. W., (1973a), "The Creep of Sapphire
    Filament with Orientations Close to the C-Axis," J. Mater.
    Sci., 8, 1238.
Gooch, D. J. and Groves, G. W., (1973b), "Non-basal Slip in Sap-
    phire," Phil. Mag., 28, 623.
Gordon, R., (1973), "Mass Transport in the Diffusional Creep of
    Ionic Solids," J. Am. Ceram. Soc. 56, 147.

Gordon, R., (1974), "Ambipolar Diffusion and Its Application to Diffusion Creep," in Proc. of the 9th Univ. Conf. on Ceramic Science, ed. A. Cooper and A. H. Heuer, Case Western Reserve University, Cleveland, Ohio.

Gordon, R. S. and Terwilliger, G. R., (1972), "Transient Creep in Fe-Doped Polycrystalline MgO," J. Am. Ceram. Soc., 55, 450.

Govorkov, V. G., Kozlovskaya, E. P. and Voinova, N. N., (1972), "On the Thermoactive Nature of Plastic Deformation of α-Al₂O₃ Crystals," Phys. Stat. Sol. (a), 11, 411.

Green, H. W., (1970), "Diffusional Flow in Polycrystalline Materials," J. Appl. Phys., 41, 3899.

Groves, G. W. and Kelly, A., (1969), "Change of Shape due to Dislocation Climb," Phil. Mag. 19, 977.

Herring, C., (1950), "Diffusional Creep Viscosity of a Polycrystalline Solid," J. Appl. Phys., 21, 437.

Heuer, A. H., Cannon, R. M. and Tighe, N. J., (1970), "Plastic Deformation in Fine Grain Ceramics," in Ultrafine-Grain Ceramics, ed. J. J. Burke, N. L. Reed, R. Weiss, Syracuse Univ. Press, p. 339.

Heuer, A. H., Firestone, R. F., Snow, J. D. and Tullis, J. D., (1970), "Non-basal Slip in Aluminum Oxide," in 2nd Inter. Conf. on the Strength of Metals and Alloys, ASM. p. 1165.

Heuer, A. H., Firestone, R. F., Snow, J. D. and Tullis, J. D., (1971), "Non-basal Slip in Alumina at High Temperatures and Pressures," in Ceramics in Severe Environments, Mater. Sci. Res. V5, ed. W. W. Kriegel and H. Palmour III, Plenum Press, New York, p. 331.

Hewson, C. W. and Kingery, W. D., (1967), "Effect of MgO and MgTiO₃ Doping on Diffusion-Controlled Creep of Polycrystalline Aluminum Dioxide," J. Am. Ceram. Soc. 50, 218.

Hollenberg, G. W. and Gordon, R. S., (1973), "Effect of Oxygen Partial Pressure on the Creep of Polycrystalline Al₂O₃ Doped with Cr, Fe, or Ti," J. Am. Ceram. Soc., 56, 140.

Hull, D. and Rimmer, D. E., (1959), "The Growth of Grain Boundary Voids under Stress," Phil. Mag. 4, 673.

Johnson, D. L. and Berrin, L., (1967), "Grain Boundary Diffusion in the Sintering of Oxides," in Sintering and Related Phenomena, ed. G. C. Kuczynski, N. A. Hooton, and C. G. Gibbon, Gordon and Breach, New York, p. 445.

Jones, H., (1969), "A Comparison of Theroy with Experiment for Diffusion Creep in Metal Foils and Wires," Mat. Sci. Eng. 4, 106.

Jones, T. P., Coble, R. L. and Mogab, C. J., (1969), "Defect Diffusion in Single Crystal Aluminum Oxide," J. Am. Ceram. Soc., 52, 331.

Jorgenson, P. J., (1965), "Modification of Sintering Kinetics by Solute Segregation in Al₂O₃," J. Am. Ceram. Soc. 48, 207.

Kingery, W. D., (1974), "Plausible Concepts Necessary and Sufficient for Interpretation of Ceramic Grain-Boundary Phenomena I: J. Am. Ceram. Soc., 57, 1.

Kitazawa, K. and Coble, R. L., (1974a), "Electrical Conduction in Single Crystal and Polycrystalline Al₂O₃ at High Temperatures," J. Am. Ceram. Soc., 57, 245.

Kitazawa, K. and Coble, R. L., (1974b), "Chemical Diffusion in
    Polycrystalline Al_2O_3 as Determined from Electrical Conduc-
    tivity Measurements," J. Am. Ceram. Soc. <u>57</u>, 250.

Kronberg, M. L., (1962), "Dynamical Flow Properties of Single
    Crystals of Sapphire, I," J. Am. Ceram. Soc., <u>45</u>, 274.

Lifshitz, I. M., (1963), "On the Theory of Diffusion-Viscous Flow
    of Polycrystalline Bodies," Sov. Phys. JETP, <u>17</u>, 909.

Lifshitz, I. M. and Geguzin, Ya. E., (1965), "Surface Phenomena in
    Ionic Crystals," Sov. Phys. Sol. St., <u>7</u>, 44.

Lifshitz, I. M., Kossevich, A. M. and Geguzin, Ya. E., (1967),
    "Surface Phenomena and Diffusion Mechanism of the Movement of
    Defects in Ionic Crystals., J. Phys. Chem. Sol., <u>28</u>, 783.

Lifshitz, I.M. and Shikin, V. B., (1965), "The Theory of Dif-
    fusional Viscous Flow of Polycrystalline Solids," Sov. Phys.
    Sol. St., <u>6</u>, 2211.

Mistler, R. E., (1967), "Grain Boundary Diffusion, Widths and
    Migration Kinetics in Aluminum Oxide, Sodium Chloride, and
    Silver," Sc.D. Thesis, M.I.T., Cambridge, Mass.

Mocellin, A. and Kingery, W. D., (1971), "Creep Deformation in MgO
    Saturated Large Grain Size Al_2O_3," J. Am. Ceram. Soc. <u>54</u>, 339.

Nabarro, F. R. N., (1948), "Deformation of Crystals by the Motion of
    Single Ions," Report of a Conference on Strength of Solids,
    Physical Society, London, p. 75.

Nabarro, F. R. N., (1967), "Steady State Diffusional Creep," Phil.
    Mag., <u>16</u>, 231.

Nix, W. D., Gasca-Neri, R. and Hirth, J. P., (1971), "A Contribution
    to the Theory of Dislocation Climb." Phil. Mag., <u>23</u>. 1339.

Oishi, Y. and Kingery, W. D., (1960), "Self-Diffusion of Oxygen in
    Single Crystal and Polycrystalline Aluminum Oxide," J. Chem.
    Phys., <u>33</u>, 480.

Paladino, A. E. and Coble, R. L., (1963), "Effect of Grain Bounda-
    ries on Diffusion Controlled Processes in Aluminum Oxide, J.
    Am. Ceram. Soc., <u>46</u>, 133.

Paladine, A. E. and Kingery, W. D., (1962), "Aluminum Ion Diffusion
    in Aluminum Oxide," J. Chem. Phys. <u>37</u>, 957.

Palmour, H. III, et al, (1968), "Grain Boundary Sliding in Alumina
    Bicrystals," Progress Report No. ORO-3328-9, Contract AT(40-1)
    3328.

Pasco, R. T. and Hay, K. A., (1973), "Theory of Diffusion Creep in
    Pure, Impure and Non-Stoichiometric Ionic Materials," Phil.
    Mag. <u>27</u>, 897.

Passmore, E. M., Duff, R. N. and Vasilos, T., (1966), "Creep of
    Dense, Polycrystalline Magnesium Oxide," J. Am. Ceram. Soc.,
    <u>49</u>, 594.

Passmore, E. M. and Vasilos, T., (1966), "Creep of Dense, Pure,
    Fine Grained Aluminum Oxide," J. Am. Ceram. Soc., <u>49</u>. 166.

Radford, K. C. and Pratt, P. L., (1970), "The Mechanical Proper-
    ties of Impurity-doped Alumina Single Crystals," Proc. Brit.
    Ceram. Soc., No. 15, 185.

Raj, R. and Ashby, M. F., (1971), "On Grain Boundary Sliding and Diffusional Creep," Met. Trans, $\underline{2}$, 1113.

Rao, W. R. and Cutler, I. B., (1973), "Effect of Iron Oxide on the Sintering Kinetics of Al₂O₃," J. Am. Ceram. Soc. $\underline{56}$, 588.

Readey, D. W., (1966a), "Chemical Potentials and Initial Sintering in Pure Metals and Ionic Compounds," J. Appl. Phys., $\underline{37}$, 2309.

Readey, D. W., (1966b), "Mass Transport and Sintering in Impure Ionic Solids," J. Am. Ceram. Soc., $\underline{49}$, 366.

Reijnen, P. J. L., (1970), "Nonstoichiometry and Sintering in Ionic Solids," in Problems of Nonstoichiometry, ed. A. Rabenau, North-Holland, Amsterdam, p. 219.

Robertson, W. M., (1969), "Surface Diffusion of Oxides," J. Nucl. Mater. $\underline{30}$, 36.

Robertson, W. M. and Chang, R., (1966), "The Kinetics of Grain Boundary Groove Growth on Alumina Surfaces," in The Role of Grain Boundaries and Surfaces in Ceramics, Mater. Sci. Res., V3, ed. W. W. Kriegel and H. Palmour III, Plenum Press, New York, p. 49.

Robertson, W. M. and Ekstrom, F. E., (1969), "Impurity Effects in Surface Diffusion on Aluminum Oxide," in Kinetics of Reactions In Ionic Systems. Mater. Sci. Res., V4, ed. T. J. Grey and V. D. Frechette, Plenum Press, New York, p. 273.

Roy, S. K. and Coble, R. L., (1968), "Solubilities of Magnesia, Titania, and Magnesium Titanate in Aluminum Oxide," J. Am. Ceram. Soc., $\underline{51}$, 1.

Ruoff, A. L., (1965), "Mass Transfer Problems in Ionic Crystals with Charge Neutrality," J. Appl. Phys. $\underline{36}$, 2903.

Seltzer, M. S., Perrin, J. S., Clauer, A. H. and Wilcox, B. A., (1971), "A Review of Creep Behavior of Ceramic Nuclear Fuels," Reactor Tech., $\underline{14}$, 99.

Seltzer, M. S. and Talty, P. K., (1974), "High Temperature Creep of yttria-Rare Earth Stabilized Zirconia," in Symposium on Deformation of Ceramics Materials, ed. R. E. Tressler and R. C. Bradt, Plemum Press, New York.

Shackelford, J. F. and Scott, W. A., (1968), "Relative Energies of [1100] Tilt Boundaries in Aluminum Oxide," J. Am. Ceram. Soc. $\underline{51}$, 688.

Snow, J. D. (1972), "High Temperature Plastic Deformation of Poly-crystalline Aluminum Oxide," M.S. Thesis, Case Western Reserve University, Cleveland, Ohio.

Stevens, R. N., (1972), "Grain Boundary Sliding and Diffusional Creep," Surf. Sci. $\underline{31}$, 543.

Stuart, J., (1964), "Internal Friction and Creep in Polycrystalline Alumina," in Studies of the Brittle Behavior of Ceramic Materials, N. M. Parikh, ASD-TR-61-628. Part III, p.103.

Sugita, T, and Pask, J. A., (1970), "Creep of Doped Polycrystalline Al₂O₃," J. Am. Ceram. Soc. $\underline{53}$. 609.

Terwilliger, G. R., Bowen, H. K. and Gordon, R. S., (1970), "Creep of Polycrystalline MgO and MgO-Fe₂O₃ Solid Solutions at High

Temperature," J. Am. Ceram. Soc., 53, 241.

Tigai, N. and Zisner, T., (1968), "High Temperature Creep of Polycrystalline Magnesia, J. A. Ceram. Soc., 51, 303.

Tressler, R. E. and Barber, D. J., (1974), "Yielding and Flow of C-Axis Sapphire Filaments," J. Am. Ceram. Soc., 57, 13.

Tressler, R. E. and Michael, D. J., (1974), "Dynamics of Flow in C-Axis Sapphire," in Symposium on Deformation of Ceramic Materials," ed. R. E. Tressler and R. C. Bradt, Plenum Press, New York.

Vasilos, T. and Passmore, E. M., (1968), "Effect of microstructure on Deformation of Ceramics," in Ceramic Microstructires, ed R. M. Fulrath and J. A. Pask, John Wiley, New York, p. 406.

Warshaw, S. I. and Norton, F. H., (1962), "Deformation Behavior of Polycrystalline Aluminum Oxide," J. Am. Ceram. Soc., 45, 479.

Weertman, J., (1968), "Dislocation Climb Theory of Steady State Creep," Trans. ASM, 61, 681.

Weertman, J., (1974), "Theory of High Temperature Intercrystalline Fracture Under Static or Fatigue Loads, With or Without Irradiation Damage," Met. Trans. 5, 1743.

Yen, C. F. and Coble, R. L., (1972), "Spheroidization of Tubular Voids in Al_2O_3 Crystals at High Temperatures," J. Am. Ceram. Soc. 55, 507.

GRAIN BOUNDARY DEFORMATION PROCESSES

Terence G. Langdon

Department of Materials Science, University of

Southern California, Los Angeles, California 90007

The grain boundaries of a polycrystal play an important role in deformation processes, particularly under conditions of high-temperature creep. Firstly, the boundaries act as sources and sinks for vacancies, so that diffusional creep may occur by vacancy flow. Secondly, grain boundary sliding may occur under conditions where diffusional creep is relatively unimportant.

The various grain boundary deformation processes are reviewed, and the experimental procedures available for measurements of grain boundary sliding are described in detail. The few direct measurements reported for sliding in polycrystalline non-metals are summarized, and it is demonstrated that there is a considerable need for further work in this area. A new form of deformation mechanism map is presented which is simple to construct and which graphically illustrates the occurrence of grain boundary deformation processes.

INTRODUCTION

The deformation mechanisms occurring in a polycrystalline ceramic at high temperatures divide naturally into two distinct groups. Firstly, there are those mechanisms which take place independent of the presence of grain boundaries, and are thus equally likely to be observed in both polycrystalline material and single crystals: these are designated "lattice mechanisms." Secondly, there are those mechanisms which depend on the presence of grain boundaries, and which are therefore only relevant to the deformation of polycrystals: these are termed "boundary mechanisms."

The purpose of this review is threefold. Firstly, to summarize the various deformation processes which constitute the latter class of mechanisms. Secondly, to tabulate the few direct measurements presently available for grain boundary sliding in non-metallic materials. Thirdly, to present a new form of deformation mechanism map which directly illustrates the experimental conditions under which grain boundary deformation processes dominate the creep behavior.

THE EQUATION FOR STEADY-STATE CREEP

Creep is a thermally-activated process, and many of the mechanisms occurring in high-temperature creep have a steady-state or secondary creep rate, $\dot{\varepsilon}$, which is given by an equation of the form

$$\dot{\varepsilon} = \frac{ADGb}{kT} \left(\frac{b}{d} \right)^m \left(\frac{\sigma}{G} \right)^n \tag{1}$$

where A is a dimensionless constant, D is a diffusion coefficient (= D_0 exp (-Q/RT) where D_0 is a frequency factor, Q is the activation energy for creep, R is the gas constant, and T is the absolute temperature), G is the shear modulus, b is the Burgers vector, k is Boltzmann's constant, d is the grain size, σ is the applied stress, and m [= $-(\partial \ln \dot{\varepsilon}/\partial \ln d)_{\sigma,T}$] and n [= $(\partial \ln \dot{\varepsilon}/\partial \ln \sigma)_{d,T}$] are the exponents for the inverse grain size and stress, respectively.

An examination of Eq. (1) shows that, at constant stress, temperature, and grain size, there are only four variables in the rate equation: these are the three constants (A, m, and n) and the diffusion coefficient, D [since the value of this term depends on the particular diffusion path; whether through the lattice (D_ℓ), along the grain boundaries (D_{gb}), as pipe diffusion along the dislocation cores (D_p), or through a second phase at the grain boundaries (D_{ph})]. In practice, the precise value of A is not well known for many of the deformation mechanisms, so that agreement between experimental and theoretical creep rates is seldom achieved; but the various processes may be characterized fairly well through the predicted values for m, n, and D. Table I summarizes the relevant values for several typical lattice mechanisms and all known boundary mechanisms.

Two points arise from an examination of Table I. Firstly, since lattice mechanisms are, by definition, intragranular in nature, they have no dependence on grain size and m = 0; whereas boundary mechanisms depend on the number of grain boundaries present in the material and they therefore have a non-zero value for m. Secondly, the lattice mechanisms depend on dislocation processes within the

TABLE I

Examples of Values for m, n, and D

	$\underline{m}$	$\underline{n}$	$\underline{D}$
(1) Lattice Mechanisms			
Dislocation glide/climb controlled by climb	0	4.5	D_ℓ
Dislocation glide/climb controlled by glide	0	3	D_ℓ
Dissolution of dislocation loops	0	4	D_ℓ
Dislocation climb without glide	0	3	D_ℓ
Dislocation climb by pipe diffusion	0	5	D_p
(2) Boundary Mechanisms			
(a) Diffusional creep:			
Nabarro-Herring	2	1	D_ℓ
Coble	3	1	D_{gb}
(b) Grain boundary sliding:			
With liquid phase	1	1	D_{ph}
Without liquid phase:			
Theory	1	2	D_ℓ or D_{gb}
Experiment	~1[†]	~2.4[¶]	D_ℓ or D_{gb}

[†]*Value estimated for MgO[1] and a metal[2]*

[¶]*Value estimated for a metal;[2] no value available for a ceramic*

grains and the stress exponent, n, is then fairly high (typically, $n \geq 3$); whereas, with only one exception, the boundary mechanisms are Newtonian viscous in nature and n = 1. The exception arises for grain boundary sliding in the absence of a boundary liquid phase, where experiments on a metal (Mg - 0.78 wt % Al) indicate that $n \sim 2.4$;[2] no experimental value is presently available for a ceramic. It is not known at present whether this value of n represents a genuine value relating to the nature of the sliding process (for example, if sliding occurs by glide and climb of dislocations at the boundary[3]), or whether it represents an intermediate value between Newtonian viscous sliding with n = 1 and the accommodation of sliding at triple points and grain boundary irregularities by a lattice mechanism with $n \geq 3$. As indicated in Table I, it is also not known whether the relevant diffusion coefficient for this process is D_ℓ, D_{gb}, or an intermediate value.

DIFFUSIONAL CREEP

The grain boundaries of a polycrystal represent possible sources and sinks for vacancies. Under the action of an imposed shear stress, there is an excess vacancy concentration at boundaries under tension and a depletion of vacancies at boundaries under compression, so that a vacancy flow is created to restore equilibrium. This flow may take place either through the grains (Nabarro-Herring creep)[4,5] or along the grain boundaries (Coble creep).[6] Both of these processes are understood fairly well, but it is not always appreciated that a diffusive flux of vacancies in a polycrystalline material also requires the relative movement of adjacent grains to maintain specimen coherency.

The need for this movement may be understood by considering four grains in an hexagonal array, as shown in Fig. 1(a). The circles within the grains represent the diffusion centers, and these are the points where the vacancy concentration gradient is zero. Since diffusional creep leads to no change in the internal angles, no generality is lost by taking the point B and the lines BA and BC as fixed. Under the action of a tensile stress, there is a flow of vacancies from the boundaries inclined at 60° to the stress axis to those boundaries lying parallel to the axis. Within grain (3), material is therefore added to the boundaries BC and BA during diffusion creep, thereby giving velocity vectors to the diffusion center of this grain of V_1 and V_2, respectively. The diffusion center of grain (3) therefore moves away from point B, parallel to the stress axis in this symmetric case, with a velocity of $\overline{V}$. Since the array is symmetrical, similar movement occurs in the other grains, and, in the absence of any relative grain movement, the grains become elongated as indicated in Fig. 1(b). The new positions of the diffusion centers are now marked by crosses. For this situation, a longitudinal marker line, such as XYZ, remains in the same relative position, so that there is no offset in the marker line at the points where it crosses the grain boundaries.

In practice, however, Fig. 1(b) is an unrealistic situation, because coherency has been lost at the longitudinal boundaries with the concomitant formation of internal cracks aligned parallel to the tensile axis. Since coherency is maintained during diffusional creep, it is clear that the grains must move relative to each other to give the configuration shown in Fig. 1(c). In this case, the longitudinal marker line is divided into two segments, XY' and Y"Z, so that there is an offset in the marker at the point where it crosses the boundary A'B'. It is important to note that the appearance of this offset is essentially identical to that normally observed after grain boundary sliding, although these two processes are mechanistically different;[7] whereas the behavior depicted in Fig. 1 may occur in the *absence* of crystalline slip, normal grain

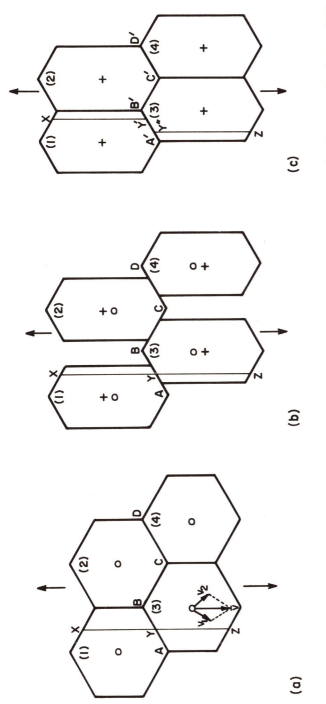

Fig. 1 (a) Four grains in an hexagonal array; the diffusion centers are marked with circles. XYZ is a marker line parallel to the tensile axis. (b) The same grains after diffusion creep in the absence of any relative grain movement; the new diffusion centers are marked with crosses. XYZ is in the same relative position. (c) The same grains after diffusion creep in the presence of relative grain movement. The marker line is now divided into two segments, XY' and Y"Z.

boundary sliding appears to occur in the *presence* of slip within the grains.[8] It is possible to regard diffusional creep as "diffusion creep with relative grain movement"[7] or as "grain boundary sliding with diffusional accommodation,"[9] whereas normal grain boundary sliding probably represents sliding with accommodation by a lattice mechanism.

GRAIN BOUNDARY SLIDING

Grain boundary sliding occurs by the movement of grains along, or in a zone adjacent to, their common boundary. This process is of considerable importance in structural materials under high-temperature creep conditions, since it leads to stress concentrations at triple points and boundary irregularities, and to the formation of cracks and cavities. The occurrence of sliding generally has a deleterious effect on the life-time of a material in a structural situation.

As indicated in Table I, grain boundary sliding may occur either in the presence of a liquid or liquid-like phase at the boundaries or in the absence of a liquid phase when the material is entirely crystalline. No experimental results are presently available on ceramic materials to unequivocally support the occurrence of sliding with a liquid phase, and it is reasonable to conclude that sliding occurs in most materials in the absence of any non-crystalline phase at the boundaries.

The phenomenon of grain boundary sliding is illustrated schematically in Fig. 2, where S represents the sliding vector in the boundary plane between the two grains. This vector may be resolved into three mutually perpendicular components: a component u along the tensile axis, and components v and w perpendicular to the tensile axis and either perpendicular to, or in the plane of, the specimen surface, respectively. Two angles are also defined in Fig. 2: an angle θ between the boundary and the tensile axis in the plane of the surface, and an angle ψ between the boundary and the surface measured on a longitudinal section cut perpendicular to the surface.

If grain boundary sliding occurs during creep, it is important that accurate and consistent procedures are established to determine the strain contributed by sliding, ε_{gbs}, to the total strain of the specimen, ε_t. Unfortunately, many of the early investigations of grain boundary sliding in metals used incorrect measuring procedures, so that it is necessary to apply corrections to the reported data.[10] More recently, four mutually consistent procedures have been established which provide a reasonably accurate measure of ε_{gbs}.[11,12]

The total strain experienced by the specimen may be expressed as

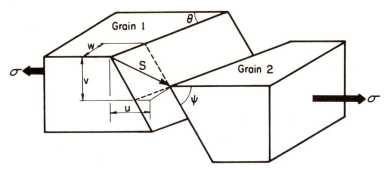

Fig. 2 Schematic representation of sliding at a grain boundary
between two adjacent grains. S is the sliding vector.

$$\varepsilon_t = \varepsilon_g + \varepsilon_{gbs} + \varepsilon_{diff(\ell)} + \varepsilon_{diff(gb)} \tag{2}$$

where ε_g is the strain due to all of the lattice mechanisms operating
intragranularly, and $\varepsilon_{diff(\ell)}$ and $\varepsilon_{diff(gb)}$ are the strains due to
diffusional creep when the vacancies move either through the lattice
or along the grain boundaries, respectively. Since diffusional creep
leads to offsets in marker lines which are not directly distinguishable from those due to grain boundary sliding (Fig. 1), it is
important to choose experimental conditions so that $\varepsilon_{diff(\ell)} \simeq$
$\varepsilon_{diff(gb)} \simeq 0$. In metals, this may be achieved fairly readily by
directly calculating the predicted creep rates for the Nabarro-
Herring and Coble processes; in ceramic materials, uncertainties in
the diffusion coefficients make these calculations more difficult,
but the magnitude of diffusional creep may be inferred indirectly
from the measured experimental values for the exponents m and n.

When the strain contributed by diffusional creep is negligible,
it follows from Eq. (2) that ε_{gbs} may be calculated either directly
or through a measure of $(\varepsilon_t - \varepsilon_g)$. Table II summarizes the four
methods available for measurements of ε_{gbs}, and these are now
examined in more detail:

(1) The intragranular strain, ε_g, may be determined directly
from the strain associated with the distortion of a grid contained
entirely within the grains. If a square grid within a grain is
distorted during deformation so that the length and breadth at a
total strain, ε_t, are L and B, respectively, the intragranular
strain associated with that grain is given by

$$\varepsilon_g = (L/B)^{2/3} - 1 \tag{3}$$

By taking a large number of measurements, it is possible to
calculate an average value of ε_g and hence, via Eq. (2), of ε_{gbs}.

TABLE II

Methods of Measuring ε_{gbs}

(1) Use an <u>intragranular</u> grid to measure ε_g:

obtain ε_{gbs} via $\varepsilon_t = \varepsilon_g + \varepsilon_{gbs}$

(2) Use a <u>longitudinal</u> marker line to measure w and θ:

$$\varepsilon_{gbs} = 2 n_\ell \overline{(w/\tan \theta)}_\ell$$

(3) Use a <u>longitudinal</u> marker line to measure w only:

$$\varepsilon_{gbs} = \zeta n_\ell \overline{w}_\ell$$

where $\zeta \simeq 1.5$

(4) Use interferometry or a calibrated microscope to measure v:

$$\varepsilon_{gbs} = \phi n_\ell \overline{v}_r$$

where $\phi \simeq 1.1$ for a polished surface at $\varepsilon_t \leq 0.05$

$\phi \simeq 1.4$ for an annealed surface or a polished surface

at $\varepsilon_t \gtrsim 0.05$

Attemps are often made to determine ε_g through Eq. (3) using the distortion of the individual grains in place of an intragranular grid. However, it has been directly demonstrated that the values of ε_g obtained in this way are often erroneously low,[13] probably due to the mobility of the grain boundaries under creep conditions; this leads to an overestimation of ε_{gbs}, as illustrated by the very high values reported for $\varepsilon_{gbs}/\varepsilon_t$ from grain shape measurements (typically > 70%) in many metal systems.[14]

(2) The most convenient methods of determining ε_{gbs} are from measurements of the offsets occurring at the grain boundaries; but several problems arise in taking these measurements, and these may be appreciated by reference to Fig. 3.

Consider a grain boundary having a trace XX on the specimen surface, oriented in the surface plane at an angle of θ to the tensile axis. Figure 3(a) shows a longitudinal marker line AA' and a transverse marker line BB' which intersect at the boundary. Sliding may occur during high-temperature creep, but at the same time the boundary will generally migrate to a new position such as X'X' in Fig. 3(b). The most obvious method of measuring ε_{gbs} is

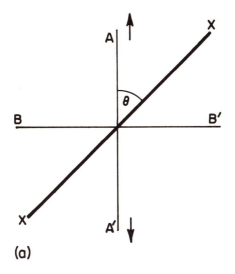

(a)

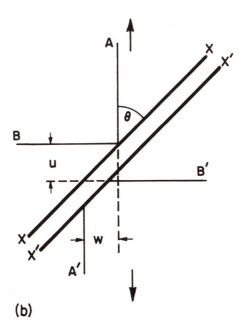

(b)

Fig. 3 (a) A grain boundary XX intersected by two marker lines
AA' and BB'. (b) The same area after the occurrence of
grain boundary sliding and migration. The new position
of the boundary is X'X'.

by a summation of the u offsets along a longitudinal marker line,
since

$$\varepsilon_{gbs} = \frac{1}{\ell} \sum_{o}^{\ell} u \tag{4}$$

where ℓ is the length of the marker line. However, Fig. 3(b)
demonstrates that it is not practical to utilize Eq. (4), since it
is experimentally difficult to measure the separation between the
end-points of a broken marker, and, furthermore, the end-point on
one side of the boundary is invariably obscured by migration.

These difficulties may initially suggest that it is preferable
to take measurements of u along a transverse marker line since, as
shown in Fig. 3(b), these offsets are unaffected by the occurrence
of migration. In practice, however, a transverse marker
preferentially intersects boundaries which are more nearly parallel
to the stress axis, and this leads to difficulties in computation
since the distribution of u with respect to θ is asymmetric.[15] The
problem is avoided by taking measurements of both w and θ along a
longitudinal marker line, since the measurement of w is then
unaffected by migration, and determining the average value of
(w/tan θ). ε_{gbs} is calculated by substituting into the relation[11]

$$\varepsilon_{gbs} = 2 n_\ell \overline{(w/\tan \theta)}_\ell \tag{5}$$

where n is the number of grains per unit length at zero strain, the
subscript ℓ denotes measurements taken along a longitudinal traverse,
and the bar signifies an average value. The factor of 2 in Eq. (5)
arises because w represents only one of the two components of
sliding (w and v) perpendicular to the stress axis.

(3) It is often fairly tedious to take a large number of
measurements of the angle θ, but in practice Eq. (5) may be
approximated for use with measurements of w only, so that

$$\varepsilon_{gbs} = \zeta n_\ell \overline{w}_\ell \tag{6}$$

where ζ is a constant. The value of ζ has been estimated
theoretically as ~ 1.6 and determined experimentally [by measuring w
and θ and using Eq. (5)] as ~ 1.4. These results suggest that a
reasonable value will be obtained for ε_{gbs} by using Eq. (6) with
$\zeta \simeq 1.5$. It should be emphasized, however, that Eq. (6) represents
an approximation, and that it is preferable, whenever facilities
are available for good angular measurements, to determine ε_{gbs}
through a measure of both w and θ and use of Eq. (5).

(4) It is not generally feasible to take accurate measurements
of offsets in marker lines when the average offset is less than
∿0.8 μm. However, the offsets perpendicular to the specimen
surface (the v component) may be determined very accurately using
interferometry, even when the average offset is of the order of
∿0.1 μm. Using this technique, the accuracy on individual measurements of v is typically ±0.06 μm. Alternative procedures are also
available for measuring v, including by use of a calibrated
microscope to directly focus on either side of the boundary and
through a surface profile device whereby a diamond stylus, connected
to a transducer, amplifier, and strip chart recorder, is moved
mechanically over the specimen surface so that the surface profile
is recorded with considerable magnification (∿5,000 – 10,000 times)
in the direction perpendicular to the surface. The former technique
is easy to perform but typically has an accuracy on each reading of
only ±0.5 μm; whereas the latter technique has an accuracy in
principle of ±0.025 μm, but is often difficult to interpret because
of the problem of distinguishing between surface irregularities and
genuine steps at the grain boundaries.

Using measurements of the v component, the value of ε_{gbs} is
calculated from the expression

$$\varepsilon_{gbs} = \phi\, n_\ell\, \overline{v}_r \qquad\qquad (7)$$

where ϕ is a constant and the subscript r indicates that, for
interferometric measurements or when using direct focusing, it is
a standard procedure to take readings at randomly selected boundaries
rather than following a linear traverse.

Unfortunately, the precise value of ϕ depends on the nature of
the specimen surface, since $\overline{v}$ is related to the angular distribution
of boundaries with the surface. Typical distributions of the
internal angle ψ are illustrated schematically in Fig. 4.[16] For a
polished surface, the angular distribution on a longitudinal section
is approximately sinusoidal, and the average angle, according to
theory and generally supported by experiment, is 57.3°. For an
annealed surface, in which the specimen is given a high-temperature
anneal without further polishing, the boundaries tend to lie more
nearly perpendicular to the surface and the average angle is
typically ∿75°; a representative distribution curve for an annealed
surface is shown in Fig. 4. A further complexity also arises
because a polished surface is relatively unstable, and the boundaries
gradually migrate during high-temperature creep so as to increase the
value of ψ. Thus the configuration of a polished surface gradually
changes during creep to an angular distribution which is more nearly
typical of an annealed surface.

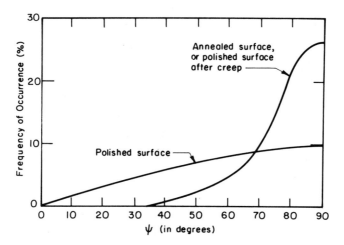

Fig. 4   Typical distributions of the internal angle ψ on polished
         and annealed surfaces.

For a specimen tested with a polished surface, a series of
detailed measurements has established that $\phi \sim 1.1$ at low total
strains; but at high strains, and at all strains for specimens with
an annealed surface, $\phi \sim 1.4$.[13] For a polished surface, the strain
at which ϕ begins to deviate significantly from a value of 1.1 is
not well defined, and depends on the experimental conditions;
however, as a general rule, it is reasonable to take $\phi = 1.1$ for
polished specimens at strains less than ~ 0.05, and, if required, a
more accurate value may be established at higher strains by
measuring both $\overline{w}_\ell$ and $\overline{v}_r$ and using the relation[11]

$$\phi = 1.4 \; (\overline{w}_\ell / \overline{v}_r) \tag{8}$$

In summary, interferometry provides a very accurate and simple
method for measuring grain boundary offsets after creep testing,
and, with the exception of polished specimens at strains in excess
of ~ 0.05, a reasonably accurate value of ε_{gbs} may be readily
estimated by putting $\phi = 1.1$ in Eq. (7). In practice, it should be
noted that a large number of individual readings of v are required
in order to obtain a statistically meaningful result; for example,
the average value of $\overline{v}$ determined from 300 individual offset measurements is typically $\pm 10\%$ at the 95% confidence level.

MEASUREMENTS OF GRAIN BOUNDARY SLIDING IN NON-METALS

Although the occurrence of grain boundary sliding has been inferred in several investigations using polycrystalline non-metals,[12] very few systematic measurements of ε_{gbs} have so far been reported. Only five sets of data are available at present, and these are summarized in Table III for Al_2O_3,[17] $CaCO_3$,[18] MgO,[1,19,20] and a U-Pu carbide.[21] For each material, the experimental values of stress, temperature, and grain size are recorded, together with the method used to measure sliding and the reported range for $\varepsilon_{gbs}/\varepsilon_t$.

An inspection of Table III reveals that two of the sets of data, for MgO[19,20] and the U-Pu carbide,[21] were obtained by measuring ε_g through the change in shape of individual grains and use of Eq. (3). As already noted, this procedure invariably yields erroneously high values for $\varepsilon_{gbs}/\varepsilon_t$, and this is confirmed by the tabulated values which are in the range of 80-100%. Table III also indicates that the tests on $CaCO_3$ were performed at a constant rate of strain rather than the usual creep conditions of constant stress or load.[18] Since a constant strain rate corresponds to a continuously increasing stress, it is unfortunately not possible to directly utilize the values reported for $\varepsilon_{gbs}/\varepsilon_t$ in terms of an equivalent stress level.

Only two of the sets of data documented in Table III appear to represent reliable values for $\varepsilon_{gbs}/\varepsilon_t$, and these both involve a measure of the v component of sliding. In Al_2O_3, this was achieved using a surface profile device;[17] in MgO, measurements were taken using interferometry.[1] These experimental results are logarithmically plotted as $\varepsilon_{gbs}/\varepsilon_t$ versus normalized stress, σ/G, in Fig. 5. The higher values of $\varepsilon_{gbs}/\varepsilon_t$ reported for Al_2O_3 may not be significant, because the magnitude of sliding depends critically on the precise experimental conditions of stress, temperature, and grain size.

Despite the obvious lack of extensive data for $\varepsilon_{gbs}/\varepsilon_t$ in non-metals, the few results presently available confirm two trends previously observed in metals.[14] Firstly, under conditions of constant stress, sliding increases in importance with a decrease in grain size; this trend is confirmed by the results on both Al_2O_3 and MgO. Secondly, under conditions of constant grain size, sliding increases in importance with a decrease in stress; this is demonstrated by the results on MgO, although the highest value recorded for $\varepsilon_{gbs}/\varepsilon_t$ at the smallest grain size and lowest stress level was only $\sim$20%.[1] It is expected on theoretical grounds that both of these trends would be reversed at very low stress levels

TABLE III

MEASUREMENTS OF $\epsilon_{gbs}/\epsilon_t$ IN NON-METALS

Material	σ, MN m^{-2}	T, K	d, μm	$\epsilon_{gbs}/\epsilon_t$	Method	Reference
Al$_2$O$_3$	82.7	1923	25-65	40-56%	Surface profile	Cannon (1971)
CaCO$_3$	$-$ †	773-1073	200	10-20%	Marker lines	Heard and Raleigh (1972)
MgO	6.9-41.4	1473-1773	13-68	100%	Grain shape	Hensler and Cullen (1967a,b)
MgO	34.4-103.3	1473	33-52	4-20%	Interferometry	Langdon (1974)
U$_{0.79}$Pu$_{0.21}$C$_{1.02}$	27.6-41.3	1673-1773	25	80-100%	Grain shape	Tokar (1973)

†Tests performed at constant strain rate rather than constant stress

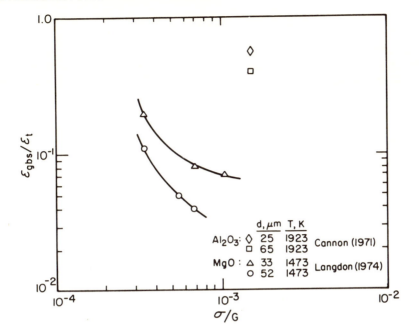

Fig. 5 $\varepsilon_{gbs}/\varepsilon_t$ versus normalized stress, σ/G, for Al_2O_3 and MgO.

and small grain sizes, because of the increasing importance of diffusional creep under these conditions.[3] However, since diffusional creep also gives rise to boundary offsets which are similar in appearance to those due to grain boundary sliding (Fig. 1), it would then be difficult to obtain meaningful values for $\varepsilon_{gbs}/\varepsilon_t$.

The sparcity of results in Fig. 5, especially when compared to a comparable plot for metals (e.g. Fig. 5-26 of ref. 22), demonstrates the urgent need for detailed and systematic measurements of grain boundary sliding in non-metals, using the experimental procedures outlined in the preceding section. In the absence of this quantitative information, it is not yet possible to reach definitive conclusions concerning the overall role of sliding in non-metals under different experimental conditions.

DEFORMATION MECHANISM MAPS

Considerable progress has been made in recent years in identifying the various mechanisms which contribute towards plastic deformation during high-temperature creep. By developing the best possible constitutive equation to describe each of the mechanisms, either theoretically or by an empirical examination of the creep data, it is possible to construct deformation mechanism maps which graphically illustrate the dominance of different mechanisms under various creep conditions. Traditionally, these maps plot normalized stress, σ/G, as a function of homologous temperature, T/T_m, where T_m is the melting point of the material, and the map is then divided into fields of stress/temperature space within which a single mechanism dominates the creep behavior.[23] However, a drawback of this form of map is that it must be constructed for one specific grain size, and, in addition, the various fields are separated by curved lines which can only be calculated in any detail by using a computer to solve the constitutive equations at a large number of points in stress/temperature space. To circumvent these two difficulties, a new form of deformation mechanism map has been introduced which is simple to construct without recourse to a computer and which is ideally suited to illustrate the influence of grain boundary deformation processes: this map plots normalized grain size, d/b, versus normalized stress, σ/G, for one particular temperature.[24]

In order to develop a map for a typical ceramic, such as MgO at a temperature of 1200°C, it is first of all necessary to derive a constitutive equation for the creep behavior of the material when a lattice mechanism dominates. Since lattice mechanisms have no dependence on grain size ($m = 0$), Eq. (1) shows that the data may be logarithmically plotted in the form of the normalized creep rate, $\dot{\varepsilon}kT/DGb$, versus the normalized stress, σ/G. Three materials were selected (Al_2O_3,[17] MgO,[25] and UC[26]) where the creep results were documented in detail, the tests were all conducted in compression (thereby avoiding the problems associated with the interpretation of bending data), and the stress exponent in the lattice mechanism regime was known to be close to 3.

The lattice diffusion coefficient was taken for the anion, using published diffusion data. The shear modulus at a selected temperature, T, was estimated from the relationship

$$G = G_o - (\Delta G)T \tag{9}$$

where G_o is the value of G obtained by a linear extrapolation from high temperatures to absolute zero and ΔG is the variation of G per degree Kelvin. The values taken for D_ℓ, G_o, and ΔG, and the appropriate values of b, are summarized in Table IV.

TABLE IV

Values of D, G, and b for Al_2O_3, MgO, and UC

Al_2O_3: $D_\ell(O^{2-})$ = 2.0 exp (-110,000/RT)   (a)

G_o = 1.71 × 10^5 MN m^{-2}
ΔG = 23.4 MN m^{-2} (b)

b = 4.75 × 10^{-8} cm[†]

†*For slip on {0001}<11$\overline{2}$0> basal system*

MgO: $D_\ell(O^{2-})$ = 2.5 × 10^{-6} exp (-62,400/RT)   (c)

G_o = 1.387 × 10^5 MN m^{-2}
ΔG = 26.2 MN m^{-2} (d)

b = 2.98 × 10^{-8} cm

UC: $D_\ell(U^{4+})$ = 7.5 × 10^{-5} exp (-81,000/RT)   (e)

G_o = 2.058 × 10^5 MN m^{-2}
ΔG = 16.1 MN m^{-2} (f)

b = 3.51 × 10^{-8} cm

References

(a) Y. Oishi and W.D. Kingery, *J. Chem. Phys.* **33**, 480–86 (1960).
(b) D.H. Chung and G. Simmons, *J. Appl. Phys.* **39**, 5316–26 (1968).
(c) Y. Oishi and W.D. Kingery, *J. Chem. Phys.* **33**, 905–06 (1960).
(d) N. Soga and O.L. Anderson, *J. Amer. Ceram. Soc.* **49**, 355–59 (1966).
(e) P. Villaine, *Diffusion Data* **3**, 515 (1968).
(f) A. Padel and Ch. de Novion, *J. Nucl. Mater.* **33**, 40–51 (1969).

Figure 6 shows the experimental data for these three materials plotted in the normalized form. It is apparent that the results for Al_2O_3 and MgO are in excellent agreement, within a factor of two, despite significant differences in testing temperature (1600°C, 1650°C, and 1700°C for Al_2O_3; 1200°C for MgO) and grain size (65 μm for Al_2O_3; 12 μm, 33 μm, and 52 μm for MgO). The results for UC are slightly higher, and this difference may arise due to uncertainties in $D_\ell(U^{4+})$. Although the data plotted in Fig. 6 support a stress exponent of n ∿ 3, it should be noted that a similar analysis of

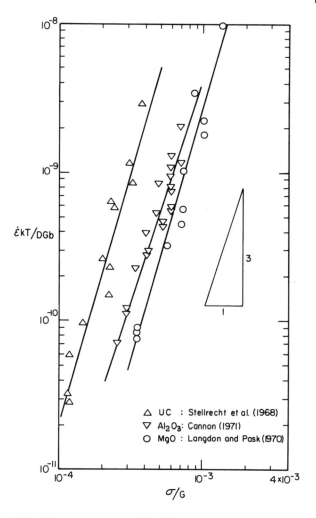

Fig. 6   Normalized creep rate, $\dot{\varepsilon}kT/DGb$, versus normalized stress, σ/G, for Al_2O_3, MgO, and UC.

creep results obtained on polycrystalline alkali halides gives n ∿ 5.[27] The reason for this difference is not known at present, but it has been suggested that the ratio of the anion to cation radius may be important (ratios > 2 giving n ∿ 3 and ratios < 2 giving n ∿ 5),[28] or that it may arise due to differences in the nature of the atomic bonds.[29]

   Using the data in Fig. 6, it is possible to estimate an empirical equation for the lattice mechanism applicable to ceramic oxides and carbides. This mechanism, here designated "dislocation creep" since the precise deformation process is not known, obeys a

TABLE V

Deformation Mechanisms in MgO

(1)   Dislocation creep:

$$\dot{\varepsilon} = 3 \frac{D_\ell Gb}{kT} \left(\frac{\sigma}{G} \right)^3$$

(2)   Nabarro-Herring:

$$\dot{\varepsilon} = 28 \frac{D_\ell Gb}{kT} \left(\frac{b}{d} \right)^2 \left(\frac{\sigma}{G} \right)$$

(3)   Coble:

$$\dot{\varepsilon} = 66.8 \frac{D_{gb} Gb}{kT} \left(\frac{b}{d} \right)^3 \left(\frac{\sigma}{G} \right)$$

relationship of the form

$$\dot{\varepsilon} = 3 \frac{D_\ell Gb}{kT} \left(\frac{\sigma}{G} \right)^3 \tag{10}$$

The equations for the three primary deformation processes in MgO are summarized in Table V. For Nabarro-Herring creep, the theoretical equation is given by[4,5]

$$\dot{\varepsilon} = \frac{B\Omega\sigma D_\ell}{d^2 kT} \tag{11}$$

where B is a constant and Ω is the atomic volume. The value of B depends on grain shape and load distribution, but is in the range of ~12 – 40.[30] The equation shown in Table V was obtained by putting $\Omega = 0.7 \, b^3$, and, following a recent review of experimental data for a number of metals tested in tension,[31] taking B = 40. For Coble creep, the theoretical equation is given by[6]

$$\dot{\varepsilon} = \frac{150}{\pi} \frac{\Omega\sigma\delta D_{gb}}{d^3 kT} \tag{12}$$

where δ is the effective width of the boundary for enhanced diffusivity. The equation in Table V was obtained by assuming $\delta \approx 2b$; however, there is some evidence that δ may be substantially larger than 2b in ceramic systems,[32] so that this value must be

regarded as a lower limit. In the absence of definitive data for
grain boundary diffusion in MgO, it was assumed in this analysis
that, to a first approximation, the activation energy for boundary
diffusion is 0.6 of the activation energy for lattice diffusion,
and that the pre-exponential factor is the same.

Using the three constitutive equations shown in Table V,
Fig. 7 illustrates the method of constructing a deformation
mechanism map for MgO at 1200°C, where the normalized grain size,
d/b, is plotted as a function of the normalized stress, σ/G. The
range of d/b from 10^3 to 10^8 is equivalent to grain sizes of
~ 0.3 μm to 3 cm, so that the map covers all grain sizes which may
be of interest. Laboratory experiments are generally conducted at
normalized stresses in the range of $\sim 10^{-4} - 10^{-3}$, but for
structural applications the range of $\sim 10^{-6} - 10^{-4}$ is of major
interest.

The map may be constructed very easily by the following three-
step procedure:

(1) The position of the boundary between the Nabarro–Herring
and Coble mechanisms is calculated by putting

$$\frac{d}{b} = \frac{A_{Co}}{A_{NH}} \left(\frac{D_{gb}}{D_{\ell}} \right) \tag{13}$$

where A_{Co} and A_{NH} are the values of the dimensionless constant A in
the Coble and Nabarro–Herring equations, respectively, and D_{gb} and
D_{ℓ} are calculated for a temperature of 1200°C. Since both of these
processes have the same dependence on stress, the boundary between
them is represented by a horizontal line passing through the point
given by Eq. (13).

(2) The value of the normalized stress at which this horizontal
line passes into the field for dislocation creep is given by

$$\frac{\sigma}{G} = \left(\frac{A_{NH}^3}{A_{Co}^2 \, A_c} \right)^{1/(n_c - 1)} \left(\frac{D_{\ell}}{D_{gb}} \right)^{2/(n_c - 1)} \tag{14}$$

where A_c is the value of the constant A in the equation for
dislocation creep and n_c is the stress exponent for the dislocation
mechanism. Equation (14) therefore indicates the total length of
the line.

(3) The slope of the line separating any two mechanisms is
given by the difference in stress exponent divided by the

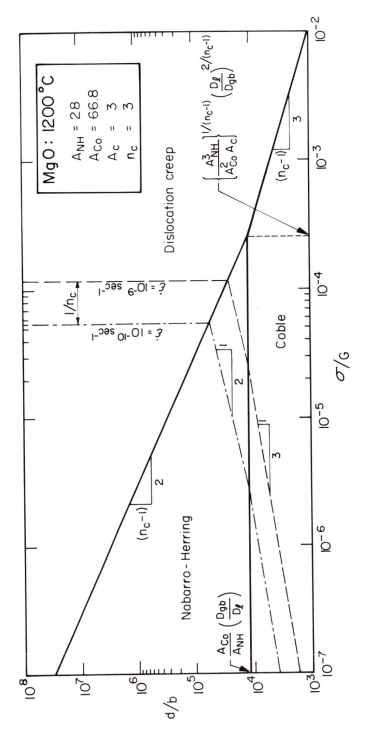

Fig. 7 Method of constructing a deformation mechanism map, illustrated for MgO at 1200°C. The values used for the various dimensionless constants are indicated.

difference in the exponent for the inverse grain size. Thus, the line between the Nabarro-Herring and dislocation creep fields has a slope of $(n_c-1)/2$, and the line between the Coble and dislocation creep fields has a slope of $(n_c-1)/3$. Lines having these slopes are therefore drawn from the point on the horizontal line given by Eq. (14). The map for MgO at 1200°C is now complete.

Having constructed the map for one particular temperature, a new map may be obtained very easily for a different temperature ($\geq 0.5\ T_m$) by noting that the position of the boundary between the Nabarro-Herring and dislocation creep mechanisms is independent of temperature, since both processes are governed by lattice diffusion. A map for a different temperature therefore requires only the solving of Eq. (13) for the temperature of interest and the drawing of a horizontal line through the appropriate value of d/b to intersect the Nabarro-Herring/dislocation creep boundary. The slope (but not the position) of the boundary between the Coble and dislocation creep mechanisms remains unchanged.

Contours of constant strain rate may be inserted on the map by solving one of the three constitutive equations within an established field. For example, by setting $\dot{\varepsilon} = 10^{-10}\ \text{sec}^{-1}$ in Eq. (10), it is possible to solve for the appropriate value of σ/G. Within the dislocation creep field, the contour for this strain rate is therefore given by a vertical line at this value of σ/G. Within the other fields, the slope of the contour is given by n/m, which corresponds to 1/2 for Nabarro-Herring and 1/3 for Coble. Furthermore, additional contours may be added for other strain rates since an order of magnitude change in $\dot{\varepsilon}$ displaces the contour by a factor of 1/n when measured parallel to the stress axis. It is therefore a simple task to establish the position of the complete contour for any strain rate of interest.

The final form of the map for MgO at 1200°C is shown in Fig. 8: a more detailed presentation of maps for other ceramic materials is given elsewhere.[33] Figure 8 also indicates the range of stress and grain size used for the MgO creep data in Fig. 6.[25] This map is identical to that shown in Fig. 7 with the exception that the strain rate contours are curved in the vicinity of the boundaries between two adjacent mechanisms. This difference arises because the simplified procedure of drawing straight lines up to the boundary ignores the contribution of other mechanisms in adjacent fields, and this becomes increasingly important near the boundaries between two fields. The contours shown in Fig. 8 were calculated by computer and are exact for the constitutive equations chosen to construct the map; as indicated, these contours deviate only slightly from those obtained by the approximate method of drawing straight lines within any field.

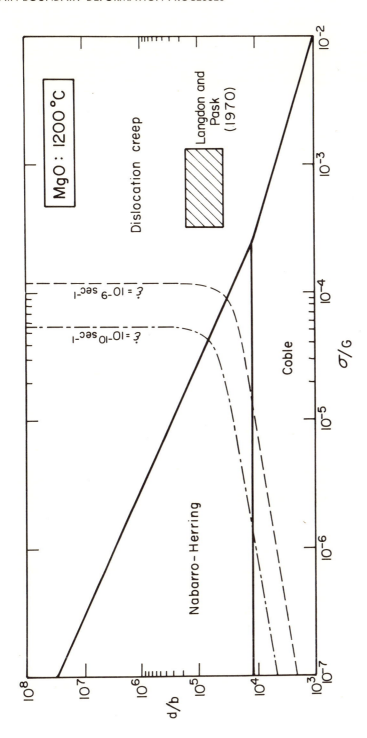

Fig. 8   A deformation mechanism map for MgO at 1200°C.   The range of stress and grain size used for the MgO creep data in Fig. 6 is indicated.

Deformation mechanism maps of the type shown in Fig. 8 may be constructed very easily for a wide range of materials, and graphically illustrate the importance of various mechanisms under different conditions of stress and grain size. However, it is important to emphasize two points regarding their use. Firstly, the accuracy of a map is limited by the accuracy of the constitutive equations used to construct it; this point is particularly important with ceramics, because of the uncertainties associated with grain boundary, and to some extent lattice, diffusion coefficients. Secondly, it is only possible to include those mechanisms for which a reasonably accurate constitutive equation may be derived, so that some mechanisms, which may be of primary importance under certain experimental conditions, are necessarily excluded from the map. For example, grain boundary sliding has not been included as a deformation mechanism in the construction of the map shown in Fig. 8, due to the uncertainties associated with both the form of the relevant constitutive equation and the nature of the accommodation of sliding by lattice mechanisms; a simple method of including sliding on deformation mechanism maps, which tends to overestimate the sliding field, is given elsewhere.[34] Despite these limitations, deformation mechanism maps of this type provide a good pictorial representation of the deformation processes and a useful indication of anticipated behavior.

SUMMARY AND CONCLUSIONS

1. The deformation mechanisms occurring at high temperatures may be divided into "lattice mechanisms," which are entirely intragranular, and "boundary mechanisms," which are directly associated with the presence of grain boundaries. By definition, lattice mechanisms are independent of grain size.

2. Boundary mechanisms divide into two types: diffusional creep (Nabarro-Herring, Coble) and grain boundary sliding (with or without the presence of a liquid phase).

3. Diffusional creep leads to offsets in marker lines which are similar in appearance to those due to grain boundary sliding.

4. Four experimental methods are available to measure the strain due to grain boundary sliding, ε_{gbs}: these methods are described in detail.

5. The few direct measurements of grain boundary sliding in polycrystalline non-metals are summarized, and it is demonstrated that there is a considerable need for further work in this area.

6. A new form of deformation mechanism map is presented which

is simple to construct without recourse to a computer; the method of construction is outlined, and a map is presented for MgO at 1200°C.

Acknowledgments

This work was supported in part by the National Science Foundation under Grant No. GH-36129 and in part by the United States Atomic Energy Commission under Contract AT(04-3)-113 PA-26.

REFERENCES

1. T.G. Langdon, "Grain Boundary Sliding during Creep of MgO," submitted for publication in *J. Amer. Ceram. Soc.*
2. T.G. Langdon, *The Microstructure and Design of Alloys*, Proc. Third Intl. Conf. on Strength of Metals and Alloys, The Institute of Metals and The Iron and Steel Institute, London, Vol. 1, pp. 222-26, 1973.
3. T.G. Langdon, *Phil. Mag.* 22, 689-700 (1970).
4. F.R.N. Nabarro, *Report of a Conference on Strength of Solids*, The Physical Society, London, pp. 75-90, 1948.
5. C. Herring, *J. Appl. Phys.* 21, 437-45 (1950).
6. R.L. Coble, *J. Appl. Phys.* 34, 1679-82 (1963).
7. R.C. Gifkins and T.G. Langdon, *Scripta Met.* 4, 563-66 (1970).
8. T.G. Langdon, *Rate Processes in Plastic Deformation*, Proc. John E. Dorn Memorial Symposium, Cleveland, Ohio, 1972 (in press).
9. R. Raj and M.F. Ashby, *Met. Trans.* 2, 1113-27 (1971).
10. R.L. Bell, C. Graeme-Barber and T.G. Langdon, *Trans. TMS-AIME* 239, 1821-24 (1967).
11. T.G. Langdon, *Met. Trans.* 3, 797-801 (1972).
12. T.G. Langdon, *J. Amer. Ceram. Soc.* 55, 430-31 (1972).
13. T.G. Langdon and R.L. Bell, *Trans. TMS-AIME* 242, 2479-84 (1968).
14. R.L. Bell and T.G. Langdon, *Interfaces Conference*, R.C. Gifkins, ed., Butterworths, Sydney, pp. 115-37, 1969.
15. R.C. Gifkins, A. Gittins, R.L. Bell and T.G. Langdon, *J. Mater. Sci.* 3, 306-13 (1968).
16. R.L. Bell and T.G. Langdon, *J. Mater. Sci.* 2, 313-23 (1967).
17. W.R. Cannon, Ph.D. thesis, Department of Materials Science, Stanford University, 1971.
18. H.C. Heard and C.B. Raleigh, *Geol. Soc. Amer. Bull.* 83, 935-56 (1972).
19. J.H. Hensler and G.V. Cullen, *J. Amer. Ceram. Soc.* 50, 584-85 (1967).
20. J.H. Hensler and G.V. Cullen, Final Technical Report, AAEC Research Contract No. 64/D/13, Industrial Research Section, Department of Metallurgy, University of Melbourne, 1967.
21. M. Tokar, *J. Amer. Ceram. Soc.* 56, 173-77 (1973).

22. F. Garofalo, *Fundamentals of Creep and Creep-Rupture in Metals*, Macmillan Company, New York, 1965.
23. M.F. Ashby, *Acta Met.* 20, 887-97 (1972).
24. F.A. Mohamed and T.G. Langdon, "Deformation Mechanism Maps Based on Grain Size," *Met. Trans.* (in press).
25. T.G. Langdon and J.A. Pask, *Acta Met.* 18, 505-10 (1970).
26. D.E. Stellrecht, M.S. Farkas and D.P. Moak, *J. Amer. Ceram. Soc.* 51, 455-58 (1968).
27. F.A. Mohamed and T.G. Langdon, *J. Appl. Phys.* 45, 1965-67 (1974).
28. W.R. Cannon and O.D. Sherby, *J. Amer. Ceram. Soc.* 56, 157-60 (1973).
29. S.H. Kirby and C.B. Raleigh, *Tectonophysics* 19, 165-94 (1973).
30. G.B. Gibbs, *Mém. Sci. Rev. Mét.* 62, 781-86 (1965).
31. J.E. Harris, *Met. Sci. J.* 7, 1-6 (1973).
32. L.W. Barr, I.M. Hoodless, J.A. Morrison and R. Rudham, *Trans. Faraday Soc.* 56, 697-708 (1960).
33. F.A. Mohamed and T.G. Langdon, to be published.
34. T.G. Langdon and F.A. Mohamed, *Proc. Fourth Bolton Landing Conf. on Grain Boundaries in Engineering Materials*, Lake George, N.Y., 1974 (in press).

THERMALLY ACTIVATED DISLOCATION MOTION IN CERAMIC MATERIALS

A. G. Evans

Institute for Materials Research
National Bureau of Standards
Washington, D. C.   20234

ABSTRACT

The current role of thermal activation studies in ceramic systems is examined, identifying three research areas. The theory of thermal activation is then summarized and the types of experiments needed to achieve the research objectives are identified. Finally, the data analysis techniques required for effective barrier analysis are described and some existing data are evaluated.

1.   INTRODUCTION

The primary intent of this review is to examine the current role of thermally activated dislocation motion studies in ceramic systems, and to identify the experiments needed to impliment the emergent objectives. It is not intended to comprehensively review the field of thermally activated dislocation motion; because this has been the subject of a very excellent and extensive recent review by Kocks, Argon and Ashby [1] and our readers are urged to consult this article if they require a detailed appreciation of the thermal activation phenomenon. The basic experimental techniques for studying thermally activated dislocation motion have also been reviewed several times in recent years. [2] Hence, these will only be cursorily examined in this paper; more emphasis will be placed on a critique of the various experimental approaches, as applied to ceramic systems.

2. MOTIVATION

2.1 Application of Deformation Studies

Studies of plastic deformation in ceramic materials have application in three related areas:

(a) The characterization and understanding of the steady state (creep) and cyclic (fatigue) deformation of ceramic materials that occurs prior to fracture, usually at elevated temperatures and at low stress levels.

(b) The evaluation of the deformation response for rapid, monotonically varying, stresses.

(c) The determination of the role of crack tip plasticity on fracture toughness and the onset of macroscopic ductility.

The first study can be readily justified, both as a means for predicting the in-service dimensional changes that occur in components subjected to slowly varying (or steady) states of stress, and as an aid in the development of materials with a maximum resistance to deformation. Exemplary schemes of this type are contained in the work of Ashby. [3] The modes of deformation typically encountered under these conditions are diffusion related, as discussed in detail in this volume by Cannon and Coble [4] and by Langdon. [5]

Significant plastic deformation under more rapid loading rates is rarely observed in ceramic polycrystals, because the deformation must then occur primarily by dislocation motion. This lack of dislocation ductility limits the ability of structural ceramic materials to alleviate the stresses imposed under, for example, impact and thermal shock conditions. There is a strong motivation, therefore, to study dislocation activity in structural ceramics in an attempt to develop polycrystalline materials with superior ductility. A substantial effort, over a 10 year period between 1960 and 1970, was devoted to the development of a 'ductile ceramic,' and reached a largely unsuccessful culmination.* But, some very admirable research was conducted during this period which has enabled us to appreciate the magnitude of the problem, and to specify the research areas which need further development. Realistically, we must consider this development as a long term exercise that requires good fundamental research.

---

*The best results were achieved with the ionic materials, e. g. NaCl [17], CaF_2 [18] which are ductile above $\sim$300°C.

Conversely, the use of polycrystalline ceramics as laser windows requires that the extent of the dislocation motion be minimized in order to maintain optimum transparency. The initial deleterious deformation occurs on the primary slip systems, and the resistance to dislocation motion on this system can be increased by selective alloy additions. The first definitive current role of thermal activation studies is thus apparent. These studies can be utilized to identify alloy additions that increase the resistance to primary slip (using single crystals). Thermal activation studies are not the only approach that can be utilized, but these studies are capable of generating very useful quantitative detail, as we shall see later.

The deformation that occurs in the vicinity of a crack tip strongly influences the fracture toughness of a material; in general, the more extensive the deformation the higher the toughness. Since the strain-rates that pertain to the crack tip region are usually large, deformation in this region normally occurs by dislocation activity. Consequently, crack tip plasticity in ceramic materials is generally rather restricted and this class of materials exhibits low toughness. Again, therefore, there is a strong motivation for developing a structural ceramic material which can deform extensively by dislocation motion. The problem is comparable in form to the ductile ceramic development, but it is less ambitious in scope and has more justifiable current research possibilities. We shall thus examine crack tip dislocation processes in order to identify the role of thermal activation studies in the development of a higher toughness ceramic.

2.2 Crack Tip Dislocation Activity

2.2.1 <u>Restricted Primary Slip</u>. For polycrystalline ceramics, the detailed features of crack tip dislocation activity are not well formulated. There have been a few preliminary studies, using transmission electron microscopy [6] and scanning electron microscopy [7] (in conjunction with etch pitting). But our comprehension should advance substantially during the next few years as this work progresses. We do not, however, require a detailed knowledge of crack tip dislocation morphologies to determine the initial role of thermal activation studies; a general appreciation of the dislocation behavior is sufficient.

Dislocation motion near a crack is activated by the shear stresses, τ, which increase in magnitude as the crack tip is approached; [8]

$$\tau = \frac{K_I}{\sqrt{2\pi r}}\ \sin\frac{\theta}{2}\ \cos\frac{\theta}{2}\ \cos\frac{3\theta}{2} \tag{1}$$

where K_I is the stress intensity factor, r is the distance from
the crack tip and θ is an angular dependence. Dislocation motion
commences when the shear stress exceeds a critical value, τ_c,
at a nearby source. In polycrystals, dislocation sources are
expected to occur primarily at grain boundaries. Hence, we
envisage a grain boundary source, S, emitting dislocations
which tend to be transmitted across the grain, as depicted in
figure 1. The shear stress will, of course, vary in a rather

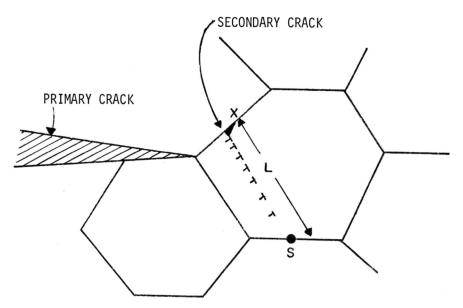

Figure 1. Dislocation pile-up and microcrack formation at a
primary crack.

complex manner across the slip plane. The stress can, however,
be mapped out for each potential slip plane and source location
to show that the dislocation transmission often proceeds to the
adjacent grain boundary, thereby generating a grain boundary pile-
up. [9] This pile-up results in microcrack initiation, at the
grain boundary, if the normal stress at the tip of the pile-up
exceeds the theoretical strength of the boundary. [10] An analysis
of this process, using a simple model, suggests that microcrack
formation occurs when; [17]

$$(K_I)_m \approx \beta \left\{ \sqrt{2x}\ \tau_o + \left[\frac{2\zeta E^3 b}{(1+\phi)^2} \right]^{1/4} \right\} \tag{2}$$

where x is the distance of the pile-up from the crack tip, τ_o
is the lattice resistance to dislocation motion on the operative
slip system, E is Young's modulus, b is the Burger's vector, ζ
is the strain-energy release rate for grain boundary fracture,
ϕ is Poisson's ratio, $(K_I)_m$ is the stress intensity factor for
microcrack initiation and β is a constant, $\approx\sqrt{\pi}$. The microcrack
extends to a length roughly equal to the pile-up length; hence,
microcrack formation constitutes an increment in the length of
the primary crack when x<ℓ where ℓ is the length of the pile-up.
The stress intensity factor for crack propagation assisted by
limited crack tip dislocation activity is thus;

$$K_I \approx \beta \left\{ \sqrt{2\ell}\ \tau_o + \left[\frac{2\zeta E^3 b}{(1+\phi)^2}\right]^{1/4} \right\} ,\qquad (3)$$

   This process competes with bond rupture, which occurs at a
stress intensity factor, $(K_I)_b$, as a mode of crack propagation.
Typically, as the temperature is increased, τ_o decreases. We expect,
therefore, that dislocation assisted crack propagation to commence
above a critical temperature, T_c causing K_I to decrease as the
temperature is further increased (figure 2)--as observed in Al_2O_3
[12, 13] and SiC [14] for example. Inserting typical values for

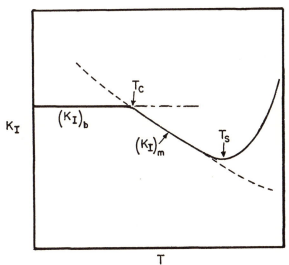

Figure 2. A schematic of the temperature dependence of K_I: $(K_I)_b$
          is the stress intensity factor for bond rupture at
          the primary crack, and $(K_I)_m$ refers to microcrack
          formation by limited dislocation activity at the
          primary crack. The increase in $(K_I)_m$ above T_s is due
          to an increase in slip homogeneity, etc.

$(K_I)_b$, E, b, and ℓ (which we can reasonably equate to the grain size) for ceramic systems into eqn. (3) gives a value for τ_o of ~ 100 MNm^{-2}.$\dagger$

Hence, in situation where significant slip homogeneity can not be induced, crack tip dislocation activity should be <u>discouraged</u> (a similar conclusion has been reached for single crystal crack propagation [15]) and the lattice resistance to primary dislocation motion, τ_o, should be maintained above ~ 100 MNm^{-2} at the use temperature for the material.

Another important role for thermally activated dislocation motion studies has thus emerged. These techniques can be used as one means of studying the lattice resistance to dislocation motion. The primary purpose of such a study is to identify alloying additions which increase τ_o on the primary slip system, without significantly diminishing the grain boundary fracture energy.

2.1.2 <u>Cross Slip Effects</u>. When dislocation cross-glide can occur, the grain boundary pile-up will not be restricted to a single plane; a slip band will form and the stress at the tip of the pile-up will be less intense. [16] Microcrack initiation will thus require many more dislocations in the pile-up and the $(K_I)_m$ value might become substantially larger than the value predicted by eqn. (3). At this stage, the crack tip deformation can become beneficial. The benefit is realized when the stress reduction at the primary crack tip, caused by the interaction with the super-dislocation self-stress field, results in a significant increase in $(K_I)_b$ before microcrack formation intervenes, i.e., when

$$(K_I)_m > (K_I)_b > (K_I)_{b,i}$$

where $(K_I)_{b,i}$ is the stress intensity factor for crack propagation due to bond rupture at the primary crack tip, in the absence of dislocation activity.

$\dagger$More homogeneous dislocation activity would clearly reduce the propensity for microcrack formation and dislocation activity at stresses below ~ 100 MNm^{-2} might then be tolerable.

$*$ Grain boundary fracture studies, using bi-crystals for example, should, therefore, constitute an important part of the total study.

The onset of substantial cross glide can thus signify the
onset of some benefit from crack tip dislocation activity. A
study of the cross glide phenomenon is single crystals is thus
an important input to the problem of enhancing fracture toughness.
Cross glide can be studied using microscopic techniques, but
more fundamental studies may be more informative. For example,
we know that extensive cross glide is encouraged by reducing
the difference between the dislocation flow stress on the primary
and secondary slip systems. [17, 18] Hence, coordinated studies
of dislocation motion on the primary and secondary slip systems,
directed at identifying alloying additions that reduce this flow
stress difference, seem especially appropriate. Thermal activation
studies would constitute an integral part of this investigation.

2.1.3 <u>Homogeneous Slip</u>. Several additional conditions must
be satisfied before a general deformation (leading to a toughness
comparable to metals) can occur. Slip must be possible on at
least five independent systems. [19] This may not be achievable
in some materials at reasonable stress levels; for these materials
high toughness can never be realized.* When sufficient systems
are available, all systems must operate at comparable shear stress
levels to avert local stress concentrations and premature microcrack
formation. A general deformation also involves intersecting
slip, [17] so that slip band penetration must proceed without
excessive impediment (and it is not entirely clear at this time
whether effective slip band interpretation occurs before the
primary and secondary slip stress equivalence condition is achieved).
Hence, two important fundamental studies are emerging. The first
is the study of secondary slip, as outlined in the preceeding
sub-section. The second is the study of intersecting slip.
Again, thermal activation techniques are not the only techniques
that can be used to address the problem. More macroscopic 'latent
hardening' studies can generate very useful information. [21]
But, thermal activation studies can provide additional valuable
detail to enhance our appreciation of the problem.

2.2 The Role of Thermal Activation Studies

The importance of thermal activation studies to materials
development, for laser window and structural ceramic applications,
has been outlined in the preceeding section. A summary of the
proposed studies is presented in this section for ease of reference.
Three primary studies have emerged all of which can be conducted
on single crystals.

---

*Except perhaps as a matrix phase in a composite. [20]

(a)   The study of the rate limiting lattice process on the primary slip system, directed primarily at identifying alloy additions which increase the resistance to dislocation motion. This study has application to both laser window and structural ceramic problems.

(b)   The investigation of the relative lattice resistance on the primary, τ_p, and secondary, τ_s, slip systems, directed toward the identification of alloy systems that tend to reduce the ratio τ_s/τ_p to unity.

(c)   The study of intersecting slip, directed at both the evaluation of the 'latent hardening' effect and (if possible) the elucidation of approaches for reducing the 'latent hardening' to zero.

3.   THEORY OF THERMAL ACTIVATION

3.1   Thermodynamics

A typical resistance profile for dislocation motion along a slip plane is shown in figure 3.   When a shear stress, τ_a, is

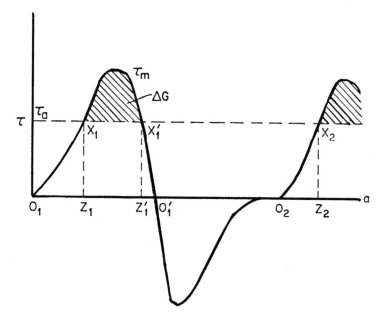

Figure 3. A barrier profile showing the variation of the barrier resistance, τ, as the dislocation sweeps through an area, a, of slip plane.

applied across the slip plane, the dislocation moves reversibly
to position X_1 (figure 3) on the barrier. At this stage, the applied
stress has performed work given by the area $O_1X_1Z_1$. If a thermal
fluctuation of sufficient magnitude is now made available to the
restrained dislocation segment the barrier can be surmounted and
the dislocation will move to a new position, X_2, at the next barrier.
The energy that must be supplied thermally to induce this forward
motion of the dislocation is given by the cross-hatched area in
the figure. Work is also done by the applied stress <u>during</u> the
activation, given by the area $X_1Z_1Z_1'X_1'$. After the activation,
additional work is performed by the applied stress before it reaches
its next equilibrium position, given by the sum of the areas
$Z_1'X_1'O_1'$ and $O_2X_2Z_2$.

     The thermal energy is the activation free energy, ΔG, and
it is apparent from fig. 3 that this is given by;

$$\Delta G = b \int_{\tau_m}^{\tau_m} \Delta a d\tau \tag{4}$$

where Δa is the true activation area and τ_m is the peak barrier
resistance. The activation free energy is related to the activation
enthalpy, ΔH, and the activation entropy, ΔS, by;

$$\Delta G = \Delta H - T \Delta S$$

where T is the absolute temperature. Note that ΔH is usually temperature independent for a single dominant barrier, ΔS is also relatively
temperature independent, while ΔG is temperature dependent.

3.2 Kinetics

     The rate of forward motion of a dislocation by thermal acti-
vation, v, can be described by a rate equation of the Arrhenius type;

$$v = \nu \exp \left[\frac{-\Delta G}{kT} \right] \tag{6}$$

where ν is a frequency factor, $\stackrel{\sim}{<}10^{-11}s^{-1}$. This relation is expected
to apply when $\Delta G \gg kT$, and if the probability of reverse motion
over the obstacle is negligible. Substituting for ΔG from eqn.
(5) gives;

$$v = \nu \exp \left(\frac{\Delta S}{k} \right) \exp \left(\frac{-\Delta H}{kT} \right) \tag{7}$$

The group velocity of the mobile dislocations in a macroscopically
deforming body is related directly to the rate of strain, $\dot{\gamma}$, by;

$$\dot{\gamma} = \rho_m b \bar{v} \tag{8}$$

where ρ_m is the density of mobile dislocations and $\bar{v}$ is the average dislocation velocity. The quantity, $\bar{v}$ is related to velocity of the individual dislocation segments by; [1]

$$\bar{v} = \frac{1}{V\rho_m} \int_{0}^{V\rho_m} v \, d\xi \tag{9}$$

where V is the total volume of the body. Combining eqns. (7) and (8) gives;

$$\dot{\gamma} = \rho_m b \left[\nu \, \exp\left(\frac{\Delta S}{k}\right) \exp\left(\frac{-\Delta H}{kT}\right) \right]_{av} \tag{10}$$

4. EXPERIMENTAL MEASUREMENTS

4.1 Theoretical Development

It is immediately apparent from eqns. (7) and (10) that measurements of the dislocation velocity or the macroscopic strain-rate at different temperatures should generate important information about the nature of the obstacles impeding dislocation motion, through the thermodynamic quantities ΔG, ΔH, ΔS, ν and ρ_m. In this section we shall inspect the rate equations and identify the pertinent experimental measurements.

(a) <u>Direct Dislocation Velocity Studies</u>. Expressing eqn. (7) in logarithmic form and differentiating with respect to reciprocal temperature, at constant applied stress, gives

$$\left.\frac{\partial \ell n \nu}{\partial (1/T)}\right|_{\tau_a} = \left.\frac{\partial \nu}{(1/T)}\right|_{\tau_a} + \frac{1}{k} \left.\frac{\partial \Delta S}{\partial (1/T)}\right|_{\tau_a} - \frac{\Delta H}{k} \tag{11}$$

The activation entropy is usually temperature invariant, and the temperature dependence of the frequency factor is small, e.g., for a linear barrier $\nu \propto 1/\sqrt{T}$. [1] Hence, we frequently expect eqn. (11) to reduce to:

$$\Delta H = - k \left.\frac{\partial \ell n \nu}{\partial (1/T)}\right|_{\tau_a} \tag{12}$$

and a plot of $\ell n\ v$ vs. $1/T$, at constant τ_a, should frequently yield a straight line, with the slope proportional to ΔH, as depicted in fig. 4. When the experimental data indicate a significant curvature,

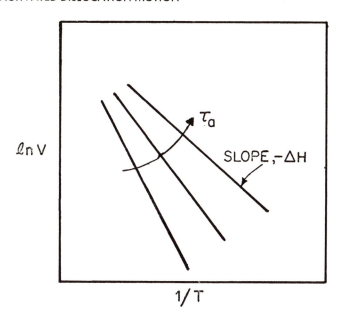

Figure 4. A schematic of a plot of ℓnv vs. $1/T$ at constant τ_a for a single dominant barrier with a stress invariant activation enthalpy.

several alternative conclusions are possible: (a) the data is in-
valid (and should be retaken), (b) there is a substancial dependence
of ΔS or v on temperature, (c) the barrier is changing with temperature from one dominant type to another, or (d) the barrier has a
characteristic temperature dependent activation enthalpy. We are
not aware of any single barrier with a significant temperature de-
depence of v and ΔS, although this possibility should be comprehensively explored before discarding it. Conclusions (c) or (d) must
be consistent with the theoretical analysis of the dominant barrier,
e.g., the constriction of an extended dislocation during intersection or cross slip can give a temperature dependent ΔH. [22]

The evaluation of the activation enthalpy at a single stress
is not sufficient for barrier analysis, because ΔH is stress dependent.
The stress dependent characteristics of the dominant barrier are
directly related to ΔG, as described by eqn. (4), and for detailed
barrier analysis, the stress dependence of ΔG, rather than ΔH,
is preferred. A phenomenological relation for this stress dependence,

which has sufficient than ΔH, is preferred. A phenomenological
relation for this stress dependence, which has sufficient versatility
to describe essentially all barrier profiles, is;

$$\Delta G = F_o \left[1 - \left(\frac{\tau_a}{\tau_m} \right)^z \right]^y$$ (13)

where F_o is the Helmholtz free energy and z and y are constants:
usually $z \sim 1/2$ and $y \sim 3/2$. For most barriers, F_o is expected
to be a function of the Burger's vector b and the shear modulus,
μ, specifically; [1]

$$F_o \propto \mu b^3$$

and the glide resistance, τ_m is expected to be porportional to
the shear modulus, [1]

$$\tau_m \propto \mu$$

suggesting that the following normalized variables might be used
to obtain ΔG;

$$\frac{\partial \ell n v}{\partial [\mu b^3 / T]} \bigg|_{\tau_a / \mu} = \frac{-k \Delta G}{\mu b^3}$$ (14)

Hence, a plot of $\ell n v$ vs. $\mu b^3 / T$ at constant τ_a / μ (fig. 5) should

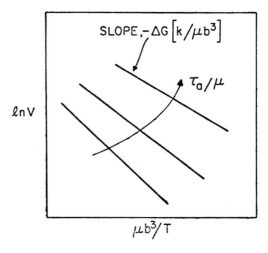

Figure 5. A schematic of a plot to obtain ΔG using normalized
          variables.

generate a straight line with a slope related to ΔG, if the temperature dependencies of F_o and τ_m have been properly anticipated.
When this procedure generates a linear plot, then ΔG has been
identified and it is possible to proceed. Alternatively, if some
substantial curvature is observed, it is clear that an incorrect
assignment of the temperature dependencies of F_o and τ_m has been
made, and this, by itself, provides some important information
about the character of the dominant barrier. Theoretical analysis
of barriers with different functional forms should then be initiated.

Values for ΔG should be obtained for a range of τ_a at several
temperatures to conduct a comprehensive barrier analysis. The
data can be treated in several ways. Values of z, y and F_o can
be obtained directly using eqn. (13), and the resultant energy
profile constructed. This profile may then be compared with theorti-
cal estimates to identify the dominant barrier. Alternatively,
the results can be expressed in terms of an apparent activation
area, Δa^*, defined by:

$$b\Delta a^* = \left.\frac{\partial \Delta G}{\partial \tau_a}\right|_T \equiv kT\left[\left.\frac{\partial \ell n \nu}{\partial \tau_a}\right|_T - \left.\frac{\partial \ell n \nu}{\partial \tau_a}\right|_T\right] \tag{15}$$

This is not the true activation area, Δa, defined in eqn. (4)
unless the barrier profile is pressure independent. In general,
[1]

$$\Delta a^* = \Delta a - \int_{\tau_a}^{\tau_m} \left.\frac{\partial \Delta_a}{\partial \tau_a}\right|_T \, d\tau \tag{16}$$

and frequently,

$$\Delta a^* \propto \Delta a$$

The apparent activation area can be used to construct stress profiles as shown in fig. 6. These profiles can then be compared with
theoretically computed profiles to achieve barrier identification.
(Note the difference between this stress profile and the resistance
profile shown in fig. 3). Clearly, the stress and energy profiles
are directly related, and differentiation of eqn. (13) shows that,

$$b\Delta a^* = F_o zy \left(\frac{\tau_a}{\tau_m}\right)^{z-1}\left[1 - \left(\frac{\tau_a}{\tau_m}\right)^z\right]^{y-1} \tag{17}$$

The stress profile does not, therefore, add any new information;
it is merely an alternative description of the barrier profile,
although it might be regarded by some as physically more satisfying.

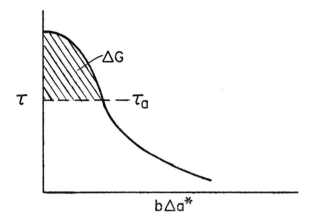

Figure 6. A schematic of a stress profile

Finally, a quantity m defined by

$$m = \frac{\partial \ell nv}{\partial \ell n\tau_a}\bigg|_T \left(\frac{\partial \ell nv}{\partial \ell n\tau_a}\bigg|_T - \left(\frac{1}{kT}\right) \frac{\partial \Delta G}{\partial \ell n\tau_a}\bigg|_T \right) \tag{18}$$

may be obtained directly from the stress dependence of the dislocation velocity at constant temperature. This clearly does not
add any additional information, but it is introduced here because
m values have been frequently quoted in thermal activation studies,
[1] and it is useful to idnetify its relation to the obstacle
profile. From eqns. (15) and (18), we obtain;

$$m = \frac{b\tau_a}{kT} \Delta a^* + \tau_a \left|\frac{\partial \ell nv}{\partial \tau_a}\right|_T \tag{19}$$

Hence, if the stress dependence of v is small, m gives a direct
measure of the apparent activation area;

$$b\Delta a^* = \frac{mkT}{\tau_a} \tag{20}$$

In fact, the stress dependence of v can be found by comparing
Δa^* obtained from m with Δa^* obtained directly from ΔG.

In summary, dislocation velocity data should be treated in
the following sequence to generate information about the dominant
barrier. (a) Plot ℓnv vs. $1/T$ at constant applied stress to

obtain the activation enthalpy, ΔH; if ΔH is temperature independent, a single well-defined barrier is dominant; but if ΔH is temperature dependent, it can usually be concluded that either two competing dominant barriers (with different temperature dependencies) are operating or that a barrier with a characteristic temperature dependent ΔH is dominant. (b) Plot the normalized variables $\ell n v$ vs. $\mu b^3/T$, at constant τ_a/μ, to obtain the activation free energy; a temperature independent slope indicates a valid normalization and identifies ΔG; a significant curvature suggests several possibilities, including further theoretical development. (c) Finally, the stress dependence of ΔG is used to delineate the barrier profile, which can then be compared with theoretically determined profiles to determine the characteristics of the dominant barrier.

(b) <u>Strain-Rate Studies</u>. The same procedure outlined above for the dislocation velocity studies can be used to obtain information about the dominant barrier from strain-rate studies; but the interpretation sequence is more complex and frequently, the conclusions are less specific.

The principal problem with the interpretation of strain-rate studies can be appreciated by differentiating the logarithmic form of eqn. (10);

$$\left.\frac{\partial \ell n \dot{\gamma}}{\partial (1/T)}\right|_{\tau_a} = \left.\frac{\partial \ell n \rho_m}{\partial (1/T)}\right|_{\tau_a} + \left.\frac{\partial \ell n b v}{\partial (1/T)}\right|_{\tau_a} + \left.\frac{1}{k}\frac{\partial \Delta S}{\partial (1/T)}\right|_{\tau_a} - \frac{\Delta H}{k} \quad (21)$$

It is immediately apparent that direct information about ΔH can only be obtained if the temperature dependence of the mobile dislocation density is well characterized. Similarly, information about the barrier profile can only be elucidated if the stress dependence of the mobile dislocation density is also well characterized.

One potential approach to this problem is the use of very rapid incremental changes in stress or temperature which, hopefully, would not substantially change the mobile dislocation density; thereby, permitting the term containing this density to be neglected. Unfortunately, the 'instantaneous' reponse of the load to an instantaneous increment (or decrement) in strain-rate (or the strain-rate response to a load change) are not unambiguously delineated. Rather, a gradual curvature leading to a new steady-state is usually observed (fig. 7). There is ample opportunity for the mobile dislocation density to change before the steady-state is attained (it has generally been the convention to use the steady-state condition). It is not reasonable, therefore, to implicitly assume that the mobile dislocation density term can be neglected; some independent verification is needed.

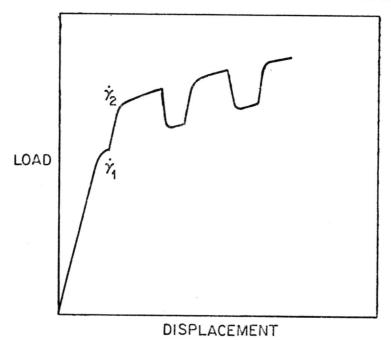

Figure 7. A typical load response to incremental and decremental
strain-rate changes.

The difference between the stress differential of the dislocation velocity (eqn. 7) and the stress differential of the strain-
rate (eqn. 10) is clearly the stress differential of the mobile
dislocation density. Inspection of these stress differentials
indicates that when Δa^* (and hence m) is large, the stress dependence
of ρ_m should not be as dominant an effect as in situations where
Δa^* is small. It might well be reasonable, therefore, to neglect
effects due to changes in ρ_m when Δa^* is large. This general result
is largely confirmed by experiments which compare m values obtained
from dislocation velocity measurements (eqn. 18) with m values
obtained from strain-rate cycling measurements,† extrapolated
to zero strain. Any discrepancy between the m values may indicate
a significant stress dependence of ρ_m.**

$$\dagger \quad m = \frac{\partial \ell n \dot{\gamma}}{\partial (1/T)} \bigg|_T = \frac{\partial \ell n \rho_m}{\partial \ell n \tau_a} \bigg|_T + \frac{\partial \ell n \nu b}{\partial \ell n_a} \bigg|_T - \frac{1}{kT} \frac{\partial \Delta G}{\partial \ell n \tau_a} \bigg|_T$$

**Discrepancies due to the development of an 'internal stress'
may also be possible.

Data for relatively impure NaCl and CaF$_2$ with large m values
[23] (>17) gave reasonably good m correlations, but data for
a 'pure' NaCl [22] with a much smaller m ($\sim$3-6) indicated major
discrepancies. More data of this type is evidently needed to
examine the generality of this result. But, in the interim,
it seems reasonable to obtain barrier profiles directly from
strain-rate data when m is large, say >20, (using the procedures
outlined above for the velocity data). When m is smaller, models
for the stress and temperature dependence of ρ_m must be developed
for effective barrier profile analysis. Some models of this
type are discussed by Kocks et al. [1]

4.2 Experimental Techniques

    The preceeding section has emphasized the interpretive advantages
afforded by direct dislocation velocity studies compared to macroscopic strain-rate studies. But, dislocation studies are experimentally much more difficult to perform; and it is extremely
time consuming to amass the substantial quantities of data needed
for a comprehensive barrier profile analysis. In many cases,
therefore, it might be possible to cautiously combine limited
dislocation velocity data with more comprehensive strain-rate
data. We shall, thus, briefly describe dislocation velocity
and strain-rate techniques.

    (a) Dislocation Velocities Studies. Most direct dislocation
velocity data has been obtained using etch pitting techniques,
following the original work by Gilman and Johnson. [24] This
technique gives a measure of the average velocity of isolated
dislocation segments moving near the surface of the material.
The velocity data for given external conditions (of stress and
temperature) exhibit a statistical variation, which for 'pure'
NaCl approximated a Poisson function. [22] The median velocity
of many dislocation segments, obtained under isothermal constant
stress conditions for a range of stresses and temperatures, may
be conveniently plotted on $\ell n v$, τ_a (or $\ell n \tau_a$) axes for each temperature, as shown schematically in fig. 8. A least square fit to
the data can then be used to compute the functional temperature
(at constant stress) and stress (at constant temperature) dependencies needed for barrier profile analysis.

    A possible limitation of the etch pit technique is that it
only examines the near surface motion of dislocations and, hence,
is susceptible to environmental influences; and the work by Westwood
and colleagues [25] has clearly indicated the important role
of the environment on dislocation motion in many materials.
It is important, therefore, that the experiments are performed
in non-active environments. The problem of potential dislocation/

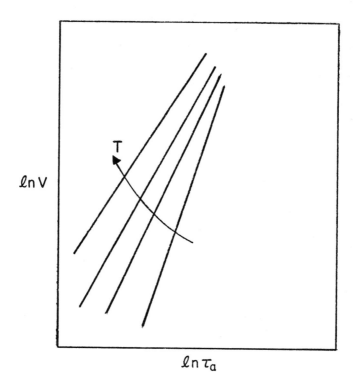

Figure 8. A schematic of the stress dependence of the dislocation velocity, showing the effect of temperature.

surface interactions still exists, however, and it would be useful
to have available a technique that can investigate the internal
motion of dislocations, for comparison purposes. A technique
with this potential is x-ray topography. [26] It is a difficult
technique to use and is confined to materials with a very low
dislocation density. But, where it can be applied it has the
basic prerequisites for effective dislocation velocity characterization.

(b) Strain Rate Studies. Macroscopic strain-rate studies
are much more readily performed than the dislocation velocity
studies. The experiments can be performed in either of two modes:
constant displacement rate or constant load. Usually, the experi-
mennts will be performed using incremental changes in the external
conditions (of temperature, load, or displacement rate) until
the new 'steady state' is achieved: alternate increments and
decrements are preferred.

5. DOMINANT BARRIERS IN CERAMIC MATERIALS

5.1 Theoretical Models

A comprehensive list of potential dominant barriers in ceramic materials was presented in a previous publication. [27] More recently, Gilman [28] has proposed another potential impurity barrier, which can be added to this list of possibilities. His model derives its dislocation resistance from the increase in electrostatic energy that occurs when an aliovalent impurity complex is intersected by a dislocation.

Detailed treatments of several barriers are available in the literature in a form which permit a proper comparison with experimental barrier profile analysis, see for example, ref. 1. Many more barrier types are inadequately modelled, or the models are not sufficiently distinctive to permit a definitive identification based on an experimental profile. The adequately modelled barriers include several of the 'point' barriers, such as solutes and precipitates, and a formal treatment of these can be found in the review by Kocks et al. [1] The potential for empirical distinction of the various point barriers is, thus, rather good although some additional model development will probably be required. Since examination of alloying effects has been suggested as one of the important objectives of thermal activation studies in ceramic materials, the existing status of the theoretical models is satisfying. The linear barriers, such as the Peierls barrier, cross slip etc., are much more difficult to model; and the available models do not exhibit sufficiently characteristic profiles for an unambiguous experimental identification, using barrier profile analysis. For these cases, the formal thermal activation analysis is of questionable merit at this time. Instead, indirect inference might be used to signify the dominant barrier. For example, cross slip only occurs at screw dislocations; and hence, separate dislocation velocity studies on edge and screw components might reveal different behaviors that could be attributed to cross slip control at the screw dislocations.

5.2 Data Analysis

The only available data at the time of writing that is sufficiently extensive to attempt a detailed barrier profile analysis is the data on 'pure' NaCl obtained by Argon and Padawar. [22] The analytical sequence recommended in the preceeding section terminates, however, after the first step; because it is found that the activation enthalpy exhibits a strong temperature dependence. From this point barrier analysis procedes less formally, involving some inference (see below), and suggests that forest intersection is the dominant barrier (with the temperature dependent ΔH being attributed to constriction of the dislocation core prior to inter-section). It is unfortunate that there are no equivalent data for a comprehensively modelled dominant barrier to demonstrate the total profile analysis.

A substantial amount of uncoordinated dislocation velocity and strain-rate data are available for ceramic materials, but these are only amenable to a partial analysis. The first useful result comes from isothermal strain-rate cycling studies on a range of different ceramic materials: NaCl, [22, 23] MgO, [29] AgCl. [36] The stress increment, $\Delta\tau_a$, during a strain-rate incre-ment, $\Delta\dot{\gamma}$, is independent of strain in all cases, except for the 'pure' NaCl where $\Delta\tau_a$ increases with strain. If we approximate the increments by a differential to obtain (from eqns. 10 and 15:

$$\left.\frac{\partial\ell n\dot{\gamma}}{\partial\tau_a}\right|_T = b\Delta a^* + \left.\frac{\partial\ell n\rho_m}{\partial\tau_a}\right|_T + \left.\frac{\partial\ell nb\nu}{\partial\tau_a}\right|_T , \tag{22}$$

we note that, if the stress dependencies of ρ_m and ν are essentially strain independent, the strain independence of $\Delta\tau_a$ implies a strain independent apparent activation area. A possible inference, therefore, is that the barrier density (reflected in Δa^*) is strain independent when $\Delta\tau_a$ is strain independenct. This is not, of course, an unambiguous conclusion; but it does suggest that a strain dependent dislocation barrier (such as dislocation intersection) may be dominant in the 'pure' NaCl; [22, 27] whereas, a non-dislocation barrier (impurity interaction, Peierls, etc.) may be dominant for the other materials. [23, 27, 28, 29] The 'pure' NaCl is also unique in that it exhibits a relatively small m value at low temperatures, which tends to reinforce the conclusion that a different barrier is dominant in this material.

The nature of the strain independent barrier is uncertain; but, for NaCl at least, it is unlikely to be a Peierl's barrier; because this is inconsistent with the transition to a strain dependent barrier in the 'pure' material. Unfortunately, the temperature range of the available velocity data is not extensive enough to obtain valid barrier profile characteristics from a prior determination of ΔG. But, some limited information about the barrier profile can be obtained from the isothermal velocity data, if we <u>assume</u> that ν is relatively stress independent; because we can then use eqn. (15) to obtain Δa^*. The velocity/stress data usually exhibit a good empirical fit to a logarithmic plot over a wide velocity range, i.e., m is constant. Hence, from eqn. (20) we note that the following approximate proportionality applies;

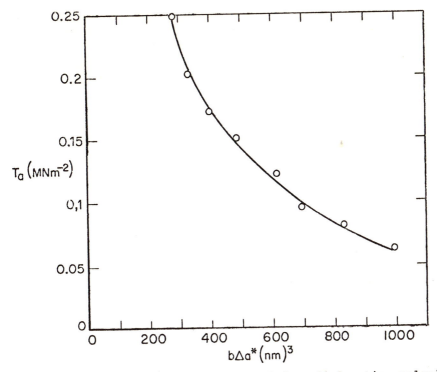

Figure 9. A stress profile obtained from dislocation velocity
          data for 'impure' NaCl at -29°C, indicating that only
          the central portion of the profile is defined by the
          data.

$$\Delta a* \propto \tau_a^{-1}$$

This relation is plotted in fig. 9 using m values for an impure
NaCl. It is apparent that only the central zone of the barrier
profile can be obtained from this data. Additional data at higher
and lower temperatures are needed to assess the characteristics
of the lower and upper portions of the profile. Then, some reasonable conclusions concerning the character of the dominant barrier
might be possible.

REFERENCES

1. U. F. Kocks, A. S. Argon and M. F. Ashby, "Progress in
    Materials Science, in press.

2. See for example, A. G. Evans and R. D. Rawlings, Phys.
    Stat. Sol, 34, 9 (1969).

3. M. F. Ashby, Acta Met., $\underline{20}$, 887 (1972).

4. R. M. Cannon and R. L. Coble, this volume.

5. T. G. Langdon, this volume.

6. S. M. Wiederhorn, B. J. Hockey and D. E. Roberts, Phil. Mag., $\underline{23}$, 783 (1973).

7. A. G. Evans, Phil Mag., $\underline{22}$, 841 (1970).

8. P. C. Paris and G. C. Sih, ASTM STP, $\underline{381}$, (1967).

9. R. M. Thomson, private communication.

10. A. N. Stroh, Advan. Phys., $\underline{6}$, 418 (1957).

11. A. G. Evans, "Ceramics for High Performance Applications," proceedings of a conference held at Hyannis, Mass. in Nov. 1973, in press.

12. P. L. Gutshall and C. E. Cross, Eng. Frac. Mech. $\underline{1}$, 463 (1969).

13. A. G. Evans, M. Linzer and L. R. Russell, Mat. Sci. Eng. $\underline{15}$, 253 (1974).

14. F. F. Lange, in ref. 11.

15. C. N. Alquist, Acta Met. in press.

16. E. Smith, Acta Met. $\underline{16}$, 313 (1968).

17. R. J. Stokes, Proc. Brit. Ceram. Soc., $\underline{6}$, 189 (1966).

18. A. G. Evans, C. Roy and P. L. Pratt, Proc. Brit. Ceram. Soc., $\underline{6}$, 173 (1966).

19. G. W. Groves and A. Kelly, Phil. Mag. $\underline{8}$, 877 (1963).

20. D. C. Phillips and A. S. Tetelman, Composites, $\underline{3}$, 216 (1972).

21. T. H. Alden, Trans. Metall. Soc. AIME, $\underline{230}$, 649 (1966).

22. A. S. Argon and G. E. Padawar, Phil. Mag., $\underline{25}$, 1073 (1972).

23. A. G. Evans and P. L. Pratt, Phil. Mag., $\underline{21}$, 951 (1970).

24. J. J. Gilman and W. G. Johnson, Solid State Physics, $\underline{13}$, 147 (1962).

25. A.R.C. Westwood, P. L. Goldheim and R. G. Lye, Phil. Mag. 17, 951 (1968).

26. A. R. Lang, Acta Cryst., 12, 249 (1959).

27. A. G. Evans, Proc. Brit. Ceram. Soc., 15, 113 (1970).

28. J. J. Gilman, Jnl. Appl. Phys., 45, (1974).

29. M. N. Sinha, D. J. Lloyd and K. Tangri, Phil. Mag., 28, 1341 (1973).

30. D. J. Lloyd and K. Tangri, Phil. Mag., 26, 665 (1972).

RHOMBOHEDRAL TWINNING IN ALUMINUM OXIDE

William D. Scott

Ceramic Engineering Division
University of Washington
Seattle, Washington 98195

ABSTRACT

A model is presented for the atom movements and structure of rhombohedral $\{\bar{1}101\}$ type twins in aluminum oxide. The oxygen lattice twinning is the same as $\{11\bar{2}2\}$ type twins in hcp metals. The aluminum lattice has the correct twin orientation but is faulted at the boundary and is not a true reflection across the twin plane.

INTRODUCTION

Rhombohedral twinning is an important mode of deformation in aluminum oxide. It has been observed in uniaxial compression of single crystals between 1100°C and 1500°C by Stofel and Conrad (1) and Conrad et al. (2) and by Bertolotti and Scott (3) between 1500°C and 1700°C. In low temperature compression tests, Auh and Scott (4) have observed substantial large-scale twinning at 600°C at a resolved shear stress as low as 14.7 MN/m^2 (2137 psi).

Rhombohedral twins also form under conditions other than uniaxial compression. Heuer (5) observed twins in bend test specimens after fracture at -196°C. Twins also commonly form as a result of thermal shock as reported by Kronberg (6) and Barber and Tighe (7). Recently Hockey (8) observed rhombohedral twins in the electron microscope which originated from room temperature grinding of the

This work supported by NASA under Grant NGL 48-002-004 to the University of Washington.

surface of alumina. Twins also play an important role in fracture because twin-twin and twin-grain boundary intersections often nucleate cracks (3,5,8).

This paper presents a model for the atom motions involved in rhombohedral twin formation in aluminum oxide and a model for the twin boundary structure.

CRYSTALLOGRAPHY OF RHOMBOHEDRAL TWINNING

Heuer (5) determined the crystallographic elements of rhombo-hedral twinning in Al_2O_3 as follows:*

$$K_1 = (10\bar{1}1), \quad K_2 = (\bar{1}012), \quad \eta_1 = [10\bar{1}2], \quad \eta_2 = [10\bar{1}1], \quad s = 0.202.$$

The first undistorted plane, K_1 is the twin composition plane and contains the glide direction, η_1. The second undistorted plane is K_2^0 before twinning and K_2 after twinning. It contains the direction η_2, the reciprocal twinning direction. The plane perpendicular to K_1 and K_2 containing η_1 and η_2 is called the plane of shear (9,10). In Al_2O_3 this is the $\{11\bar{2}0\}$ type plane. The shear is defined by the distance s through which a point moves that lies a unit distance from K_1. The angle between K_2^0 and K_2 is given by

$$\Theta = 2 \arctan \frac{s}{2}. \tag{1}$$

There are three K_1 planes of the type $\{10\bar{1}1\}$ in aluminum oxide. As pointed out by Schmid and Boas (11) and Frank and Thompson (12), complete twinning shear by compression can only occur if the compressive stress is applied in crystallographic directions between K_1 and K_2^0. It is possible, therefore, to delineate directions in the crystal where tensile or compressive stresses can produce twinning on one, two, or three twinning systems. Figure 1 shows a stereographic projection of these regions as seen from the c-axis (13,14).

The poles of the three twin composition planes are designated r_I, r_{II}, and r_{III}, and the traces of the appropriate K_1 and K_2^0 planes are shown. The Roman numerals in the various areas of the projection indicate those twin systems which will produce contraction, i.e., operate under compressive stresses applied in the region shown. The operation of a twin system not designated in a region of the projection will cause extension.

---

*Miller-Bravais hexagonal indices based on the morphological unit cell, c/a 1.365 are used throughout this paper when referring to Al_2O_3. See Refs. 13,14,15 for discussion of Al_2O_3 structure.

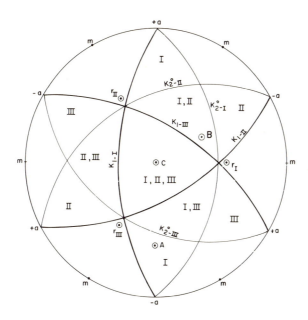

Figure 1. Stereographic projection of alumina showing the three
rhombohedral twinning systems. See text for full explanation of
the symbols.

The point marked A in the lower part of Fig. 1 is the location of the compression axis in the specimens tested by Stofel and
Conrad (1). From their specimen orientation and photographs one
can confirm that twin system I was the operating system with a
Schmid Factor (resolved stress factor) of only 0.07. In contrast,
system r_{II} had a Schmid factor of 0.33 but did not operate under
this compressive loading because the shear direction could only
produce extension.

The point marked B in the upper right quadrant of Fig. 1 is
the compression axis position of a crystal deformed by Bertolotti
and Scott (3) at 1500°C and 4000 psi compressive stress. Of the
two possible systems, I and II, only system I operated, although
their respective Schmid factors were 0.216 and 0.235.

There are several ways to describe the crystallography of
twinning. For example, one can describe the twin by 180° rotation about the twin composition plane normal, by reflection across
the twin composition plane or by the usual twinning elements and
shear as determined by Heuer (5). The actual atom movements required to produce the twin are not necessarily the same as the
crystallographic translations.

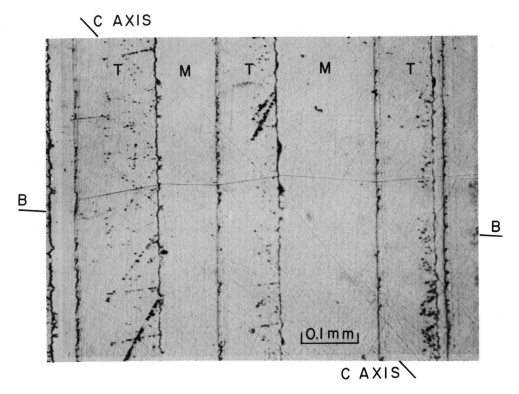

Figure 2.   Twinning shear as shown by the offset of a polishing
scratch. T designates a twinned area, and M the matrix. The
compression axis is marked B as in Fig. 1, and the face shown is
an m plane.

        The surface tilt due to twinning shear has been measured by
Heuer (5), and the present author has measured twinning shear
from scratch offsets such as that shown in Fig. 2.   The measurements of the latter specimens were not exact because of simultaneous dislocation shear, but the results in both investigations
gave shear offsets between 8° and 12.5°.   It was concluded
that the crystallographic shear angle of 11.5° was the true atom
displacement in twinning.

                          ATOM MOVEMENTS IN TWINNING

A.   Structure of Aluminum Oxide

        In discussing the relationships of twinning and structure,

we will follow the notation and the excellent structural diagrams
published by Kronberg (15). The basal plane arrangement and the
axis system for Al_2O_3 are shown in Fig. 3. Atom movements and
twinning shear are viewed perpendicular to the plane of shear,
and in Fig. 3 this is the (11$\bar{2}$0) plane shown. It is chosen here
to pass through the centers of the aluminum ion sites, and it is
the plane which will be viewed in later figures.

The arrangements of aluminum ions and vacancies in the
morphological unit cell are shown in Fig. 4. Two of the bounding
planes in the diagram are a (11$\bar{2}$0) plane of shear and a (1$\bar{1}$01)
twinning plane. In the actual structure, the aluminum ion sheets
are not planar but are puckered because Al ions adjacent in the
c-direction move vertically away from one another toward the empty
site above and below them. This slight deviation from planar
positions will not be shown in the diagrams.

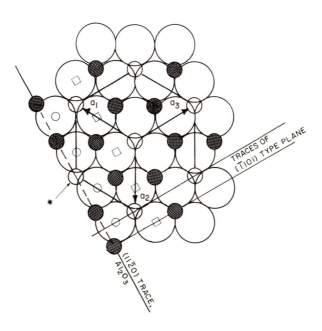

Figure 3. Basal plane of alumina. The large circles are oxygen
ions, and the round and square open symbols conform to symbols
used in later diagrams. The aluminum ions are the cross-hatched
circles, and empty octahedral sites are open circles. The empty
site marked with an asterisk is a reference point for later
diagrams. The (11$\bar{2}$0) plane is perpendicular to the paper, while
the {$\bar{1}$101} type planes are at 57.6° to the basal plane.

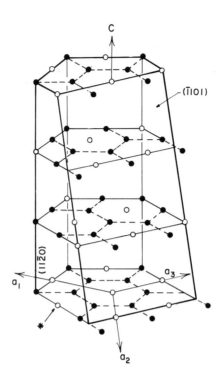

Figure 4.   Arrangement of aluminum ions and vacancies in the
morphological unit cell.   The basic diagram is after Kronberg (15).

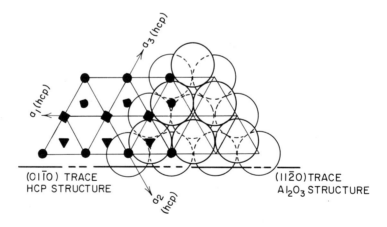

Figure 5.   Hexagonal close packed oxygen structure showing two
layers in the basal plane and the symbols which will be used to
identify particular atoms.   The difference between the normal hcp
and the Al_2O_3 axis system is discussed in Ref. 15.

The oxygen ions in Al_2O_3 are in hexagonal close packed arrangement. Figure 5 shows two repeat hcp basal plane layers and the symbol notation which will be used to identify particular atoms. Note that the conventional axis system in hcp is rotated 30° from that adopted from Al_2O_3 (Fig. 4). All notations in this paper will be based on the Al_2O_3 morphological unit cell unless they are identified as hcp.

Figure 6 shows the proposed twinned structure for Al_2O_3 as viewed perpendicular to the plane of shear, $(11\bar{2}0)$. The atom movements for twinning will be considered separately for the oxygen lattice and the aluminum lattice. Symbols for the oxygen packing are shown in the lower part of Fig. 6.

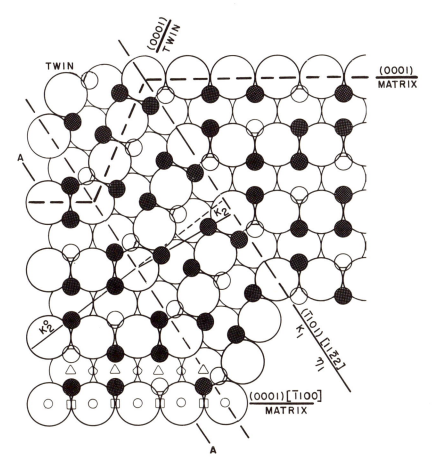

Figure 6. The proposed rhombohedral twin structure in Al_2O_3. The plane of the paper is $(11\bar{2}0)$ and the twin boundary, $(\bar{1}10\bar{1})$, is perpendicular to the plane of the paper.

B.   Atom Movements in the Oxygen Lattice

     The oxygen ion lattice is shown in Fig. 7 with the first four
(11$\bar{2}$0) planes indicated by different symbols in conformance with
Figs. 3, 5 and 6.   For the oxygen lattice, the ($\bar{1}$101) twin plane
in Al$_2$O$_3$ is the same as the (11$\bar{2}$2) twin plane in hcp metals because of the 30° rotation of the axis system in Al$_2$O$_3$.   Kronberg
first pointed out the relationship of these two twin composition
planes, and his unpublished work on rhombohedral twinning in
alumina has been cited by Heuer (5) and Rosenbaum (16).

     Twinning has been reported on {11$\bar{2}$2} planes (hcp) in Zr (17)
and Ti (18) and the K_1, η_1, K_2, η_2, elements are given by

$$\{11\bar{2}2\} \ \langle 11\bar{2}3 \rangle \ \{11\bar{2}4\} \ \langle 22\bar{4}3 \rangle .$$

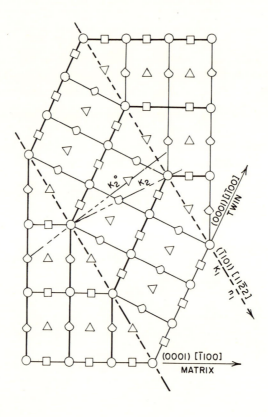

Figure 7.   Oxygen lattice as projected on the (11$\bar{2}$0) plane.   The
symbols correspond to the atom positions indicated in Figs. 3,
5, and 6.   The shear K_2^o to K_2 is produced by compression as the
twin advances from lower left to upper right.

In this twinning mode, only one-third of the atoms are sheared directly to the correct twin position (18,19). The remaining atoms acquire the twinned position by shuffles in the K_1 plane parallel and perpendicular to the η_1 direction. When the oxygen atoms in Al_2O_3 twin on the system identified by the Heuer (5) they are, relative to their own hcp packing, twinning on the same system cited above for Zr and Ti.

Figure 8 shows the motions of the oxygen atoms necessary to produce one unit of twin advance. Six ($\bar{1}101$) planes take part, but only the atoms in the third and sixth planes shear directly to their final positions (18). Other atoms must undergo some motion other than the shear to assume their final twinned positions. This extra non-shearing motion is called shuffling.

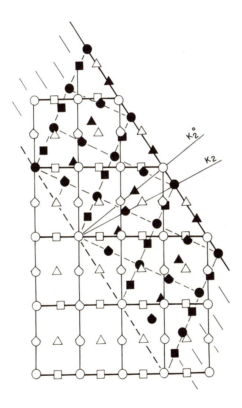

Figure 8. Relative positions of matrix (open symbols) and twin atoms (shaded symbols). The matrix is converted to the twin orientation by shear down and to the right, while the twin advances upwards from left to right. Only atoms in the third and sixth ($\bar{1}101$) planes shear direct to their final positions.

Crocker and Bevis (18) introduced the concept of representing
a lattice point in the hcp structure by motif pairs of atoms.
Figure 9 uses this motif notation to show a section of the twin
from Fig. 8. The motif pairs in the matrix are connected by a
line, and these two atoms are considered to move together. Once
again, the third and sixth planes shear direct. Other atoms must
undergo two kinds of shuffles. The first kind of shuffle is perpendicular to the plane of the paper and is indicated by the change
in symbol type. This type of shuffle occurs in all but the third
and sixth planes. It is in the opposite direction for each of the
two atoms of the motif pair and occurs between adjacent $\{11\bar{2}0\}$
planes (see symbols, Fig. 5). The second type of shuffle is parallel to the direction of shear. These shuffles are shown as vectors in Fig. 9. The length of the vector is the length of the
shuffle: i.e., shear plus or minus the shuffle equals the new
position. The shuffles in this twinning scheme obey the rules
formulated by Bilby and Crocker (20) in that the total vector addition of all the shuffles is zero so that no shear displacement
other than that given by the twinning shear occurs.

C.   Atom Movement   in the Aluminum Lattice

The selection of the twin boundary in Fig. 6 was based on
three considerations: 1) The twin boundary should pass through
the centers of the close packed oxygen to make a perfect twin in
this lattice, 2) aluminum ion sites should not be adjacent to one

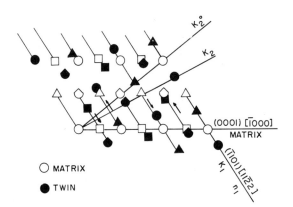

Figure 9. Movement of oxygen atoms from matrix (open symbols)
to twin (shaded symbols). The arrows indicate the direction and
magnitude of shuffles required to assume the twinned positions.
The symbols ◯ △ ▢ ◇ indicate atoms sequentially lower in the
plane of the figure. See Fig. 5.

another across the twin boundary, and 3) the shears and shuffles of the aluminum ions should be in the {Ī101} type twinning plane and not perpendicular to it. No true reflection of the aluminum lattice across the twinning plane fulfills these conditions. However, the boundary selected in Fig. 6 can be produced by shear which creates the correct twin orientation relative to the matrix. Furthermore, the repeat spacing of this boundary is the same as the six {Ī101} planes involved in the twinning shear of the oxygen lattice.

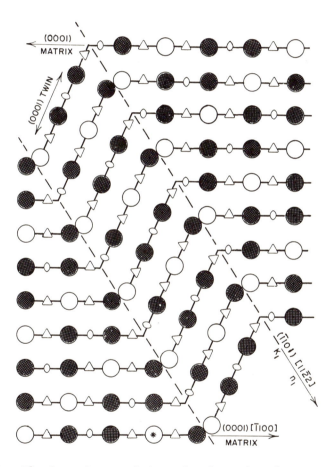

Figure 10. Aluminum ion positions in the twinned structure The large circles are in the plane of the paper and are shown cut by the (11Ī0) plane in Fig. 3. The triangles are the next layer of Al, and the ovals are empty sites in this layer.

Note in Fig. 6 that the two boundaries of the twin are not symmetrical. One the lower left boundary, the holes are in the matrix, and the first ($\bar{1}101$) planes of the twin contains Al ions, while on the upper right boundary the holes are in the twin and the Al ions are in the matrix. This asymmetry could be reversed by selecting the twin boundary along the line marked A-A.

Figure 10 shows the arrangement of the first two ($11\bar{2}0$) planes of aluminum ions which repeat through the structure. The boundary structure consisting of holes adjacent to Al ions is maintained in the second and subsequent layers.

Because of the asymmetry of the two twin boundaries, the Al ion movements during the advance of the twin are different in two directions. These movements are shown in Fig. 11. If the twin advances upwards to the right, the Al ions in the second and fifth $\{\bar{1}101\}$ planes shear direct while those in the third and fifth shuffle in the same direction as the shear and change ($11\bar{2}0$) planes perpendicular to the shear. Both shuffles are in the $\{\bar{1}101\}$ type plane.

If the twin advances downward to the left, the same two planes shear direct. However, the Al ions in the first and fourth planes must shuffle counter to the shear direction while changing ($11\bar{2}0$) planes. An alternative movement for these atoms would be a longer shuffle in the direction of shear but no change in ($11\bar{2}0$) planes perpendicular to the shear direction.

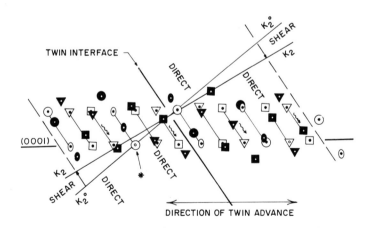

Figure 11. Aluminum ion movement during twinning. Open symbols are matrix positions, and filled symbols are twin positions. The arrows indicate shuffle directions.

It is not possible to predict from the present information
which of the two kinds of advance may be preferred. Asymmetric
behavior in twin growth has been observed by Auh and Scott (4)
where one boundary of a twin advanced while the other remained
stationary. Asymmetry has also been observed in twin removal at
high temperatures by Bhandari et al. (21). In many other cases,
however, rapid twin growth in two directions has been observed
which indicates that both types of shear are feasible or that
symmetric boundaries can be formed.

A symmetric twin may be generated by the formation of a thin
twin or faulted structure in the region A-A in Fig. 12. This
would require slip on the ($\bar{1}$101) plane with typical atom motions
shown. The entire upper right portion shears downwards to the new
position. Rosenbaum (16) has developed a detailed model for slip
on the (11$\bar{2}$2) (hcp) surface, and non-basal dislocations have been
reported in alumina (22).

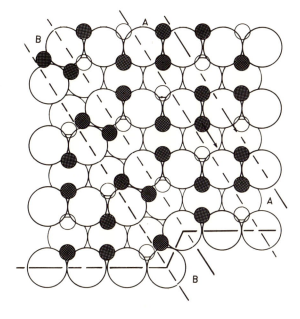

Figure 12. Production of a symmetric twin by shear faulting in
a narrow region A-A. Typical atom movements are indicated by
arrows. This shear could produce the thin symmetric twin shown
at B-B which could then advance in units of six ($\bar{1}$101) planes.

Fig. 13 compares the 6-fold packing of oxygen ions around an aluminum ion in the normal structure (A) and in the twin boundary (B). It is apparent that the ocahedral Al site in the boundary is distorted. Measurements of relative twin boundary interfacial energies (21) gave a value of 0.31 which is about 60% of the value 0.54 reported for high angle tilt boundaries in alumina (22). This high twin boundary energy is consistent with the twin model proposed here.

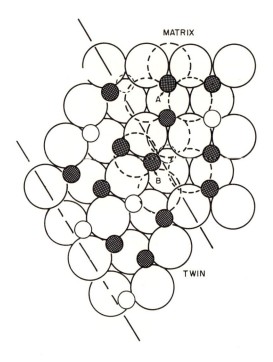

Figure 13. Comparison of oxygen coordination around an Al ion in normal packing and in the twin boundary. The dashed circles are oxygen ions in the next layer above the paper. Site A is normal octahedral coordination, and B is the distorted twin boundary site. The oxygen ions are shown overlapping in the twin boundary because the necessary dialation in the boundary was not included in the drawing.

CONCLUSIONS

A ($\bar{1}101$) twin structure in alumina has been proposed in which the oxygen ion movements are the same as the $\{11\bar{2}2\}$ (hcp) twinning system in hcp metals. An aluminum ion structure consistent with the oxygen ion motion is proposed. Although the relative orientations of the twin and matrix are correct in the crystallographic sense, the twin boundary is a faulted structure which is not a true twin reflection for the aluminum ion lattice. Furthermore the twin boundaries in this structure are asymmetric with different kinds of faulted boundaries at each of the two sides of the twin bands. Symmetric twin boundaries are produced if a thin twin fault is first produced by shear before twin growth occurs. Aluminum ions in the boundary are in distorted octahedral sites which accounts for the high energy interfacial observed in these boundaries.

ACKNOWLEDGEMENTS

The author thanks Keunho Auh and Raymond L. Bertolotti for providing experimental specimens and unpublished data for study.

REFERENCES

1. E. Stofel and H. Conrad, Trans. AIME 227 1053 (1963).

2. H. Conrad, K. Janowski and E. Stofel, Trans. AIME 233 255 (1965).

3. R. L. Bertolotti and W. D. Scott, J. Amer. Ceram. Soc. 54 [6] 286 (1971).

4. Keunho Auh and W. D. Scott, to be published.

5. A. H. Heuer, Phil. Mag. 13 [122] 379 (1966).

6. M. L. Kronberg, Mechanical Properties of Engineering Ceramics, p. 329, Interscience, New York (1961).

7. D. J. Barber and Nancy J. Tighe, Phil. Mag. 11 [111] 495 (1965).

8. B. J. Hockey, Proc. Brit. Ceram. Soc., [20] 95 (1972).

9. C. S. Barrett, Structure of Metals, p. 383, McGraw-Hill, New York (1952).

10. S. Mahajan and D. F. Williams, Intl. Metallurgical Reviews, 18 43 (1973).

11. E. Schmid and W. Boas, Kristallplastizität, Springer, Berlin, 1935 and translation, Hughes, London, 1950.

12. F. C. Frank and N. Thompson, Acta Met., 3 30 (1955).

13. Horace Winchell, Bull. Geol. Soc. Am. 57 295 (1946).

14. J. B. Wachtman, Jr., W. E. Tefft, D. G. Lam, Jr., and R. P. Stinchfield, J. Res. Natl. Bur. Stds., 64A [3] 213 (1960).

15. M. L. Kronberg, Acta Met., 5 507 (1957).

16. H. S. Rosenbaum, Deformation Twinning, pp. 43-76,

17. E. J. Rapperport and C. S. Hartley, Trans. AIME 218 869 (1960).

18. A. G. Crocker and M. Bevis, pp. 453-58 The Science, Technology and Application of Titanium, R. I. Jaffee and N. E. Promisel, Ed., Pergammon, 1969.

19. E. J. Rapperport, Acta Met., 7 254 (1959).

20. B. A. Bilby and A. G. Crocker, Proc. Roy. Soc. A288 240 (1965).

21. Om P. Bhandari, R. L. Bertolotti and W. D. Scott, Acta Met. 21 1515 (1973).

22. J. D. Snow and A. H. Heuer, J. Amer. Ceram. Soc. 56 [3] 153 (1973).

23. J. F. Shakelford and W. D. Scott, J. Amer. Ceram. Soc., 51 [12] 688 (1968).

PYRAMIDAL SLIP ON $\{11\bar{2}3\}$ $<\bar{1}100>$ AND BASAL TWINNING IN Al_2O_3

B. J. Hockey

Institute for Materials Research
National Bureau of Standards
Washington, D. C. 20234

ABSTRACT

Plastic deformation of Al_2O_3 by slip and twinning has been investigated by examining the regions surrounding a microhardness indentation using transmission electron microscopy (TEM). The results establish: (1) the occurrence of pyramidal slip on $\{11\bar{2}3\}<\bar{1}100>$, and (2) the nature of basal twins in this material. The observations on basal twins, in particular, have led to a completely different description for the twinning process, which is briefly described.

1. INTRODUCTION

Previous observations by TEM [1] have shown that generalized plastic flow involving both slip and twinning occurs during the room temperature indentation of Al_2O_3. Indentation of various surface orientations--for example, by the placement of indentations within individual grains of polycrystals--thus provides a simple experimental means for causing and subsequently studying the possible modes of deformation by TEM.

In this paper, the general usefulness of these techniques, which are particularly applicable to "hard, brittle materials" [2], is illustrated and specific results (1) on the occurrence of pyramidal slip on the $\{11\bar{2}3\}$ $<\bar{1}100>$ system and (2) on the structure of basal microtwins in Al_2O_3 are presented.

2.   EXPERIMENTAL DESCRIPTION

The methods used in the initial specimen preparation, indentation, and the preparation of thinned electron microscope specimens
have been previously published [1]. In the present study, the
electron microscope was operated at an accelarating voltage of 200
kv and a double-tilt (± 45°) specimen holder was used to tilt the
specimen to various two-beam conditions.

All observations presented here were obtained from the region
surrounding a 200 gram Vickers hardness indentation, made at room
temperature within a single grain of polycrystalline Al_2O_3 (Lucalox*). Fig. 1 is a sterographic projection** onto the plane of the
foil and shows that the normal to the original crystal surface was
close to the $(11\bar{2}3)$ pole. The other poles listed correspond to the
various reflections*** used in the analysis of slip and twinning
described in the following sections.

3.   RESULTS AND DISCUSSION

A.   Slip on $\{11\bar{2}3\}$ $<\bar{1}100>$

Observations on the indented specimen revealed numerous arrays
of dislocations which extended from the indentation in various directions. Fig. 2 shows the configuration of dislocation found in
one area. In this micrograph, the beam is nearly normal to the
foil, and it is clear that the region contains arrays of short dislocation segments (E) situated on inclined planes that intersect
the foil along A-B, together with long dislocation segments (S)
which are nearly parallel to A-B. The regions of dark contrast, F,
are apparently faults which also intersect the foil along A-B. Although not apparent in this projected view, stereomicroscopy revealed
that all of the isolated dislocations (S and E) lie on the same set
of parallel planes, which is also the plane of the faults, F.

The presence of curved dislocations on parallel planes strongly suggests that the observed configuration is a result of slip.

---

*General Electric product.

**Miller-Bravais indices corresponding to the structural unit cell
c/a = 2.730, are used throughout.

***Indices of corresponding reciprocal space vectors satisfy the
reflecting conditions: - h+k+ℓ=3n; hh2hℓ:(ℓ=3n); hhoℓ:h+ℓ=3n,
ℓ=2n [3].

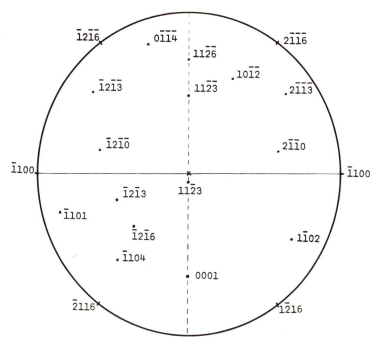

Figure 1. Stereographic projection normal to the indented specimen
surface. With the exception of 11$\bar{2}$3, all poles listed
correspond to the various reflections used in the analyses
described in Section 3.

This conclusion was confirmed by determining both the apparent slip
plane and the Burgers vector of the isolated dislocations. The
lack of direct evidence for the various slip systems proposed for
Al_2O_3 [4], make it worthwhile to consider the details of the results.

    (a) Slip Plane. With reference to Fig. 1, the intersection of
the apparent slip planes and faults, F, with the surfaces of the
foil was determined to be along [$\bar{1}$100]. Accordingly, the plane of
the dislocations must belong to the [$\bar{1}$100] zone, which is indicated
by the dotted great circle in Fig. 1. Identification of this plane
was then made by tilting the foil about [$\bar{1}$100] until the dislocations and faults were edge-on (i.e. parallel to the electron beam).
As shown in Fig. 3, this position of the foil was determined by
electron diffraction to coincide with the 11$\bar{2}$3 reflecting position.
Since 11$\bar{2}$3 lies on the [$\bar{1}$100] zone, this uniquely defines the slip
(and fault) plane as (11$\bar{2}$3).

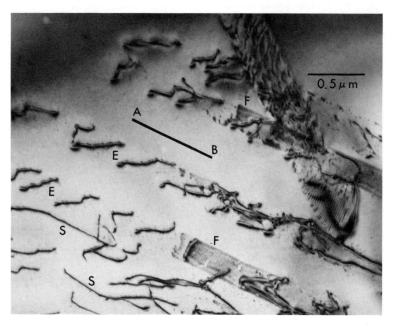

Figure 2.   Arrays of dislocations associated with 200 g Vickers
            hardness indentation which were determined to be a
            result of slip on (112̄3)[1̄100]. See text for description.

        (b) Slip Direction. Examination of the region of Fig. 2 under
various 2-beam conditions (listed in Fig. 1), showed not only that
the isolated dislocations S and E are total dislocations but that
their Burgers vector corresponds to the only expected direction of
slip on {112̄3}. The nature of the faults, however, has not yet
been established.

        The fact that there is only one expected direction of slip on
(112̄3) can be deduced from the crystal structure of Al_2O_3, which is
trigonal, (R3̄2/c[3]), and not hexagonal as implied by the use of
Miller-Bravais indices. As a consequence, it is possible to show
that since all six {112̄3} planes are crystallographically equivalent, there is but one set of equivalent directions lying on {112̄3},
namely, the <101̄0> directions. In other words, {112̄3} <1̄100>
represents the only combination of slip plane and slip (lattice)
vector which is consistent with the crystal symmetry of Al_2O_3.
Accordingly, the Burgers vector of total dislocations associated
with slip on (112̄3) is expected to be [1̄100].

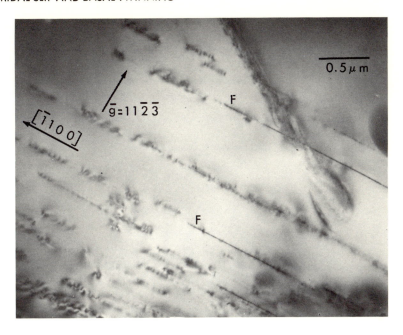

Figure 3.   Dislocations and faults are seen "edge-on" with specimen
             tilted to the 11$\bar{2}$3 reflecting position.

     In agreement, comparison of Figs. 4a and 4b with Fig. 3 shows
that the long dislocation segments (S) lying close to [$\bar{1}$100] are
either out of contrast or show only a weak residual contrast for
$\bar{g}$ = 11$\bar{2}$3 and $\bar{g}$ = 11$\bar{2}\bar{6}$. Under both conditions, $\bar{g} \cdot$[$\bar{1}$100] = 0 and it
can be concluded that these dislocations are screw or near screw
segments with $\underline{b}$ = [$\bar{1}$100]. From their line orientation, those dislocations which are steeply inclined to the foil (E) appear to
correspond to edge or near edge [$\bar{1}$100] dislocation segments. Although these dislocations are in contrast, particularly in Fig. 4b,
the nature of the line contrast is characteristic of that expected
for edge dislocations when $\bar{g} \cdot \underline{b}$ = 0, but m = 1/8$\bar{g}$ $\underline{b}$ x $\underline{u}$ $\geqslant$ 0.2, where
$\underline{u}$ is a unit vector tangent to the dislocation line (see e.g. [5]
and [6]). Since $\underline{b}$ x $\underline{u}$ is simply a vector of magnitude $|\underline{b}|$ along
the normal to the (11$\bar{2}$3) slip planes, this expression for m can be
reduced to:

$$1/8 \; |g|_{hki\ell} \quad |b| \cos \theta = 1/8 \; \frac{|b| \cos \theta}{d_{hki\ell}}$$

for pure edge dislocations, where θ is the angle between $\bar{g}_{hki\ell}$ and
the $(11\bar{2}3)$ pole. Using this expression, calculated values of m
for $(11\bar{2}3)$ $[\bar{1}100]$ edge dislocations are: m = 0.61 for $\bar{g} = 11\bar{2}\bar{6}$,
Fig. 4b, and m = 0.49 for $\bar{g} = 11\bar{2}\bar{3}$, Fig. 4a. These values are
quite large and the strong contrast observed experimentally is
expected.

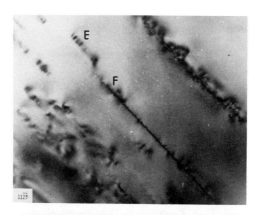

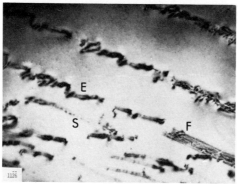

Figure 4.   Image contrast of dislocations on $(11\bar{2}\bar{3})$ planes produced
            for: a) $\bar{g} = 11\bar{2}\bar{3}$ and b) $\bar{g} = 11\bar{2}\bar{6}$. For both reflections,
            $\bar{g} \cdot \underline{b} = 0$ for $[\bar{1}100]$ dislocations. Only $[\bar{1}100]$ screw
            segments are invisible; edge segments show strong con-
            trast due to large values of m = 1/8 $\bar{g} \cdot \underline{b} \times \underline{u}$.

     In further agreement, strong double line contrast for both
screw (S) and edge (E) segments was observed in reflections for
which $\bar{g} \cdot [\bar{1}100] = 2$ at s = 0 (i.e., at the Bragg reflecting position). This is illustrated in Fig. 5, where $\bar{g} = 1\bar{1}02$.

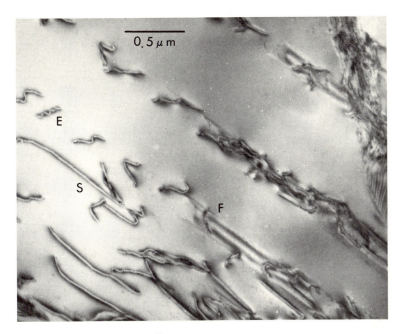

Figure 5.   Edge and screw [$\bar{1}$100] dislocation segments in double
            image contrast ($\underline{g} \cdot \underline{b}$ = 2) for g = $\bar{1}$102 at s = 0.

     Finally, a comparison was made of the observed contrast for
each of the reflections listed in Fig. 1 with that expected
assuming other values of $\underline{b}$ equal to a lattice vector. No other
possible Burger vector of reasonable magnitude was found to be
consistent with the observed results.

     The present results thus provide clear evidence that slip on
{11$\bar{2}$3} <$\bar{1}$100> occurs during the room temperature indentation of
Al_2O_3. The relative activity of this system compared to other
slip systems, however, cannot be ascertained. Conceivably, the
extremely high stresses developed during point loading may result
in the activation of certain slip systems which may never be
energetically favorable modes of bulk deformation under uni- or
bi-axial loading. The assumption then that {11$\bar{2}$3} <$\bar{1}$100> slip
occurs during the high temperature bulk deformation of Al_2O_3 cannot
be made solely on the basis of these results. However, Klassen-
Neklyudova et al. [7] have reported impurity segregation along
{11$\bar{2}$3} planes in ruby after deformation above 1850 °C. The present
results, thus, appear to justify their conclusion that these bands

are the result of slip. In addition, the present observation provide the first positive identification of glissile <$10\bar{1}0$> dislocations in Al_2O_3, which until now, could only be inferred from etch
pits or the macroscopic displacements associated with {$1\bar{2}10$} <$10\bar{1}0$>
slip [8,9,10].

B. Basal Twinning

Unlike {$11\bar{2}3$} <$\bar{1}100$> slip, there is ample evidence for deformation twinning on {0001} [11,12,13]. The macroscopic elements of
basal twinning-first determined by morphological examination and
later confirmed by X-ray diffraction-are known to be:*

$$k_1 = (0001); \quad \sigma_1 = <10\bar{1}0>$$

$$k_2 = \{10\bar{1}1\}; \quad \sigma_2 = <\bar{1}012>$$

$$S = 0.635$$

Crystallographically, twinning on (0001) can be described by a
rotation of 180° about [0001], which requires that the macroscopic
shear direction σ_1 be [$10\bar{1}0$], [$\bar{1}100$], or [$0\bar{1}10$] and not the reverse
directions.

However, as pointed out by Kronberg, [14] uniform shear displacements along the observed twinning shear direction, σ_1, cannot
possibly lead to a twinned structure. As a consequence, Kronberg
proposed a rather complicated shearing mechanism involving the
synchronized motions of "quarter-partial" twinning dislocations.

In this section, the structure of basal microtwins, as determined by TEM, is presented. The results identify the actual displacements associated with the twinned structure and have led to
the development of a completely different description of the twinning process.

Figure 1 agains specifies the orientation of the indented specimen which contained numerous thin lamallar regions that often extended far beyond the zone of intense slip deformation. The identification of these lamallae as basal microtwins was readily established
by selected area diffraction and the contrast observed under different diffracting conditions, Figs. 6a and 6b. Typically, these
twins were quite thin (≤ 2μm) and essentially wedge-shaped with
arrays of dislocations present within the twin-matrix boundary.
These features are best illustrated in Fig. 6c, which shows the
same twin viewed edge-on. Here the twin boundaries are delineated

---

*It is important to emphasize that indices based on the true struc-
tural cell are used here.

by the residual contrast associated with the boundary dislocations. Previous observations [15] on basal twins produced by indentation or surface grinding have revealed similar characteristics.

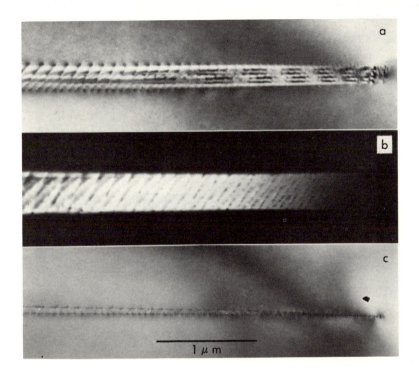

Figure 6. Basal microtwin under various diffracting conditions:
a) Bright field $\bar{g}$ = $(0\bar{1}14)_{matrix}$, b) Dark field
$\bar{g}$ = $(10\bar{1}4)_{twin}$, c) Bright field $\bar{g}$ = 000.12.

        In the present study, the peculiar nature of the twin boundaries was established by unambigously determining the Burgers vectors of the boundary dislocations by application of the $\bar{g}\cdot\underline{b}$ = 0 invisibility criterion [5]. The essential results are contained in Figs. 7a-f, which show that, for this orientation, one face of the wedge boundary contains only $\pm[10\bar{1}0]$ dislocations, while the other face contains only $\pm[0\bar{1}10]$ dislocations. Since similar contrast is observed at the points where the dislocations emerge from the foil, the dislocations within each face must be of the same sign. Although the magnitude of $\underline{b}$ was not determined, it can be shown that the boundary dislocations cannot be actual twinning partials, but instead must correspond to accommodation dislocations.

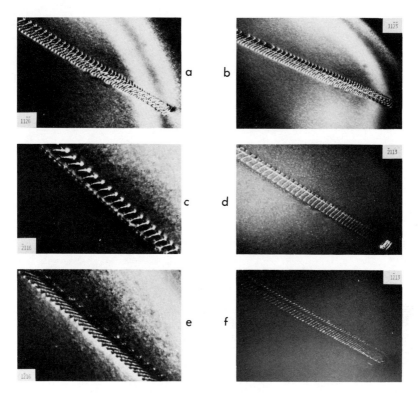

Figure 7. Basal twin boundary dislocations in different diffracting
conditions. Dislocations in both faces of wedge-shaped
twins are in contrast for a) $\bar{g}$ = 112$\bar{6}$ and b) $\bar{g}$ = 112$\bar{3}$.
Dislocations in one face are invisible for c) $\bar{g}$ = $\bar{2}$116
and d) $\bar{g}$ = $\bar{2}$113; $\bar{g}\cdot\underline{b}$ = 0 for $\underline{b}$ = x[01$\bar{1}$0]. Dislocations
in opposite face are invisible for e) $\bar{g}$ = 1$\bar{2}$16 and f)
$\bar{g}$ = 1$\bar{2}$13; $\bar{g}\cdot\underline{b}$ = 0 for $\underline{b}$ = x[10$\bar{1}$0].

For the purposes of discussion, it is assumed that the unequal
densities of dislocations present within the two boundaries simply
indicatesthat the boundaries are not equally inclined to (0001) –
the subtended angle at the tip was too small (>2° in Fig. 6c) to
allow experimental verification. Fig. 8a, thus, schematically
represents the configuration of boundary dislocations at the tip
of an ideal fully symmetric basal twin.

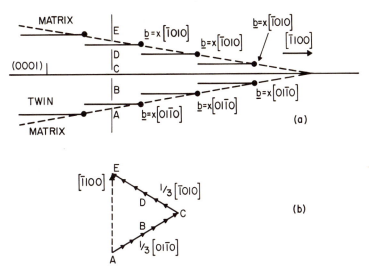

Figure 8. a) Schematic representation of configuration of boun-
              dary dislocations at the tip of a symmetric wedge-
              shaped basal twin based on present observations.
              Twin is viewed parallel to (0001) along [11$\bar{2}$0] direc-
              tion as in Fig. 6c.

           b) Vector diagram illustrating nature of the relative
              shear displacements on (0001) planes through cross-
              section of basal twin, e.g., ABCDE in (a).

     The significance of the results can be understood by noting
that, since the boundary dislocations are glissile on {0001}, each
face of the twin corresponds to a simply shear boundary. The observed structure of the twin boundaries thus directly suggests a
twinning mechanism involving uniform shear on (0001) along $\pm$[10$\bar{1}$0]
above the mid-plane of the twin and along $\pm$[0$\bar{1}$10] below the mid-
plane. Such a process is clearly imcompatible with a unidirectional
shear process normally used to describe twinning. It is also
incompatible with Kronberg's proposed dislocation model for basal
twinning [14].

The observed structure can, however, be explained on the basis
of a different and relatively simple dislocation model for basal
twinning. The details of this model are included in a formal treatment of basal twinning which is currently being prepared for publication elsewhere. Only those aspects relating to the present
results and admittedly idealized by Fig. 8a will be described here.

Basic to this model, is the realization the displacements of
$1/3[\bar{1}010]$, $1/3[01\bar{1}0]$, or $1/3[1\bar{1}00]$, i.e., opposite to the macroscopic twinning shear directions, on every other basal lattice
plane will result in a twinned structure. As a result, a symmetric
twin, whose orientation and boundary displacements are depicted in
Fig. 8a, can be described by the passage of $1/3[\bar{1}010]$ partial dislocations on every other (0001) lattice plane situated above the
mid-plane of the twin and the passage of $1/3[01\bar{1}0]$ partial dislocations on every other (0001) lattice plane below the mid-plane.
This process can be envisioned by considering the action of a
hypothetical double-ended pole dislocation source emitting
$1/3[\bar{1}010]$ dislocations at one end and $1/3[01\bar{1}0]$ dislocations at the
other end under a uniform applied stress. The resulting displacements through any cross-section of the twin, e.g. ABCDE in Fig. 8a,
will then be represented by a simple vector diagram, Fig. 8b. As
seen, the relative shear displacements above and below the mid-
plane correspond to the experimentally determined directions (the
sign of which cannot be uniquely specified). Moreover, the resultant matrix displacement across the twin is along $[\bar{1}100]$. Propagation of the twin through the crystal would thus eventually result
in a maximum surface offset along $[\bar{1}100]$, which corresponds to the
required macroscopic twinning shear direction, σ_1. Since the
passage of each $1/3[\bar{1}010]$ and $1/3[01\bar{1}0]$ dislocations produces a
net resolved displacement of $1/6[\bar{1}100]$ on every other (0001) lattice
plane, the resultant shear along $[\bar{1}100]$ is simply

$$\frac{|1/6[1100]|}{|1/6c|} = \sqrt{3}\ a/c, \text{ or } 0.634,$$

which corresponds to the experimentally verified required value.

The proposed model thus satisfies all the macroscopically
observed aspects of twinning. Unlike Kronberg's model, the syn-
chonous displacements of adjacent oxygen and aluminum lattice
planes in different direction under a uniform stress is not required.
Moreover, the resulting twin interface on (0001) is a true mirror
plane, not a shear (faulted) interface.

The author wishes to acknowledge the support of the Army
Research Office.

References

[1] B. J. Hockey, J. Am. Ceram. Soc., 54, 223 (1971).

[2] B. J. Hockey, pp. 21-50 in The Science of Hardness Testing and Its Research Applications. Edited by J. H. Westbrook and H. Conrad. American Society for Metals, Metals Park. Ohio 1973.

[3] International Tables for X-ray Crystallography, Vol. 1; No. 167, p. 275. Kynoch Press, Brimingham, England, 1962.

[4] J. D. Snow and A. H. Heuer, J. Am. Ceram. Soc., 56, 153 (1973).

[5] P. B. Hirch, A. Howie, R. B. Nicholson, D. W. Pashley, and M. J. Whelan. Electron Microscopy of Thin Crystals. Butterworth, Inc., Washington, D.C., 1965.

[6] J. M. Silcock and W. J. Tunstall, Phil. Mag. 10, 361 (1964).

[7] M. V. Klassen-Neklyudova, V. G. Govorkov, A. A. Urusovskaga, M. N. Voinova, and E. P. Kozlovskaga, Phys. Stat. Solidi, 39, 679 (1970).

[8] M. V. Klassen-Neklyudova, J. Tech. Phys. (USSR), 12, 519, 535 (1942).

[9] R. Scheuplein and P. Gibbs, J. Am. Ceram. Soc., 43, 458 (1960).

[10] D. J. Gooch and G. W. Groves, J. Am. Ceram. Soc., 55, 105 (1972).

[11] K. Veit, Neues Jahrb. Mineral., Geol. Palaeontol., Beilageband, 45, 121 (1921).

[12] E. Stofel and H. Conrad, Trans. AIME, 227, 1053 (1963).

[13] A. H. Heuer, Phil. Mag., 13, 379 (1966).

[14] M. L. Kronberg, Acta Met., 5, 507 (1957).

[15] B. J. Hockey, Proc. Brit. Ceram. Soc., 20, 95 (1972).

STRENGTHENING MECHANISMS IN SAPPHIRE

B.J. Pletka, T.E. Mitchell and A.H. Heuer

Case Western Reserve University

Cleveland, Ohio 44106

ABSTRACT

Strengthening of sapphire by work hardening, solid solution hardening and precipitation hardening is described. Examination of the dislocation structures in pure sapphire deformed by basal glide has led to the development of a semi-quantitative work hardening model; edge dipoles, formed by trapping of dislocations on parallel slip planes, are the primary obstacles to glide dislocations. As work hardening proceeds, the dipoles break up by climb into smaller loops and, when the rate of accumulation of dipoles is equal to their rate of annihilation, a region of zero work hardening is observed.

Strong solid solution hardening is observed when sapphire is doped with Cr and Ti. The relative hardening of these ions can be explained by the size difference and type of distortion produced in the lattice.

Precipitation hardening of sapphire doped with Ti is found to be no more effective a strengthening mechanism than solid solution hardening with Ti^{4+}, due to the fact that the precipitates lie in the glide plane and the dislocations can by-pass them by glide and climb with relative ease.

1. INTRODUCTION

Although pure metals often have low strengths, a variety of strengthening mechanisms are available to improve their properties so that it becomes feasible to use them in engineering applications.

Similarly in the high temperature regimes where ceramic systems
exhibit ductile behavior, their strengths often decrease rapidly
with increasing temperature; thus, they are rarely used for advanced
structural engineering applications. Particularly for this high
temperature regime, one would like to utilize proven strengthening
mechanisms to enhance mechanical performance; with this in mind,
a study of hardening processes in sapphire oriented for basal slip
has been undertaken. Experimental results and mechanisms for work
hardening, solid solution hardening, and precipitation hardening
will be presented in turn in the next three sections. Details of
the experimental procedures are available elsewhere [1,2].

2. WORK HARDENING

Figure 1 illustrates a typical stress-strain curve for pure
sapphire, deformed in compression at 1400°C with an initial
strain-rate of 1.33 x 10^{-4} sec^{-1}. On exceeding the elastic limit,
distinct upper and lower yield points are observed which are
believed to be due to a dislocation multiplication mechanism [3,4].
Just past the lower yield point, a significant initial work-
hardening rate of $\sim \mu/400$ (where μ is the shear modulus) is found,
which gradually levels off to a zero work-hardening rate at high
strains.

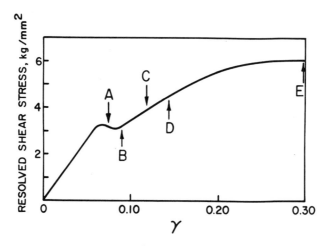

Figure 1 -- Typical shear stress-shear strain curve of sapphire
single crystals used for transmission electron microscopy. Samples identified by the letters were deformed
prior to examination in the electron microscope to the
indicated strain (T=1400°C, $\dot{\varepsilon}$=1.33 x 10^{-4} sec^{-1}).

To examine the development of the dislocation structures in
the various work hardening regions of the flow stress curve, a
series of basal plane foils was examined from specimens deformed
to the various strains indicated on the stress-strain curve in
Fig. 1, using a Hitachi HU-650B high voltage electron microscope
operating at 650 kV. Figure 2 illustrates the dislocation structure found in specimen A, which had been deformed to between the
upper and lower yield points; it is dominated by dipoles and
multipole configurations, (such as E and A-B, respectively) with
linking glide dislocations. Note that the dislocations lie predominantly along the [1$\bar{1}$00] direction and since their Burgers
vector is 1/3 [11$\bar{2}$0], they are primarily edge in character.
Through climb or cross-slip, the dipoles can pinch off at one end,
e.g. at A' and E'. Thus the glide dislocations interact with each
other to form dipole configurations, which are then linked by the
remaining portions of the glide dislocations to other dipole or
multipole formations.

At this stage of deformation, dipoles (or portions of dipoles)
with some screw character are present. However, as deformation
proceeds, the dipoles have been found to become even more predominantly edge in character. As has been discussed previously [2],

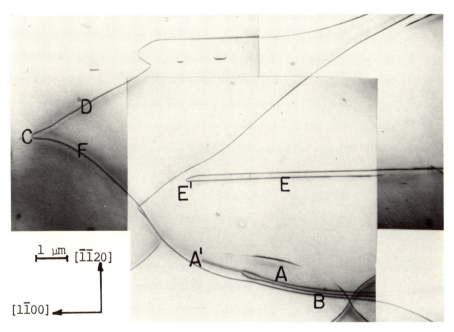

Figure 2 -- Dislocation structure of specimen deformed to between
            the upper and lower yield point (A of Fig. 1); the
            structure consists predominantly of edge dipoles (e.g.
            E) and multipole configurations (A-B). ε = 0.016
            $\vec{g}$ = (30$\bar{3}$0), basal foil.

the dipoles form by "edge trapping" of dislocations gliding on parallel slip planes; the ability of screw dislocations to annihilate by cross-slip is responsible for the dominant edge character of the dipoles.

Increasing the deformation beyond the yield point (specimen B) results in the dislocation structure shown in Fig. 3. The dipole, multipole, and linking glide dislocation structure is again apparent with virtually all the dipoles being predominantly edge in character. In addition, however, a number of small dislocation loops are present in the background. This feature of the dislocation structure, while visible at smaller strains (Fig. 2), becomes more pronounced at higher strains (Fig. 3). Such loops form through the breakup of dipoles by climb, because of the reduction in the dislocation line energy [2].

Further deformation into the work hardening region (specimen C) results mainly in an increase in dislocation density, as shown in Fig. 4. Moreover, the breakup of the dipoles has resulted in the formation of three different types of loops. One type has the same orientation as the original dipoles (along the [1$\bar{1}$00] direction) and also has the same 1/3 [11$\bar{2}$0] Burgers vector. The other two types lie at ±30° to the dipole direction. From contrast experiments in foils of various orientations, the Burgers vectors of these "transformed" dipoles have been identified as 1/3 [10$\bar{1}$0] and 1/3 [01$\bar{1}$0].

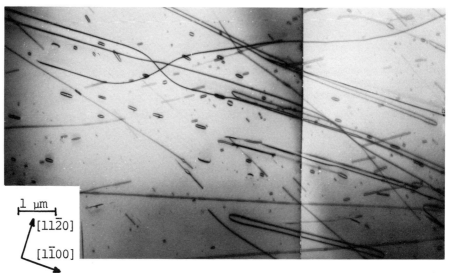

1 μm

[11$\bar{2}$0]

[1$\bar{1}$00]

Figure 3 -- Dislocation structure present at early stage in work hardening region of the flow stress curve (B of Fig. 1). In addition to the previously described structure, a number of small dislocation loops are also present. ε = 0.031, $\vec{g}$ = (30$\bar{3}$0), basal foil.

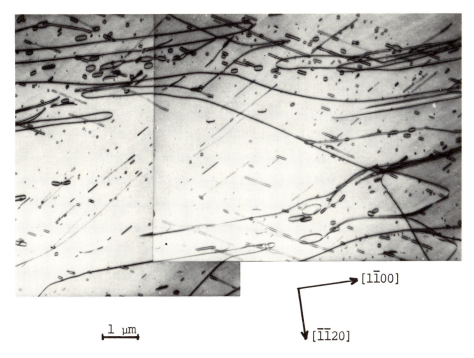

$[1\bar{1}00]$

1 μm

$[\bar{1}\bar{1}20]$

Figure 4 — Dislocation structure at intermediate stage in work
           hardening region of flow stress curve (C of Fig. 1).
           The structure is similar to specimens of lower strains.
           Note the formation of three different types of smaller
           loops; one lies along $[1\bar{1}00]$ and has the primary Burgers
           vector, while the other two lie at ±30° to the $[1\bar{1}00]$
           direction and have the Burgers vectors 1/3 $[10\bar{1}0]$ and
           1/3 $[01\bar{1}0]$. $\varepsilon = 0.055$, $\vec{g} = (0\bar{3}30)$, basal foil.

     Figure 5 shows the substructure of specimen E which has been
deformed into the region of zero work hardening and indicates that
the increase in both dipole and linking glide dislocation density
has continued with increasing strain.

     The form of the flow stress curve in Fig. 1 can be readily
correlated with the observed dislocation structures. At low strains,
the deformation structure consists of dipoles, multipoles, and
linking glide dislocations. The dipoles are relatively immobile
due to their internal stress fields and further immobilization
occurs when the dipoles form loops, since the loop ends are sessile
jogs. Thus, the dipoles serve as obstacles to other glide dislocations. As deformation proceeds, and more obstacles (dipoles and
multipoles) accumulate, it becomes harder for the glide dislocations
to move on their slip plane, resulting in the observed work hardening. At the same time, however, recovery by a decrease in obstacle

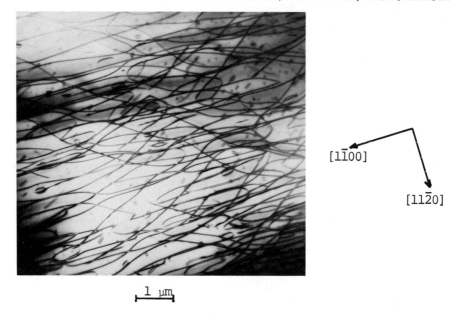

$$[1\bar{1}00]$$

$$[11\bar{2}0]$$

├─┤ 1 μm

Figure 5 -- Dislocation structure in region of zero work hardening
(E of Fig. 1). The dipole and linking glide dislocation density at this strain is greater than in specimens A-D, but no new features are present. $\varepsilon = 0.235$,
$\vec{g} = (30\bar{3}0)$, basal foil.

(dipole) density begins to be important. Dipoles can be eliminated
by two processes. Edge dipoles are only stable if their width
satisfies the relation

$$d < \frac{\mu b}{8\pi(1-\nu)(\tau-\tau_o)}$$

where ν is Poisson's ratio and τ and τ_o are the flow stress and
yield stress, respectively. As τ increases, the maximum stable
dipole width decreases; wider dipoles will break up by glide. In
addition, as indicated earlier, narrower dipoles will also tend to
break up by climb into loops to reduce their line energy. The net
effect is that the effectiveness of the dipoles as obstacles to
the glide dislocations is reduced. Thus a dynamic work-hardening-
recovery process evolves, involving the formation of edge dipoles
by edge-trapping and their breakup by climb into loops. When the
rate of accumulation of dipoles is equal to their rate of annihilation, a steady state of zero work hardening is achieved.

Dislocation densities were measured at three different strains and at two temperatures using a variation of Ham's method [5,6], in which the intersection of dislocations with randomly centered circles are counted on a net placed over the micrographs. (The thickness of each area (0.5 - 1.5 μm) was determined from stereo-micrographs.) Fig. 6 illustrates the total dislocation (i.e. dipole, linking glide, and loop) density as a function of strain for the two temperatures. The dislocation density increases with increasing strain at 1400°C but remains essentially constant with strain at 1500°C. This was surprising, in that the macroscopic work hardening rates for the two temperatures were similar. By counting separately the dipole, linking glide and loop dislocation densities, Figs. 7 and 8 were obtained. At 1400°C (Fig. 7), both the dipole and linking glide dislocation densities increase as the strain increases. In addition, the number of loops per unit volume also increases with strain. At 1500°C (Fig. 8), on the other hand, the linking glide dislocation density increases only slightly with strain while the dipole density appears to decrease slightly; the number of loops per unit volume is independent of strain. Significantly, however, the ratios of the density of glide dislocations to the density of dipoles as a function of strain is essentially the same for both temperatures. Thus it appears that this ratio controls the work hardening behavior and it is the breakup of the dipoles into loops that dominates recovery.

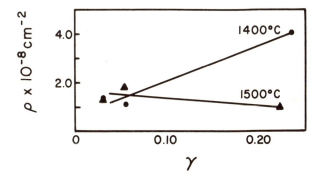

Figure 6 -- Total dislocation density (dipole, linking glide, and loop) as a function of shear strain at 1400 and 1500°C.

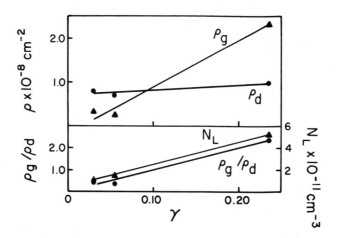

Figure 7 -- a) Dipole (ρ_d) and linking glide (ρ_g) dislocation densities as a function of strain at 1400°C. b) Number of loops per unit volume (N_L) and the ratio of glide dislocations to dipoles (ρ_g/ρ_d) as a function of strain at 1400°C.

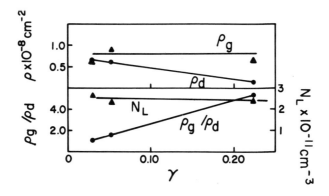

Figure 8 -- a) Dipole (ρ_d) and linking glide (ρ_g) dislocation densities as a function of strain at 1500°C. b) Number of loops per unit volume (N_L) and the ratio of glide dislocations to dipoles (ρ_g/ρ_d) as a function of strain at 1500°C.

3. SOLID SOLUTION HARDENING

To assess the effectiveness of solid solution hardening as a
strengthening mechanism for sapphire, a study was performed on
"alloys" of sapphire doped with Cr and Ti. Although other workers
[7,8,9] have examined single compositions of Cr-doped sapphire, the
present investigation involved a larger range of compositions than
has heretofore been used. The results are shown in Fig. 9, in
which the resolved shear stress for the lower yield point is plotted
as a function of temperature for various compositions. The data
for pure and 0.018 wt. % Cr_2O_3 are virtually identical and it is
not until 0.049 wt. % Cr_2O_3 is added that an increase in the lower
yield point is observed. This result suggests that the "pure"
sapphire contained some impurities which influenced its yield
strength; the effect of Cr cannot be felt until >0.018 wt. % Cr_2O_3
is added. It is also noted that in general the temperature dependence of the yield point for each alloy is similar, as is found in
metallic solid solution systems.

Plotting this data as a linear function of concentration, a
hardening rate of the order of $\mu/15$ at both 1400 and 1500°C is
obtained. In accordance with solid solution strengthening theories

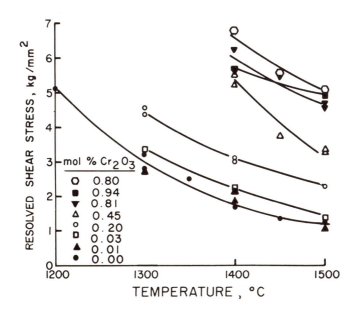

Figure 9 -- Resolved shear stress for the lower yield point vs.
        temperature for various Cr compositions.

of metallic systems, Cr can then be classified as a gradual
hardener [10]. This is not surprising since Cr in sapphire is
isovalent with Al^{3+}; no charge compensating defects are necessary
and the larger Cr ion will introduce an approximately symmetric
strain field. Hence only elastic interactions between glide dislocations and solute ions are expected. Since the solid solution
hardening rate is temperature independent, theories developed for
metallic systems can perhaps be applied. Two current theories
are Fleischer's model [10], which uses the difference in size and
modulus between solvent and solute and predicts the critical
resolved shear stress (CRSS) to be proportional to $C^{1/2}$, and
Labusch's theory [11], which uses a more rigorous statistical
treatment and gives a CRSS proportional to $C^{2/3}$, where C is the
atomic concentration of solute. Plotting our data using either
theory gives similar but only fair agreement. Moreover, the
experimental hardening rates were found to be 5 - 10 times higher
than the rates predicted by the models.

     A comparison of the effect of valence of the solute ion can
be performed in the Ti alloy system. Ti-doped crystals were
heated in a nitrogen-hydrogen atmosphere at 1300°C to reduce Ti
to the 3+ state or in air to oxidize Ti to the 4+ state. The
specimens were then deformed under identical deformation conditions
in a vacuum of 10^{-4} torr* and the CRSS's are given in Table 1.
As can be seen, Ti^{4+} is a more potent hardener than the larger
Ti^{3+} ion. This is a consequence of the fact that, when Ti^{4+} is
added to Al_2O_3, a charge compensating defect, most probably Al
vacancies, is needed to maintain charge neutrality. Thus the Ti^{4+}
ion and its associated defect introduce an asymmetric distortion
into the lattice; Ti^{3+}, like Cr^{3+}, only causes a symmetric distortion. The yield stress for the same dopant level of Cr is
also given in Table I; although Ti^{4+} is again smaller than Cr^{3+},
the asymmetric distortion of Ti^{4+} causes it to be a more potent
hardener. Comparing ions of the same valence, it is seen that the
larger Ti^{3+} ion causes a greater strengthening effect than Cr^{3+},
as expected from the large size difference between Ti^{3+} and Al^{3+}.

TABLE I

CRSS FOR PURE AND 0.26 MOLE % DOPED SAPPHIRE
DEFORMED AT 1500°C ($\dot{\varepsilon}$ = 1.75 x 10^{-4} sec^{-1})

Dopant	Ionic Radius	CRSS (kg/mm^2)
Ti^{4+}	0.60Å	7.50
Ti^{3+}	0.67Å	3.90
Cr^{3+}	0.61Å	3.22
Pure	0.54Å for Al^{3+}	1.86

     *No change in the Ti valence occurred during mechanical testing.

4. PRECIPITATION HARDENING

The Ti-doped system is also suitable for studying precipitation hardening. Fig. 10 shows an optical micrograph in which the acicular precipitates responsible for the asterism of star sapphire are shown lying along the three <10$\bar{1}$0> directions in the basal plane. Because there is a low volume fraction of the precipitate present in the crystal, x-ray techniques for determining the crystal structure of the precipitate are not possible. Using electron diffraction, however, it has been found that the set of precipitate planes perpendicular to the needle axis have the same d spacing as the {30$\bar{3}$0} planes of sapphire. Differences between particle and matrix of up to 5% are found for other planes. Thus the good fit in the <10$\bar{1}$0> direction explains the observed growth of the precipitate in this direction.

Studies of the mechanical properties of samples containing this precipitate have been hampered by the precipitate gradient illustrated in Fig. 11. Microprobe analysis has shown that a macroscopic chemical gradient is not responsible for the high density of small precipitates at the outside edge of the boule compared with the lower density of larger precipitates at the center of the boule. The observed precipitate gradient is thought to be

10μm

Figure 10 -- Optical micrograph of Ti-doped sapphire taken with partially crossed polarizers illustrating the acicular precipitates which lie along <10$\bar{1}$0> directions in the basal plane.

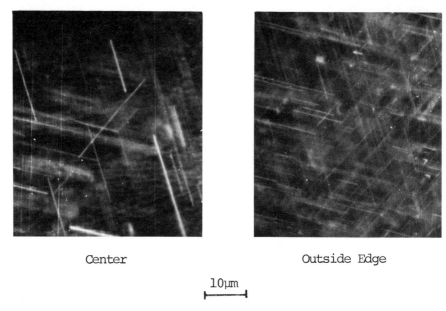

Center   Outside Edge

10µm
├──────────┤

Figure 11 -- Dark field transmission optical micrograph showing
             the precipitate gradient which exists between the
             center and outside edge of boule. Plane of thin-
             section was perpendicular to the growth axis and is
             parallel to basal plane.

due to a gradient in the density of grown-in dislocations, which
act as nucleation sites for the precipitates. This precipitate
gradient also has hindered our study of the precipitation kinetics.
However, comparison of the same area in specimens as-received, in
which Ti was primarily in the 3+ state, with specimens solution
heat treated, so that Ti was in the 4+ state, showed virtually
identical kinetics, as seen in Fig. 12. This has been interpreted
to mean that the oxidation of Ti^{3+} to Ti^{4+} is not the rate controlling
step in the precipitation process.

     Since the precipitation apparently occurs heterogeneously, it
is clearly necessary to prestrain specimens to introduce a homogeneous distribution of nucleation sites prior to precipitation in
order to control the resulting structure. Preliminary data (Fig.
13) obtained on non-prestrained specimens, however, illustrate that
there is no increase in strength due to precipitation hardening
over that due to solid solution strengthening with Ti^{4+}. This
effect is not surprising since (a) the precipitates lie in the glide
plane and (b) from the work hardening studies, extensive dislocation
climb occurs in this temperature range, and hence the dislocations
can bypass the particles with relative ease.

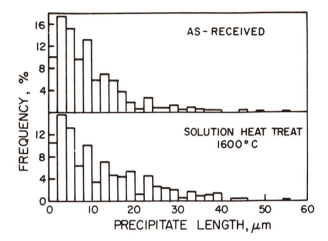

Figure 12 -- Histogram of precipitation kinetics for a) as-received
            and heat-treated and b) solution-treated at 1600°C
            for 24 hours and then heat-treated Ti-doped sapphire.
            Measurements were taken in the same area for each
            specimen.

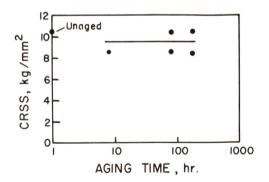

Figure 13 -- Critical resolved shear stress of Ti-doped sapphire
            crystals tested in compression at 1475°C versus aging
            time.

B. J. PLETKA, T. E. MITCHELL, AND A. H. HEUER

ACKNOWLEDGMENT

This research was supported by the United States Army Research Office, Durham, under Grant No. AROD 3112473G95.

REFERENCES

1. B.J. Pletka, M.S. Thesis, Case Western Reserve University, (1973).
2. B.J. Pletka, T.E. Mitchell, and A.H. Heuer, "Dislocation Structures in Sapphire Deformed by Basal Slip," J. Amer. Ceram. Soc., in press.
3. H. Conrad, G. Stone, and K. Janowski, "Yielding and Flow of Sapphire (α-Al$_2$O$_3$) Crystals in Tension and Compression," Trans. AIME 233 [5] 889-97 (1965).
4. R.F. Firestone and A.H. Heuer, "Yield Point of Sapphire," J. Amer. Ceram. Soc., 56 [3] 136-39 (1973).
5. R.K. Ham, "The Determination of Dislocation Densities in Thin Films," Phil. Mag., 6 [69] 1183-84 (1961).
6. P.B. Hirsch and J.S. Lally, "The Deformation of Magnesium Single Crystals," Phil. Mag., 12 [117] 595-648 (1965).
7. R. Chang, "Creep of Al$_2$O$_3$ Single Crystals," J. Appl. Phys. 31 [3] 484-87 (1960).
8. M.V. Klassen-Neklyudova, V.G. Govorkov, A.A. Urusovskaya, N.N. Voinova, and E.P. Kozlovskaya, "Plastic Deformation of Corundum Single Crystals," Phys. Status Solidi, 39 [2] 679-88 (1970).
9. K.C. Radford and P.L. Pratt, "Mechanical Properties of Impurity-Doped Alumina Single Crystals," Proc. Brit. Ceram. Sox., 1970, No. 15, pp. 185-202.
10. R.L. Fleischer, pp. 93-140 in Strengthening of Metals. Edited by D. Peckner. Reinhold, New York, 1964.
11. R. Labusch, "A Statistical Theory of Solid Solution Hardening," Phys. Status Solidi, 41 659-669 (1970).

DYNAMICS OF FLOW OF C-AXIS SAPPHIRE

R. E. Tressler[1] and D. J. Michael[2]

Material Sciences Department, The Pennsylvania State University[1], and PPG Industries, Inc., Glass Research Center, Creighton, Pa.[2]

INTRODUCTION

The extreme plastic anisotropy of α-alumina single crystals (corundum structure) was first clearly demonstrated by Wachtman and Maxwell (1) when they found that temperatures in excess of 1600°C were necessary to activate tensile creep of c-axis oriented single crystals while creep via basal slip could be detected at 900°C. Since then, detailed studies of the yielding and flow behavior of alumina single crystals via basal slip have considered the stress, strain, strain-rate, temperature and impurity content inter-relationships for plastic deformation via this slip system (2,3,4,5).

However, because of the ease with which basal slip can be activated, it is difficult to activate secondary slip systems. Therefore, relatively little work had been done until recently on the deformability of sapphire via other slip systems. Table I contains the non-basal slip systems which have been identified from macroscopically deformed specimens and those which have been established or suggested from other experiments.

Recent developments in crystal growth at Tyco Laboratories (13) and at Arthur D. Little Inc. (14) have resulted in the availability of very strong sapphire crystals in filamentary form with precise crystallographic orientations with respect to the filament axis. In particular, crystals having the c-axis within 2° of the filament axis were available from both suppliers in the "pure" form and with Ti^{3+} dopant from Arthur D. Little, Inc.

TABLE I

Non-basal Slip Systems for Sapphire

Type of Slip	Reference	Slip Systems
Prismatic	(6)(7)	$\{11\bar{2}0\}$ $<1\bar{1}00>$*
Structural Pyramidal	(8)	$\{1\bar{1}02\}$ $<10\bar{1}1>$**
Structural Pyramidal	(9)	$\{1\bar{1}02\}$
Structural Pyramidal	(10)	$\{\bar{1}012\}$ $<10\bar{1}1>$***
Structural Rhombohedral	(10)	$\{10\bar{1}1\}$ $<11\bar{2}0>$ or $<10\bar{1}1>$***
Structural Rhombohedral	(11)	$\{10\bar{1}1\}$ $<01\bar{1}1>$***
Structural Rhombohedral	(12)	$\{10\bar{1}1\}$ $<01\bar{1}1>$***

*   Identified by TEM analysis
** Tentatively identified by slip trace analysis
*** Tentatively identified by TEM analysis

The purpose of this paper is to summarize the results of deformation studies performed with these three materials, compare the stress-elongation behavior and activation parameters for the three materials, and discuss the trends in values with those found by other investigators on the same, or similar, materials.

EXPERIMENTAL PROCEDURE

The detailed characterizations of the materials studied in this investigation are given in the original reports (13,14). Briefly, the Tyco materials, $\approx 250\mu m$ diameter, were distinguished by a characteristic pore structure consisting of $\approx 1\mu m$ diameter voids lying roughly on morphological rhombohedral planes except for an outer shell ($\approx 25\mu m$) of the void free material. The total impurity content reported by the supplier was less than 50 ppm of Mo, Si, Ti, Ni, and Cr. The ADL material was void free and contained less than 3 ppm of impurity (other than Ti) in the Ti^{3+} doped filaments. The undoped filaments were not analyzed by the supplier. The Ti-dopant level in the samples tested was between 0.024 and 0.077 weight percent. The ADL sample diameters were $\approx 275\mu m$.

The pivot bearing, hypodermic needle method for sample gripping and tensile testing, as described by Tressler and Crane (15), was used in this study. The samples were heated to the testing temperature (1775-1875°C) in an internally wound Pt-40% Rh furnace with an inside diameter of 1.6 cm and a hot zone length of ≈ 2.5 cm.

The temperature was measured using a Pt40Rh–Pt20Rh thermocouple which was calibrated periodically using a National Bureau of Standards standard thermocouple of the same type. The temperature was constantly monitored with a Leeds and Northrup precision millivolt potentiometer and controlled to $\pm$ 2°C by manually adjusting the power input.

The tensile testing apparatus was a table model Instron equipped with a 50 kg tension load cell. The crosshead speeds used were 0.005, 0.01, 0.02, and 0.05 cm/min. Both differential strain-rate and differential temperature tests were performed at constant structure from 1760°C to 1875°C. Differential strain-rate tests were performed by changing the crosshead speed as the specimen deformed plastically. The differential temperature tests were performed by tensile testing a specimen at a given temperature, removing the load, lowering the temperature 25°C and restressing, a process that took approximately 10 minutes.

All samples tested at 1825°C and below were prestressed for 30 minutes at the test temperature at approximately one-third the yield stress to effect plastic deformation. The furnace was held at approximately 1000°C between tests and it took approximately one-half hour to heat the sample to the test temperature.

To guarantee that the Ti-ions in the doped specimens were all in the Ti^{4+} state the samples were oxidized using the data of Jones et al. (16). To demonstrate that the Ti^{4+} ions were in solution at the test temperatures a specimen was optimally aged at 1300°C for 14 days to produce the Ti-rich precipitates (17). The specimen was then heated at 1800°C for one hour with the result that the precipitates were completely dissolved as evidenced by microscopic examination using a petrographic microscope.

Slip traces on deformed crystals were viewed by transmitted light microscopy and scanning electron microscopy. Transmission electron microscopic examination was performed on specimens thinned by mechanical polishing and ion thinning techniques.

RESULTS AND DISCUSSION

Slip System Identification

As described by Tressler and Barber (11) scanning electron microscopic examination of slip traces on the crystal surface (Figure 1), coupled with transmission Laue X-ray diffraction verification of the orientation of the slip trace intersections, identified

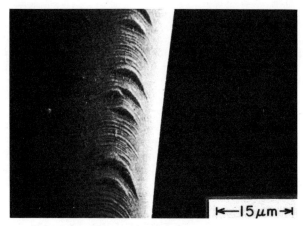

Figure 1.   Scanning electron micrograph of c-axis filament deformed
             at 1800°C showing slip traces on the crystal surface.

the active slip planes as the {10$\bar{1}$1} planes*. From stereoscopic
transmission electron microscopic examination of thinned foils it
was concluded that the elongated dislocations resulting from the
deformation lay on the {10$\bar{1}$1} planes. These results were established
by examination of the Tyco filaments. The slip trace analysis of
the deformed ADL filaments, Ti-doped and undoped, yielded the same
conclusion as above. The dislocation substructure qualitatively
appeared similar to that observed in the Tyco filaments. However,
detailed examination of the ADL filaments has not yet been done.

    As yet the Burgers vectors of the nucleated glide dislocations
have not been established unequivocally. However, from the testing
geometry (tensile load applied parallel to the c-axis) there is no
resolved shear stress in two of the three possible slip directions
(1/3 <11$\bar{2}$0> and <10$\bar{1}$0>) (18). Therefore the nucleated dislocations
most likely have Burgers vectors of the 1/3 <$\bar{1}$101>, as first proposed
by Barber and Tighe (19). Certainly, in the samples examined there
is evidence of dislocation interactions to form segments of a loose
net involving other dislocations with other Burgers vectors (Figure
2). However, the predominant slip system seems to be the {10$\bar{1}$1}
1/3 <$\bar{1}$101>.

---

*The structural unit cell, with a c/a ratio of 2.73 is used
throughout.

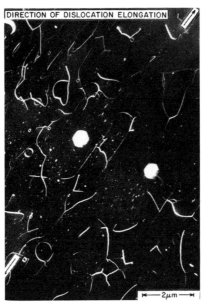

Figure 2.   Dark-field transmission electron micrograph of tilted
            foil from Tyco c-axis filament deformed at 1840°C (taken
            with $\bar{3}030$ reflection). Tilt axis is approximately
            parallel to trace of glide planes in foil.

Yielding and Flow

For the Tyco filaments gross plastic deformation was observed
from 1760 to 1875°C. Typical constant strain rate, isothermal,
stress-elongation curves are shown in Figure 3. However, the ADL
filaments fractured in a brittle fashion at temperatures below 1800°C.
Comparing Figures 4 and 5 to Figure 3 one sees that the yield drops
for the "porous" Tyco filaments are much smaller than the corresponding drops for the Ti^{4+}-doped ADL crystals tested after identical
pretest thermomechanical treatments. This difference is probably due
either to the nucleation of glide dislocations at external and
internal surfaces where the very large number of voids in the Tyco
material constitute many more nucleation sites than are available in
the void-free ADL filaments, or to a solid solution strengthening
phenomenon due to the Ti^{4+} ions. The yield drops observed for the
pure ADL crystals were substantially lower. However, insufficient
samples were available to quantitatively establish the difference
for a given pre-test treatment. Thus the fracture strength of the
ADL filaments are lower than the yield strengths at temperatures
less than 1800°C.

For all of the materials the yield stresses were irreproducible
being very dependent on the pre-test thermomechanical history.

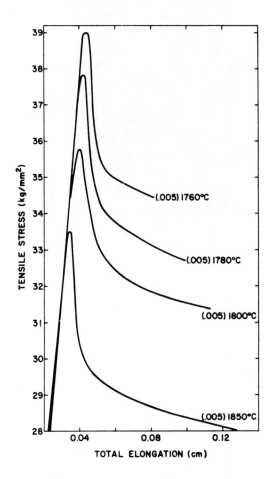

Figure 3.   Typical tensile stress vs elongation curves for Tyco c-axis
            sapphire filaments (crosshead speed is given in parentheses
            in cm/min; effective gage length ≃2.5 cm).

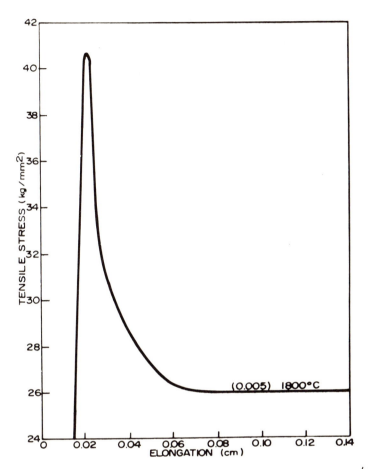

Figure 4. Tensile stress versus elongation curve for a Ti[4+]-doped
c-axis sapphire filament (ADL) tensile tested at a crosshead speed of 0.005 cm/min at 1800°C.

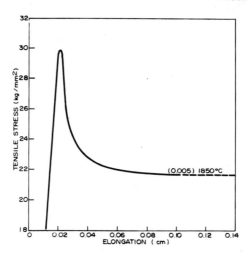

Figure 5.   Tensile stress versus elongation curve for a Ti^{4+}-doped
c-axis sapphire filament (ADL) tensile tested at a crosshead speed of 0.005 cm/min at 1850°C.

Apparently the number of mobile dislocations depends very much on
the annealing and/or prestress history, suggesting a strain ageing
effect. In Figure 6 a strain-ageing effect is verified for the Tyco
material; similar results were also obtained for both the Ti-doped
and undoped ADL crystals. The mobile dislocations are presumably
pinned by precipitates, solute ions or form equilibrium arrays
during the zero-load ageing period causing the history dependent
yield stresses. Because of the irreproducible yield-stresses the
yielding behavior could not be as well-characterized as the flow
behavior with the limited specimens available.

The flow stresses measured at given total elongation for each
material were reproducible and not affected by the pretest thermomechanical treatment. Comparing the stress-elongation curves for
the Tyco material and the ADL crystals, one finds that a steadily
decreasing flow stress is characteristic of the Tyco filaments
while the ADL crystals flowed at a nearly constant stress indicating
that the recovery process in the former is faster than in the latter.

The data from many differential strain-rate tests are constant
temperatures (Figures 7 and 8) were used to construct the flow
stress vs. temperature curves for the Tyco filaments (Figure 9) and
the Ti^{4+}-doped ADL filaments (Figure 10). It should be pointed out
at this time that the pure and doped ADL filaments demonstrated
identical flow stresses (also identical activation parameters as
discussed below). The major differences between the flow stress vs.
temperature curves for the two materials is the much higher flow

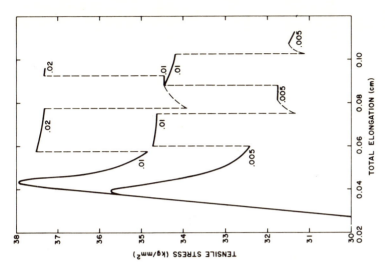

Figure 7. Representative tensile-stress-vs-elongation curves for differential strain-rate tests on Tyco crystals (T = 1800°C; crosshead speeds in cm/min are indicated on graph).

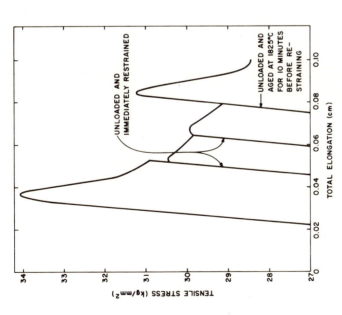

Figure 6. Tensile-stress-vs-elongation curve for Tyco c-axis sapphire filament (T = 1825°C; crosshead speed = 0.005 cm/min; effective gage length ≈2.5 cm).

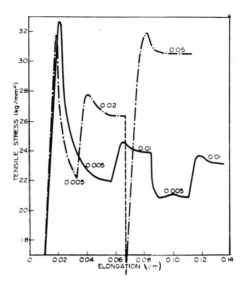

Figure 8. Typical tensile stress versus elongation curves for
differential strain-rate tests at 1850°C for Ti^{4+}-doped
c-axis sapphire filaments (ADL). The crosshead speeds
are indicated on the figure.

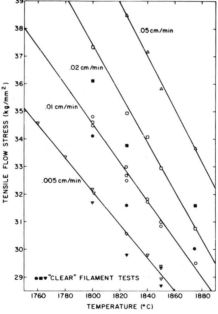

FIGURE 9. Flow stress vs temperature for Tyco c-axis filaments at
various crosshead speeds (stresses taken at 0.4 cm plastic
deformation beyond elastic strain at that stress; effective
gage length ≃2.5 cm).

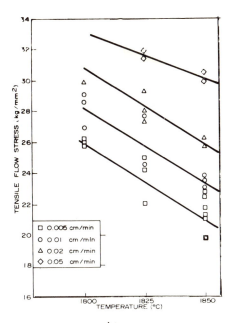

Figure 10.   Flow stresses for Ti^{4+}-doped c-axis filaments (ADL) at
              crosshead speeds ranging from 0.005 cm/min to 0.05 cm/min
              from 1800°C to 1850°C. The flow stresses were taken at
              0.08 cm from the elastic extension curve at an equivalent
              stress level.

stresses for the Tyco material at a given temperature and strain
rate indicating some additional strengthening component over the ADL
material. In addition, for the Tyco material, as the temperature was
changed at constant structure (differential temperature test) as
shown in Figure 11 the increment in flow stress did not correspond to
the change in flow stress that one derives from constant temperature
tests (Figure 9). This result was not found in the case of ADL
material where the flow stress increments were in close agreement.
In the case of the Tyco crystals, apparently there is a thermally
activated substructure variation which is superimposed on the
thermally activated rate controlling deformation mechanism; these
can only be separated via the differential temperature test.

Strain Rate Equations and Thermal Activation Parameters

Generally, the flow dynamics of sapphire via basal slip have
fit either a relation of the form

$$\dot{\gamma} = \nu \, \exp \left[\frac{-H(\tau^*)}{RT} \right] \qquad\qquad (1)$$

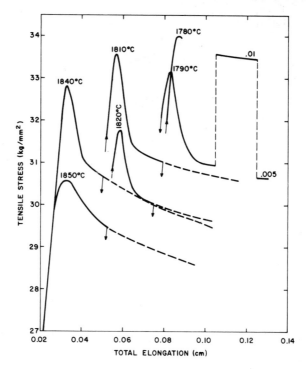

Figure 11.   Tensile stress vs total elongation for temperature-change
             tests on Tyco filaments.

indicative of the overcoming of the Peierls-Nabarro stress as being
the rate controlling mechanism (3,4) or a relation of the form

$$\dot{v} = \frac{A}{T} \; (\tau*)^n \quad \exp \; - \; \frac{H_o}{RT} \tag{2}$$

indicative of dislocation climb as the rate controlling mechanism
(5).   ($\dot{v}$ = the shear strain rate, v = a constant, H ($\tau*$) = stress
dependent activation enthalpy, R = the gas constant, T = absolute
temperature, A = a constant, $\tau*$ = the effective resolved shear stress,
n = a constant, and H_o = the activation enthalpy, relatively independent of stress).

     Assuming a deformation equation of the form of Eq. (2) and
using data from differential strain rate tests, stress exponents
were calculated from the relation

$$n = \frac{\delta \ln \dot{\varepsilon}}{\delta \ln \sigma} \approx \frac{\Delta \ln \dot{\varepsilon}}{\Delta \ln \sigma} \tag{3}$$

For the ADL material the values generally fell between 6 and 7 and
were independent of elongation, stress and temperature as illustrated
in Figure 12. The stress exponents may also be derived from the
slopes of ln ($\dot{\epsilon}$) vs ln (σ) curves at constant temperatures, which are
derived from cross-plotting the data from flow-stress vs. temperature
curves as in Figure 9. For the Tyco material n values determined in
this way (Figure 13) agrees very well with those calculated from
differential strain-rate test. However, the stress exponents for
the Tyco crystals are larger and increase dramatically with temperature in contrast to the results for the ADL crystals.

Gooch and Croves (12) also found stress exponents that were
in the range of 6 to 7 for the tensile creep of c-axis Tyco filaments
at 1600° to 1800°C at much lower strain rates. Since these values
were not calculated directly from differential strain rate tests,
they may not represent constant structure data although they are in
good agreement with the n values determined for the ADL crystals
at slightly higher temperatures.

The activation volumes may be calculated from differential
strain rate data using the relation

$$V^* = RT \frac{\partial \ln \dot{\gamma}}{\partial \tau} \simeq RT \left(\frac{\Delta \ln \dot{\epsilon}}{\Delta \sigma}\right) \frac{1}{\cos\theta\cos\phi} \tag{4}$$

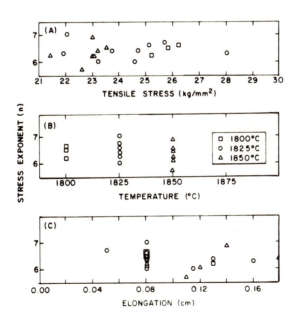

Figure 12. The stress exponent, n, as a function of tensile flow
            stress, temperature, and elongation for Ti^{4+}-doped
            c-axis filaments (ADL).

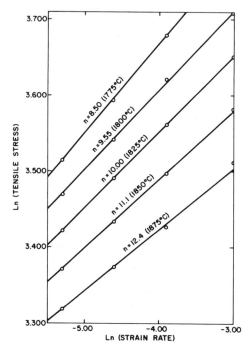

Figure 13.   Natural logarithm of flow stress vs natural logarithm
             of strain rate (crosshead speed) at various temperatures
             with respective n values for Tyco filaments.

Figure 14 illustrates that for both the Tyco material and the ADL
crystals V* is a decreasing function of stress and appears not to be
a function of temperature or elongation over the limited range
studied.   The general shape of the curves is similar to those of
previous investigators for basal slip (3,4) but the values are considerably smaller than the values determined for basal slip.   Also
to be noted is the fact that the activation volumes determined for
the ADL filaments are smaller than those for the Tyco filaments at
a given stress level.

    Activation enthalpies were calculated from the results of
differential temperature and differential strain-rate tests using
the relation

$$(\frac{\partial \ln \dot{\varepsilon}}{\partial \sigma})_T \quad (\frac{\partial \sigma}{\partial T})_{\dot{\varepsilon}} = \frac{\Delta H}{RT^2} \simeq (\frac{\Delta \ln \dot{\varepsilon}}{\Delta \sigma}) \; (\frac{\Delta \sigma}{\Delta T})_{\dot{\varepsilon}} \qquad (5)$$

This equation is only valid when the flow stress is an unique function of strain rate for a given structure as appears to be the case

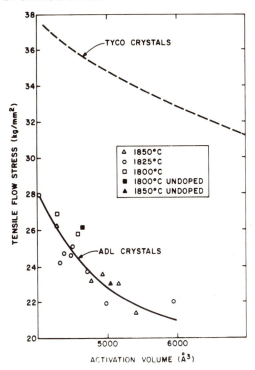

Figure 14.   Activation volume, V*, versus tensile flow stress c-axis
             filaments.

for all types of materials studied since the flow stresses were
reproducible. The enthalpies calculated for both the doped and
undoped ADL crystals fell in the range 115 kcal-170 kcal. As Figure
15 illustrates, the activation enthalpy is a decreasing function of
stress and, in Figure 16, an increasing function of temperature.
The activation enthalpies calculated for the Tyco experiments were
substantially lower being in the range of 80-90 kcal/mole. An attempt
was made to calculate, for both sets of data, the activation enthalpy
at zero effective stress, ΔH_o. This quanity is given by the
relation (20)

$$\Delta H_o = \Delta H + \int_o^{\tau^*} V^* d\tau^* \qquad\qquad (6)$$

where ΔH represents the thermal contribution to the total activation
enthalpy and the integral quantity is the mechanical contribution.

     The very approximate extrapolation of the V* vs τ^* curves to
$\tau^* = 0$ yields approximate mechanical contributions of 13.8 kcal/mole

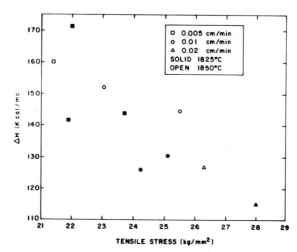

Figure 15.   Apparent activation enthalpies versus flow stress for Ti^{4+}-doped c-axis sapphire filaments (ADL).

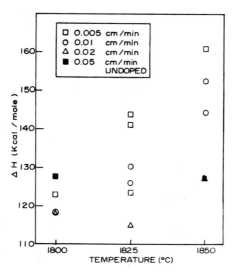

Figure 16.   Apparent activation enthalpies versus temperature for Ti^{4+}-doped and undoped c-axis sapphire filaments (ADL).

for the Tyco results and 9.3 kcal/mole for the ADL results. There-fore, there appears to be a definite difference in the ΔH_0 for the deformation processes of these two different materials. However, the difficulties in doing the integration above, and the fact that the extrapolation of the ΔH vs stress curve (Figure 15) to low

TABLE II

Flow Stresses (at a crosshead speed of 0.005 cm/min)
and Activation Parameters for the undoped, c-axis,
Arthur D. Little Sapphire Filaments.

Sample	Temp(°C)	$\sigma_{flow}(kg/mm^2)$	$V*(Å^3)$	$\Delta H(kcal/mole)$	n
1	1850	23.0	5056	127.4	6.2
2	1850	21.6	----	-----	---
3	1800	26.2	4669	127.6	6.6

stresses suggests a much larger mechanical contribution, indicate
considerable uncertainties in the ΔH_o values.

Undoped ADL Sapphire

As indicated above the results of the tests on undoped filaments
showed that the Ti-dopant did not affect the flow behavior or the
activation parameters. In Table II the data for the pure sapphire
are compiled. One must then conclude that the Ti^{4+} ion solid solution strengthening is not operative for sapphire in the concentration,
orientation, and temperature regime included in this study.

Rate Controlling Mechanisms

Unfortunately, the data collected to date have not unambiguously
identified rate controlling processes. However, significant differences in flow behavior between the "porous" Tyco filaments and the
non-porous ADL filaments have been demonstrated, which encourage one
to suggest possibilities for rate controlling processes.

For deformation limited by pure dislocation climb or dislocation
glide limited by dislocation climb the zero stress activation enthalpy
should be approximately equal to that for self-diffusion of the
slowest diffusing specie in the crystal. In sapphire this value
would most likely be that for the larger O^{2-} ion where the activation
enthalpy is 152 kcal/mole (21). The current results for the ADL
crystals, with the range of ΔH values determined, do not rule out the
possibility of an O^{2-} diffusion controlled deformation mechanism.
The activation enthalpies for the Tyco filaments are much lower than

the O^{2-} ion diffusion enthalpy being closer to the values determined
for cation diffusion in doped sapphire (16).

The stress exponents for both types of filaments are higher
than the 3 to 5 range predicted for pure dislocation climb or glide
limited by climb (22, 23). Gooch and Groves (12) found n values
comparable to those reported here for the ADL crystals in their
creep deformation studies of Tyco filaments. Based on their observations they suggested that overcoming the Peierls-Nabarro stress
was the rate controlling process in the creep deformation of c-axis
Tyco filaments at low strain rates via the $\{10\bar{1}1\}$ $<0\bar{1}11>$ system.
The n-values found for the Tyco filament flow dynamics are much
higher and not consistent with currently available theoretical models.

The observed trends in activation enthalpies with stress and
temperature for the ADL crystal deformation are consistent with that
expected for Peierls-Nabarro stress controlled flow as described
by Eq. (1). Taking the logarithm of the equation and solving for
the enthalpy yields

$$H(\tau^*) = RT\ln(\nu/\dot{\gamma}) \tag{7}$$

Therefore, $H(\tau^*)$ should be an increasing function of T at constant
ν and $\dot{\gamma}$ as demonstrated in this study. Also the apparent enthalpy
is a decreasing function of stress due to the increasing mechanical
contribution to the total enthalpy; the ADL filament deformation
results also follow this trend. On the other hand for dislocation
climb controlled processes (Eq. 2) the activation enthalpy is
relatively independent of stress. In the case of the Tyco filaments
there is insufficient data to make these sorts of statements.

The activation volumes calculated from the ADL filament results
are independent of strain as should be the case for a Peierls-
Nabarro stress rate limited process (20). Also the magnitudes
observed are not inconsistent with this type of rate-controlling
mechanism.

In the case of the Ti^{4+}-doped sapphire the predominant defect
introduced would very likely be cation vacancies (16). If the rate-
controlling mechanism were cation-defect diffusion-controlled climb
for the flow of the ADL crystals, the aluminum-ion vacancies introduced would have greatly increased the strain rates for a given
flow stress (or greatly lowered the flow stress for a given strain
rate) for the doped samples compared to the undoped samples as
observed by Hollenberg and Gordon (24) for Ti^{4+}-doped polycrystalline Al_2O_3. The fact that the dopant ions did not affect the flow
behavior further suggests a Peierls-Nabarro rate-controlling mechanism or O^{2-} controlled climb for these sapphire crystals.

The Tyco crystals show a sizeable increment of strengthening when compared to the ADL crystals. If the flow of the ADL crystals is rate limited by the overcoming of the intrinsic lattice resistance, then one must conclude that barriers to dislocation glide are present in the Tyco crystals. Further, the observed activation enthalpies suggests that the rate limiting process in overcoming these barriers may be cation defect diffusion. One must then decide how the O^{2-} ions diffuse more rapidly than the cation defects; this diffusion may occur via the pore structure in some as yet ill-defined manner.

These hypotheses are at odds with the suggestion of Gooch and Groves (12) of creep of Tyco filaments controlled by overcoming of the Peierls-Nabarro stress unless such a region of deformation rate control exists at much lower stresses and strain rages. Without more definitive data over a wider range of strain-rates and temperatures, the differences between the flow behavior of these two types of materials cannot be resolved.

SUMMARY

The yielding and flow behavior of three different types of c-axis sapphire - undoped, "porous" Tyco filaments, void-free undoped and Ti^{4+}-doped ADL filaments - have been studied by differential strain-rate and differential temperature experiments.

For the ADL filaments the Ti^{4+}-dopant had no measurable effect on the flow stresses or the activation parameters. From consideration of this result and the stress and strain dependencies of the activation volumes, the constancy and magnitude of the stress exponents, and the stress and temperature effects on the apparent activation enthalpy, it is suggested that the rate-limiting process in the flow of these crystals is the overcoming of the Peierls-Nabarro stress.

In the case of the Tyco materials where significant strengthening was observed when compared to the ADL crystals, a different rate controlling mechanism may be operative. The much lower apparent activation enthalpies are closer to the values observed for cation defect diffusion. The presence of closely spaced voids in the crystals may provide a means for rapid diffusion of O^{2-} ions permitting the rate control by cation defect diffusion, presumably, of dislocations around the barriers causing the strengthening. This process is clearly not rate-limiting in the ADL crystals. The data are not complete enough to unequivocally establish the rate controlling mechanisms.

The direct evidence indicates that slip is occurring on the $\{10\bar{1}1\}$ type planes with the probable Burgers vector, based on the testing geometry, being the $1/3 <\bar{1}101>$.

REFERENCES

1. J. B. Wachtman, Jr., and L. H. Maxwell, J. Amer. Ceram. Soc., 37 (7) 291-299 (1954).

2. H. Conrad, J. Amer. Ceram. Soc., 48 (4) 195-201 (1965).

3. H. Conrad, G. Stone, and K. Janowski, Trans. AIME, 233 (5) 889-897 (1965).

4. R. L. Bertolotti and W. D. Scott, J. Amer. Ceram. Soc., 54 (6) 286-291 (1971).

5. R. Chang, J. Appl. Phys., 31 (3) 484-487 (1960).

6. D. J. Gooch and G. W. Groves, J. Amer. Ceram. Soc., 55 (2) 105 (1972).

7. D. J. Gooch and G. W. Groves, Phil. Mag., 28 (3) 623-637 (1973).

.8. P. D. Bayer and R. E. Cooper, J. Mater. Sci., 2 301 (1967).

9. A. H. Heuer, R. F. Firestone, J. D. Snow, and J. Tullis, Proc. Conf. on Ceramics in Severe Environments, Raleigh, N.C., 331-340 (1970).

10. B. J. Hockey, J. Amer. Ceram. Soc., 54 (5) 223-231 (1971).

11. R. E. Tressler and D. J. Barber, J. Amer. Ceram. Soc., 57 (1) (1974).

12. D. J. Gooch and G. W. Groves, J. Mater. Sci., 8 1238-1246 (1973).

13. J. T. A. Pollock, J. Mater. Sci., 7 631-648 (1972).

14. J. S. Haggerty, Air Force Materials Laboratory Technical Report, AFML-TR-2, (1973).

15. R. E. Tressler and R. L. Crane, in Advanced Materials Composites and Carbon, A symposium of the Amer. Ceram. Soc., April (1971) Chicago, Ill., 59-65.

16. T. P. Jones, R. L. Coble and C. J. Mogab, J. Amer. Ceram. Soc. 52 (6) 331-334 (1969).

17. R. J. Bratton, J. Appl. Phys., 42 (1) 211-216 (1971).

18. J. D. Snow and A. H. Heuer, J. Amer. Ceram. Soc., 56 (3) 153-157 (1973).

19. D. J. Barber and N. J. Tighe, Phil. Mag., 14 531–544 (1966).

20. A. G. Evans and R. D. Rawlings, Phys. Stat. Sol., 39 9–31
     (1969).

21. Y. Oishi and W. D. Kingery, J. Chem. Phys., 33 (2) 480–486
     (1960).

22. J. Weertman, J. Appl. Phys., 28 (3) 362–364 (1957).

23. F. R. N. Nabarro, Phil. Mag. 16 231–237 (1967).

24. G. W. Hollenberg and R. S. Gordon, J. Amer. Ceram. Soc. 56
     (2) 109 (1973).

DRAG MECHANISMS CONTROLLING DYNAMIC DISLOCATION BEHAVIOR IN MgO

SINGLE CRYSTALS

R. N. Singh*

Department of Metallurgy and Materials Science
Massachusetts Institute of Technology
Cambridge, Massachusetts   02139

ABSTRACT

Edge and screw dislocation velocities in "pure" and iron-doped magnesium oxide single crystals, with and without dislocation dipoles, have been measured as a function of stress, temperature, and valence state of the iron impurities to identify the rate-controlling drag mechanisms for dislocation mobility. Edge dislocations have been observed to move faster than screw dislocations in the valence states of iron impurities over the stress and temperature regimes investigated. The edge and screw dislocations move faster in reduced (Fe^{+2}) samples than in oxidized (Fe^{+3}) samples. From the analysis of the data for the edge and screw dislocation velocities in terms of the activation parameters (activation volume, activation enthalpy, total activation enthalpy, and the stress exponent of dislocation velocity), it is suggested that the edge and screw dislocation mobilities in "pure" MgO single crystals in the reduced state are controlled by Peierls mechanism, with thermally activated double-kink nucleation as the rate-limiting step. The calculated values of the Peierls stress for edge and screw dislocation mobilities in "pure" MgO crystals in the reduced state are 0.6×10^8 and 1.7×10^8 Nm^{-2}, respectively.

In MgO single crystals containing 150 ppm Fe^{+3} dopants, the edge and screw dislocation mobilities are controlled by the interaction of dislocations with the noncentrosymmetric distortions due to ($Fe^{\cdot}_{Mg}-V^{''}_{Mg}$) defects. The strain $\Delta\varepsilon$ owing to such defects has been calculated and is related to the observed hardening of iron-doped (Fe^{+3}) MgO single crystals.

*Presently employed at Argonne National Laboratory, Materials Science
 Division, Argonne, Illinois 60439

The screw dislocation mobility in MgO single crystals containing dislocation dipoles and 90 ppm iron in the oxidized state is controlled by the interaction of dislocations with the noncentrosymmetric defects ($Fe^{\cdot}_{Mg}-V''_{Mg}$). However, the edge dislocation mobility in the oxidized state as well as the edge and screw dislocation mobilities in the reduced state are suggested to be controlled by the interaction of these dislocations with dislocation dipoles.

INTRODUCTION

The yield stress of single-crystal magnesium oxide has been shown to increase significantly with an increase in iron content,[1-4] and to be significantly affected by the valence state of the iron present.[3] These behaviors result from the defect species present in the crystal and their interaction with the dislocations. Recently, we have identified the defect species, through the interpretation of dislocation-dynamics data in "pure" and iron-doped magnesium oxide single crystals.[5,6] In addition to impurity drag, MgO single crystals acquire large concentrations of dislocation dipoles after low-temperature plastic deformation,[7] which provides a further resistance to the dislocation motion. We have also identified[8] the defect species that controls dislocation mobility in iron-doped MgO single crystals containing dislocation dipoles.

Magnesium oxide samples containing iron offer a rather unique opportunity to change the valence state of the iron by oxidation/reduction treatments while otherwise holding constant the additional purity/impurity and dislocation-dipole density. This provides an opportunity to change the character of the point defects associated with iron from centrosymmetric with iron in the reduced state (Fe^{+2}) to noncentrosymmetric ($Fe^{\cdot}_{Mg}-V''_{Mg}$) defects[1] when iron is present in the oxidized state (Fe^{+3}). From now on the notation ($Fe^{\cdot}_{Mg}-V''_{Mg}$) represents the association of defects but not necessarily the exact nature of the association. Because of the significant difference in the drag effects on the edge and screw components of centrosymmetric and noncentrosymmetric defects, the different models for the control of dislocation mobility can be definitely tested while holding other impurities at a constant level.

Measurements of the independent dynamics for the edge and screw components of dislocations are required to determine which of the components is rate controlling in plastic deformation. The results of these investigations will allow the application of specific models to the rate-controlling drag mechanisms. Therefore, edge and screw dislocation velocities have been measured as a function of stress, temperature, and state of ionization of the iron in "pure" MgO, iron-doped MgO, and iron-doped MgO single crystals containing

dislocation dipoles. The objectives of the present paper are to
summarize and discuss the results of the above investigations.[5,6,8]

EXPERIMENTAL PROCEDURE

Material Characterization

Magnesium oxide single crystals were obtained from the Oak
Ridge National Laboratory and the Norton Company. Both "pure" and
doped crystals were given a homogenization treatment at 1350°C in
air for a minimum of 40 h. This ensured that no precipitates were
present, and the iron was stabilized in the trivant state. Specimens were cooled rapidly to room temperature to avoid precipitation.
Another set of similar specimens were given a heat treatment under
reducing conditions (P_{O_2} < 10^{-15} atm) at 1300°C for one week to reduce the iron to the divalent state. The Fe^{+3} concentration in
both the oxidized and reduced samples was determined by an optical
absorption technique.[5] The results and the chemical analysis are
given in Table I.

The single-crystal samples were also characterized by measuring their yield-stress values in the oxidized state, as indicated
in Table I. We observed that the Ox-90 crystals had room-temperature

Table I

Impurities in ppm for Crystals used in Present Study

MgO Crystal	State	Room-temp. yield stress (Nm^{-2})	Fe Content	Total cation impurities	Fe^{+3} concen- tration
"Pure" (Oak Ridge)	Oxidized (Pox)	2.5 x 10^7	20	98	19
	Reduced (Pre)	-	20	98	6
Doped (Norton)	Oxidized (Ox 150)	3.5 x 10^7	150	302	150
Doped (Norton)	Oxidized (Ox 90)	5 x 10^7	90	242	85
with dislo- cation dipoles	Reduced (Re 90)	-	90	242	8

yield stress values higher than the Ox-150 crystals containing
even higher (150-ppm Fe^{+3}) iron impurity. This appears contra-
dictory because normally the yield stress should be lower for
crystals that contain smaller concentrations of impurities. To
resolve this anomaly, specimens were further characterized in terms
of as-received dislocation density by an etch-pit technique using
1 part H_2SO_4: 1 part H_2O: 5 parts saturated NH_4Cl as the etching
solution.[9] All crystals had an overall dislocation density of
5 x 10^4 dislocations/cm^2; however, the nature and arrangement of
the dislocations in the Ox-90 sample were different than in samples
Pox and Ox 150. Most dislocations in sample Ox 90 were in the form
of dislocation dipoles, whereas in other crystals the dislocations
were in the form of isolated single dislocations. This suggests
that dislocation dipoles in the Ox-90 MgO single crystals may be
responsible for the higher yield stress compared with the Ox-150
crystals.

Stress and Dislocation Velocity Measurements

The stress-velocity measurements were made in a manner similar to that described elsewhere[5]; subjecting rectangular prisms
approximately 12x3x1 mm to four-point loading from 2 s to 1 h.
Temperature was controlled to better than $\pm$ 2°C, and all high-
temperature experiments were carried out in a flowing argon atmosphere. The velocity was calculated by measuring the distance between the flat- and sharp-bottom pits in a Leitz microscope and
then dividing the distance by time. Only those measurements from
dislocations with origins unambiguously identified were used.
Stress-strain curves were also obtained in a manner described
elsewhere.[6]

Results and Interpretation. To elucidate the rate-controlling
mechanisms for dislocation mobilities in "pure" and iron-doped MgO
crystals with and without dislocation dipoles, the dislocation
velocities of both the edge and screw dislocations have been
measured at room temperature (RT), 100°C, and 150°C as a function
of the resolved shear stress and the state of ionization (Fe^{+2}
and Fe^{+3}) of the iron dopant. Various activation parameters can
be obtained from such a measurement, as explained in the following
paragraphs.

The stress-velocity data for the edge and screw dislocations
in oxidized (Fe^{+3}) and reduced (Fe^{+2}) states were presented[5,6,8]
according to the power-law relation $v = A\tau^m$, where v is the dislocation velocity, and τ is the resolved shear stress. From the
slopes of such curves, the m values were obtained at several temperatures. Magnesium oxide crystals with no precipitates and low dislocation density were utilized in these investigations. Therefore,
we can assume that the athermal component is small, and τ (applied

stress) is $\sim \tau*$ (effective stress). Recently, Srinivasan and Stoebe[10]
have shown that, during the early stages of deformation (low strains)
of MgO crystals at most temperatures, the applied stress is almost
entirely due to effective stress $\tau*$. Therefore, the activation
volume ($V*$) can be expressed in terms of $\tau*$

$$V* = \frac{m*RT}{\tau*} = \frac{mRT}{\tau}. \tag{1}$$

Activation volumes were calculated at various temperatures and
stresses for edge and screw dislocations in both the oxidized and
reduced states.[5,6,8] The m values and activation volumes (at 10^7
Nm^{-2}) at room temperature, 100°C, and 150°C are given in Table II.

     The activation energies (ΔH) were calculated as a function of
stress for both the edge and screw dislocations assuming Arrhenius
behavior for the dislocation velocities. This activation energy is
stress dependent and is related to the total activation enthalpy of
the obstacle (ΔH_o)[11]

$$\Delta H_o = \Delta H + \int_o^\tau V_T^* \, d\tau. \tag{2}$$

Since the stress dependence of the activation volume is known, ΔH_o
can be calculated for various stresses. A summary of the ΔH_o dependence on the stress, for edge and screw dislocation mobilities
in "pure" and doped MgO single crystals, is given in Table II. If
ΔH_o is independent of stress, then it can be suggested that a singly
activated process is the rate-controlling mechanism. The identification of the specific mechanism was obtained through the interpretation of other activation parameters and the effects of oxidation/
reduction treatments on the edge and screw dislocation velocities.

DISCUSSION

"Pure" Magnesium Oxide Single Crystals

     To understand the principal rate-controlling mechanism for the
edge and screw dislocation mobility in "pure" MgO, it is desirable
to study the influence of the oxidation/reduction state on the stress-
velocity relationship at a fixed temperature, as shown in Fig. 1.
It is observed that both the edge and screw dislocation velocities
have decreased in the oxidized samples (Pox) compared with the reduced samples (Pre). Therefore, noncentrosymmetric defects
$(Fe_{Mg}^{\cdot}-V_{Mg}'')$ control the edge and screw dislocation mobility in oxidized (Pox) samples. At higher stresses, dislocation velocities
(both edge and screw) in oxidized samples (Pox) approached the dislocation velocities observed in reduced samples (Pre). This may be
an indication of extrinsic to intrinsic control of dislocation

Table II

Summary of Activation Parameters for MgO Single Crystals

Specimen purity and heat treatment	Dislocation character	m values $(m = \log v/\log \tau)$			Activation volumes at 10^7 Nm^{-2}			ΔH_0 dependence on stress
		RT	100°C	150°C	RT	100°C	150°C	
Pure MgO reduced (Pre)	Edge	3.14	2.7	2.79	$40b^3$	$50b^3$	$60b^3$	Independent
	Screw	4.14	5.8	2.53	$70b^3$	$110b^3$	$55b^3$	Independent
Pure MgO oxidized (Pox)	Edge	4.34	3.61	4.05	$35b^3$	$68b^3$	$90b^3$	Increases
	Screw	6.53	3.05	3.3	$75b^3$	$55b^3$	$70b^3$	Slight decrease
MgO + 150-ppm Fe^{+3} oxidized (Ox 150)	Edge	11.45	6.45	7.65	$170b^3$	$125b^3$	$165b^3$	Slight decrease
	Screw	11.06	6.02	4.5	$160b^3$	$115b^3$	$85b^3$	Decreases
MgO + 90-ppm Fe oxidized (Ox 90)	Edge	9.51	5.22	5.0	$145b^3$	$100b^3$	$110b^3$	Slight decrease
	Screw	12.5	6.23	5.6	$190b^3$	$125b^3$	$125b^3$	Decreases
MgO + 90-ppm Fe reduced (Re 90)	Edge	12.5	6.93	5.88	$185b^3$	$135b^3$	$130b^3$	Decreases
	Screw	11.4	8.28	6.52	$165b^3$	$155b^3$	$140b^3$	Slight decrease

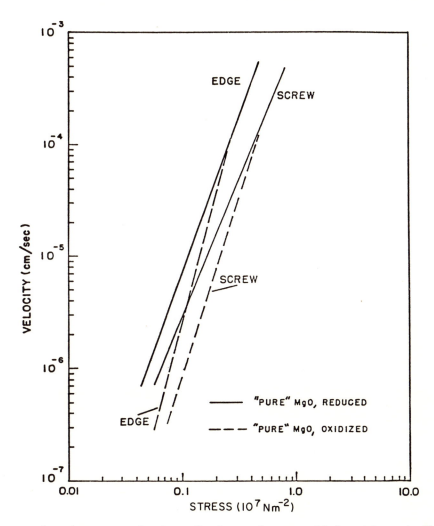

Figure 1. Average velocity of edge and screw dislocations in MgO at 150°C as a function of resolved shear stress for "Pure" crystals in oxidized (Pox) and reduced (Pre) states.

mobility in oxidized samples (Pox), or that the dislocation mobilities are controlled by an intrinsic mechanism in the reduced samples
(Pre).

In the reduced samples (Pre), the total activation enthalpy
ΔH_o is stress independent (Fig. 8 of Ref. 5) over the stress regime
investigated. This behavior strongly suggests that a singly activated mechanism controls the edge and screw dislocation mobilities.
Detailed analyses[5] of the activation parameters (ΔH_o, activation
volume and m values) have shown that Peierls mechanism is the rate-
controlling process for edge and screw dislocation mobilities in
"pure" MgO crystals in the reduced state (Pre). The rate-controlling
step in the Peierls mechanism is suggested as the nucleation of the
double kinks.[5] From the analyses, the Peierls stress values of
0.6×10^8 and 1.7×10^8 Nm^{-2} were obtained for edge and screw dislocations, respectively. The shear modulus (G) for MgO is
1.3×10^{11} Nm^{-2}. This yields Peierls stresses of 10^{-3} to 10^{-4} G,
which is reasonable.

Similar analyses[5] of the activation parameters have shown that,
for "pure" MgO single crystals in the oxidized state, the edge and
screw dislocation mobilities governed by a mixed mode consisting of
Peierls stresses and the resistance due to nonsymmetric defects
$(Fe^{\bullet}_{Mg}-V''_{Mg})$.

Iron-doped MgO Single Crystals without Dislocation Dipoles

In doped MgO single crystals containing 150-ppm Fe^{+3} impurities,
edge dislocations move faster than screw dislocations over the stress
and temperature regimes investigated. This behavior is similar to
that observed for dislocations in "pure" single crystals.[5]

The magnitudes of the total activation enthalpy ΔH_o are 38-28
and 52-31 kcal/mole, respectively, for the edge and screw dislocations. These values are such that the point-defect drag mechanism
may be suggested.[6] Better insight regarding the nature of the
species controlling the edge and screw dislocation motion in doped
MgO single crystals was gained from the temperature dependence of
m values and the magnitudes of the activation volumes (Table II).
The m values are temperature dependent for the edge and screw dislocations, therefore, Gilman's[12] point-defect drag and Fleischer's[13]
point-defect-interaction drag mechanisms may be controlling the dislocation mobilities. Since both the edge and screw dislocation
velocities decrease with an increase in Fe^{+3} concentration (Fig. 2),
Gilman's[12] point-defect drag model may be ruled out because only
edge dislocation velocities are expected to decrease as a result of
the symmetric strain fields in Gilman's[12] model. Therefore, the
edge and screw dislocation mobilities seem to be controlled by

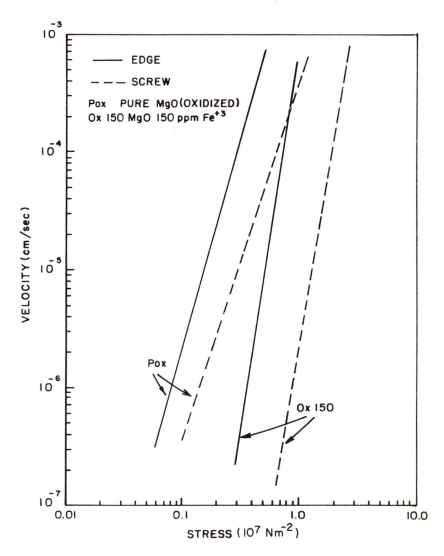

Figure 2.   Average velocity of edge and screw dislocations in MgO
at 100°C as a function of resolved shear stress for
"Pure" crystals (Pox) and crystals containing 150 ppm
Fe^{+3} impurity.

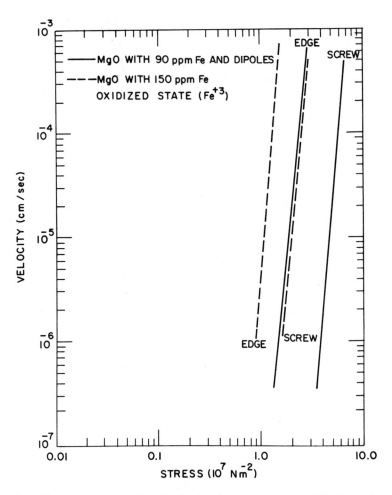

Figure 3.   The average velocity of edge and screw dislocations at
room temperature in MgO crystals containing 90 ppm iron
with dislocation dipoles and 150 ppm iron in the oxidized
state (Fe^{+3}).

Fleischer's[13,14] point-defect ($Fe_{Mg}^{\cdot}$–$V_{Mg}^{''}$) interaction mechanism.
Additional tests of the conformance of the edge and screw disloca-
tion mobilities in doped MgO single crystals were analyzed quantitatively according to Fleischer's[13,14] model (for details see
Ref. 6). From this analysis, it was concluded that both edge and
screw dislocation mobilities in MgO single crystals containing
150-ppm Fe^{+3} are controlled by the interaction of dislocations with
nonsymmetric distortions due to ($Fe_{Mg}^{\cdot}$–$V_{Mg}^{''}$) defects.

Iron-doped MgO Single Crystals Containing Dislocation Dipoles

    The important role of dislocation dipoles in determining the
strength of MgO crystals was realized when we observed[8] higher
yield-stress values for crystals containing 90-ppm iron than for
the crystals containing 150-ppm iron. The increase in flow stress
may be directly attributed to the presence of dislocation dipoles,
which can increase the athermal, thermal, or both components of the
total stress, depending on the distribution of the dipoles. The
interaction of dislocations with dipoles on parallel planes is
relatively short range; it is largely thermal if the dipole distribution is random but may be largely athermal for nonrandom arrays.

    More direct evidence of the resistance to motion due to dislo-
cation dipoles on the dislocation mobility is presented in Fig. 3.
Both the edge and screw components of dislocation have higher velocities, by an order of magnitude, in the Ox-150 crystals. Therefore,
the resistance due to dislocation dipoles on edge and screw dislocation mobilities is real and significant.

    To elucidate the principal rate-controlling mechanism for the
edge and screw dislocation mobilities in MgO crystals containing
90-ppm iron and dislocation dipoles, it is desirable to study the
influence of defect concentration on the stress-velocity relationship at a fixed temperature. This was achieved by changing the
state of ionization of the iron impurity from trivalent (Fe^{+3}) to
divalent (Fe^{+2}). The influence of such a treatment on dislocation
velocity is shown in Fig. 4. The edge and screw dislocation velocities are higher in the divalent state (Fe^{+2}) than in the trivalent
state (Fe^{+3}). This effect seems to be significantly different in
magnitude for the different components of the dislocation lines.
The screw dislocation velocities have increased much more than the
edge dislocation velocities. Another significant observation is
that the magnitudes of the edge and screw dislocation velocities
in the reduced state (Fe^{+2}) and the edge dislocation velocity in the
oxidized state (Fe^{+3}) are almost similar. This suggests that different kinds of defect species may be controlling the edge and
screw dislocation velocities in the oxidized state (Fe^{+3}); however,
in the reduced state (Fe^{+2}) only one type of defect may be the

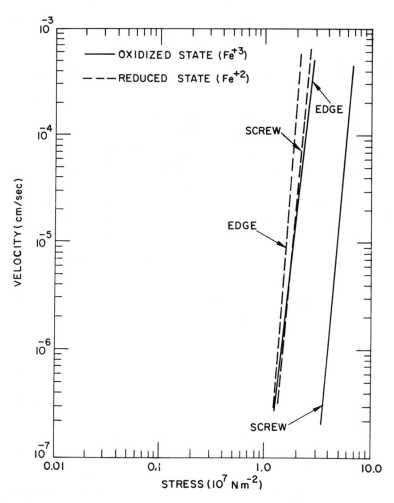

Figure 4.   The average velocity of edge and screw dislocations
            in MgO crystals at room temperature as a function of
            resolved shear stress for crystals containing 90 ppm
            iron and dislocation dipoles in oxidized (Ox 90) and
            reduced (Re 90) states.

Table III
Summary of Drag Mechanisms Controlling Edge and
Screw Dislocation Mobilities in MgO Single Crystals

Specimen purity and heat treatment	Total cation impurities (ppm)	Fe^{+3} concentration (ppm)	Rate-controlling drag mechanisms	
			Edge dislocation	Screw dislocation
Pure MgO reduced (Pre)	98	6	Peierls mechanism with thermally activated double-kink nucleation as rate-limiting step. Peierls stress = 0.6×10^8 Nm^{-2}	Same as for edge Peierls stress = 1.7×10^8 Nm^{-2}
Pure MgO oxidized (Pox)	98	19	Mixed control due to Peierls stresses and nonsymmetric defects ($Fe_{Mg}^{\cdot}$–V_{Mg}'')	Mixed control similar to edge dislocation
MgO + 150-ppm Fe oxidized (Ox 150)	302	150	Interaction of dislocations with nonsymmetric distortions due to ($Fe_{Mg}^{\cdot}$–V_{Mg}'') defects. $\Delta\varepsilon = 0.33$	Same as for edge dislocation
MgO + 90-ppm Fe and dislocation dipoles oxidized (Ox 90)	242	85	Interaction of edge dislocation with dislocation dipoles	Interaction of screw dislocation with non-centrosymmetric defects ($Fe_{Mg}^{\cdot}$–V_{Mg}''). $\Delta\varepsilon = 0.4$
MgO + 90-ppm Fe and dislocation Dipoles reduced (Re 90)	242	8	Interaction of edge dislocation with dislocation dipoles	Interaction of screw dislocation with dislocation dipoles

principal rate-controlling mechanism for both of the components
of dislocations.

To identify the species controlling the edge and screw dislocation velocities in the oxidized and reduced states, we have
further analyzed the thermal-activation parameters (for details see
Ref. 8). From the analysis it is suggested that the screw dislocation mobility, in MgO single crystals containing 90-ppm iron and
dislocation dipoles, in the oxidized state is controlled by the
interaction of dislocations with the nonsymmetric distortions due
to $(\overset{\cdot}{Fe}_{Mg}-V''_{Mg})$ defects. However, the edge dislocation mobility in
the oxidized state (Fe^{+3}) is suggested to be governed by the interaction of edge dislocations with the dislocation dipoles. The edge
and screw dislocation mobilities in the reduced state (Fe^{+2}) are
also suggested to be controlled by the interaction of these dislocations with the dislocation dipoles.

SUMMARY

The drag mechanisms controlling the edge and screw dislocation
mobilities in MgO single crystals are given in Table III. It is
apparent that care must be exercised while interpreting the results
for the identification of rate-controlling drag mechanisms in ceramic crystals. The type and the amount of impurities, their oxidation states, and the presence of other line defects will all play a
significant role in determining the rate-controlling drag mechanism
for dislocation mobilities and/or plastic deformation of ceramic
crystals.

ACKNOWLEDGMENTS

The material reported is from a thesis submitted by R. N. Singh
in partial fulfillment of the requirements for the degree of Doctor
of Science, Department of Metallurgy and Materials Science, Massachusetts Institute of Technology, Cambridge, Mass., under the guidance
of Professor R. L. Coble. This work was supported by the U. S. Atomic
Energy Commission.

REFERENCES

1. R. W. Davidge, J. Mater. Sci. $\underline{2}$, 339 (1967).
2. M. Srinivasan and T. G. Stoebe, J. Appl. Phys. $\underline{41}$, 3726 (1970).
3. G. W. Groves and M. E. Fine, J. Appl. Phys. $\underline{35}$, 3587 (1964).
4. R. L. Moon and P. L. Pratt, Proc. Brit. Ceram. Soc. $\underline{15}$, 203 (1970).
5. R. N. Singh and R. L. Coble, J. Appl. Phys. $\underline{45}$, 981 (1974).
6. R. N. Singh and R. L. Coble, J. Appl. Phys. $\underline{45}$, 990 (1974).
7. T. R. Cass and J. Washburn, Proc. Brit. Ceram. Soc. $\underline{6}$, 239 (1966).
8. R. N. Singh and R. L. Coble, submitted to J. Appl. Phys. for publication.
9. R. J. Stokes, T. L. Johnston and C. H. Li, Phil. Mag. $\underline{3}$, 718 (1958).
10. M. Srinivasan and T. G. Stoebe, J. Mater. Sci. $\underline{9}$, 121 (1974).
11. A. G. Evans and R. D. Rawlings, Phys. Stat. Sol. $\underline{34}$, 9 (1969).
12. J. J. Gilman, J. Appl. Phys. $\underline{36}$, 3195 (1965).
13. R. L. Fleischer, J. Appl. Phys. $\underline{33}$, 3504 (1962).
14. R. L. Fleischer, Acta. Metall. $\underline{15}$, 1513 (1967).

SOLUTION STRENGTHENING OF MgO CRYSTALS

C. Norman Ahlquist*

Department of Mechanical Engineering, University of
Colorado, Boulder, Colorado 80302 and *Intel Corporation,
Santa Clara, California 95051

ABSTRACT

The solution strengthening of MgO by dissolved transition
metal cations is examined. Trivalent cations, Al^{3+}, Cr^{3+} and
Fe^{3+}, harden MgO equally while divalent cations, Fe^{2+} and
Ni^{2+}, have little effect upon yield strength. The strengthening
is inconsistent with the elastic theory of Fleischer for pinning
of dislocations at tetragonal defects. The solution hardening by
Al^{3+}, Cr^{3+} and Fe^{3+} is quantitatively explained by a recent
theory by Gilman. Strengthening in this theory results from the
change in electrostatic energy that occurs when a trivalent ion
pair - cation vacancy complex is sheared by a dislocations passing through the complex.

INTRODUCTION

The experiments of Groves and Fine (1), Davidge (2),
Srinivasan and Stoebe (3) and Kruse and Fine (4) have shown that
the addition of trivalent iron, Fe^{3+}, in solid solution strongly
hardens MgO while divalent iron, Fe^{2+}, and divalent nickel,
Ni^{2+}, have little effect upon yield strength. The observed hardening in the case of trivalent cation impurities may be attributable
to the elastic pinning of dislocations by tetragonal defects.

The tetragonal defect in MgO consists of a trivalent ion
pair - cation vacancy complex. (The analogous defect in alkali-
halide crystals is the divalent ion - cation vacancy complex.)
EPR measurements indicate that Fe^{3+} and Cr^{3+} impurities form
a complex by association with Mg^{2+} vacancies (5,6). Thus the

233

requisite tetragonal defect is known to form in MgO crystals.

The Fleischer model (7) for tetragonal defect hardening predicts that the yield strength equals

$$\tau_y = \frac{1}{3} G \, \Delta\epsilon \, c^{1/2} \quad , \tag{1}$$

where G is the shear modulus of the solvent, c is the solute concentration and $\Delta\epsilon = \epsilon_1 - \epsilon_2$, the difference is principal strains due to the tetragonal defect.

The tetragonality, $\Delta\epsilon$, is expected to be a function of the relative ionic radii of the impurity cation and the divalent solvent cation, Mg^{2+} . Table I summarizes the ionic radii (8) and ionic polarizabilities (9) of the solvent Mg^{2+} and various cation impurities which form solid solutions in MgO (8). Both the divalent and trivalent impurities might be expected to produce some hardening due to size mismatch alone yet only the trivalent impurities cause significant strengthening. The tetragonality associated with the trivalent ion pair - cation vacancy complex could be responsible for the observed hardening.

The present work was undertaken to compare the solution hardening produced in MgO crystals by Al^{3+} and Cr^{3+} to the hardening previously observed by Fe^{3+} (1-4). Figure 1 shows the pseudo-binary phase diagram for MgO - Mg Cr_2 O_4 given by Alper, et al. (10). According to this diagram, an MgO crystal doped with 2% Cr_2 O_3 is single phase at 1400°C and by quenching there is the possibility of maintaining a metastable solid solution of Cr^{3+} in MgO (the preferred ionization state for Cr and Al

Table I.   Ionic Radii and Ionic Polarizabilities of Various Cations

Cation	Ionic Radius, Å (8)	Ionic Polarizability, Å³ (9)
Mg^{2+}	.720	1.8
Fe^{2+}	.770	2.8
Ni^{2+}	.700	2.7
Fe^{3+}	.645	4.7
Cr^{3+}	.615	3.9
Al^{3+}	.530	1.9

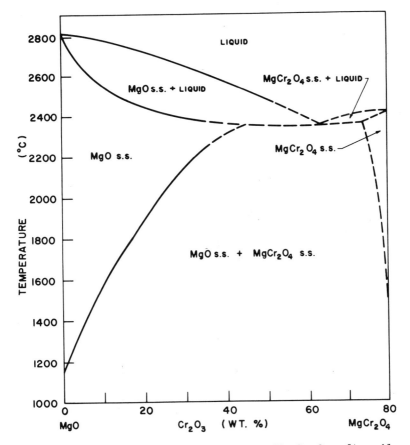

Fig. 1 Phase Diagram for the System MgO - Mg Cr_2O_4 after Alper,
et al. (10).

is +3 while Fe is observed as +2 or +3 (5,6) depending upon the
oxygen pressure in which the MgO crystals are heat treated (1,2)).
Solid solutions of Al^{3+} in MgO are formed in the same manner
(11).

If an <u>elastic</u> interaction is responsible for solution hardening of MgO , the strengthening due to Al^{3+} or Cr^{3+} should
be different than the strengthening due to Fe^{3+} due to the
difference in ionic radii and ionic polarizabilities (Table I).
An invariant feature of these impurities is the ionic charge
difference between the divalent solvent and trivalent solute cations.
If the observed hardening were <u>electrostatic</u> in origin, one would
expect the strengthening per impurity ion to be the same for
Al^{3+}, Cr^{3+} and Fe^{3+}.

TABLE II. MgO CRYSTAL COMPOSITIONS (mole ppm) AND YIELD STRENGTHS

Identification	H1	H2	C1	C2	A1	A2	A3	A4	A5	A6	A7
Al^{3+}	44	132	119	74	339	339	261	393	574	1387	1700
Cr^{3+}	–	11	262	1047	13	37	–	9	7	20	5
Fe^{3+}	57	10	199	178	83	116	231	165	165	165	116
$c=(Al^{3+}+Cr^{3+}+Fe^{3+})$	101	153	580	1299	435	492	492	567	746	1572	1821
$c^{1/2}$	10	12	24	36	21	22	22	24	27	40	43
σ_y, N/mm^2	50±8	58±9	113±9	174±25	73±8	97±7	65±2	83±6	111±5	230±20	203±20

Trace Impurities: Si < 70 ppm (Except crystal C1 which contained 530 ppm Si)

Ti < 5 ppm

Ca < 20 ppm

Mn < 10 ppm

EXPERIMENTAL METHODS

Nominally pure and doped MgO single crystals containing
Aℓ, Cr and Fe were obtained from the Norton Research Corporation
(Canada) Ltd. Spectroscopic analyses performed by Norton Research
Corporation and American Spectrographic Laboratory are listed in
Table II.

2 x 2 x 5 mm compression samples were cleaved from 5 x 5 x 10 mm
single crystals taking care to maintain parallel and flat loading
faces. The cleaved crystals were solution treated in air at 1400°C
for 20 hours and subsequently air quenched. Evidence of precipitation was not observed after quenching. The Cr doped crystals were
transparent and green while the Aℓ and Fe doped crystals were
transparent and colorless.

Compressive loads were applied in the <100> direction thru a
spherical loading seat to compensate for possible misalignment of
the loading platens. 3 - 6 tests were conducted for each composition. Only those tests where yielding clearly occurred before
fracture are included in the average value of the yield strength,
σ_y , summarized in Table II. The ± range represents the difference between the maximum and minimum measured values of σ_y. All
subsequent data are reported as the compressive yield strength,
σ_y , which is twice the critical resolved shear stress for slip
on {110} <110> .

HARDENING BY TRIVALENT COMPLEXES

Typical flow curves for the $A\ell^{3+}$, Cr^{3+} and Fe^{3+} doped MgO
single crystals tested in compression are shown in Fig. 2. The
observed increase in strength is accompanied by a marked reduction
in strain to fracture. In addition, the heavily doped crystals
exhibited more variability in their mechanical behavior than the
lower strength crystals (4), Table II.

The yield strength, σ_y, of these crystals is proportional to
$c^{1/2}$, where c is the total trivalent cation concentration
($A\ell^{3+}$ + Cr^{3+} + Fe^{3+}), as demonstrated in Fig. 3. Included in this
figure are the data of Davidge (2) and Srinivasan and Stoebe (3)
for MgO containing Fe^{3+} in solid solution. Observe the yield
strength of MgO doped with either $A\ell^{3+}$, Cr^{3+} or Fe^{3+} falls
on a common line represented by the equation:

$$\sigma_y = A + B\, c^{1/2} , \qquad\qquad (2)$$

where $A = 0.1 \times 10^9$ dynes/cm^2 and $B = 50 \times 10^9$ dynes/cm^2. The
extrapolated strength of pure MgO is represented by A which is
twice the lattice friction stress, $\tau_p = 5$ N/mm^2.

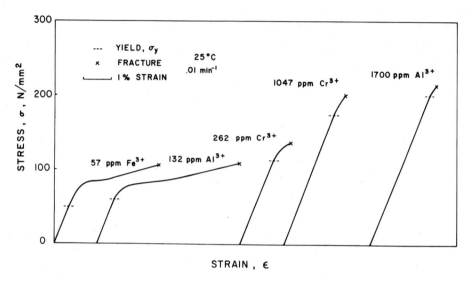

Fig. 2 Typical Stress Strain Curves for Solution Treated Crystals
    Tested in Compression.

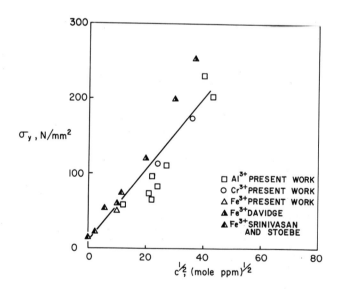

Fig. 3 Dependence of Yield Strength on Square Root of Trivalent
    Ion Concentration for MgO Crystals.

The observed concentration dependence exhibited by equation
(2) is consistent with the Fleischer theory given by equation (1).
The ionic radii and ionic polarizabilities of Al^{3+}, Cr^{3+} and Fe^{3+}
are sufficiently different to produce different values of $\Delta\varepsilon$
yet the same strengthening per cation is observed in Fig. 3. This
common behavior suggests that the elastic interaction is not responsible for the strengthening of MgO doped with trivalent cation
impurities.

THE GILMAN THEORY FOR THE SHEAR STRENGTH OF THE COMPLEX

The solution strengthening of sodium and potassium halide
crystals by dissolved alkaline earth atoms (12) was recently
quantitatively explained by Gilman (13). Strengthening is the
result of the change in electrostatic energy that occurs when a
divalent ion – vacancy complex is sheared by a dislocation passing
through the complex. The analogous complex in MgO is the trivalent
ion pair – cation vacancy complex shown schematically in Fig. 4 (5).
Gilman's theory (13) is now reformulated as it applies to the
trivalent complex in MgO.

The form of equation (2) reduces to $\sigma_y = B$ at unit concentration. Hence, B is simply twice the cohesive shear strength
of the complex shown in Fig. 4 (since the resolved shear stress
$\tau = \frac{1}{2}\sigma_y$). The hardening coefficient, B, may be estimated from
electrostatic theory as follows.

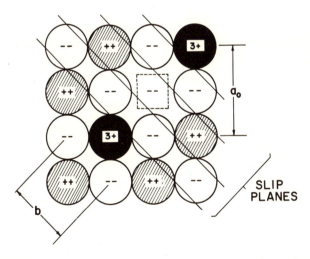

Fig. 4 Trivalent Ion Pair – Cation Vacancy Complex Shown
Schematically on the {100} Plane of the MgO Structure.

Before a dislocation with Burgers displacement b passes through the complex, the distance between the two charge centers is $b = a_o/\sqrt{2}$. After shearing it becomes $\sqrt{2}b = a_o$. If K is the static dielectric constant and e is the electron charge, the energy difference between the unsheared and sheared state is

$$\Delta U = (z^2 e^2/Ka_o)(\sqrt{2} - 1) \quad , \tag{3}$$

where $z = 1$ since each charge center has unit excess charge. The force on the complex caused by a shear stress τ is τb per unit length or $\tau b(a_o/2)$ on a segment that shears the complex. Thus the work done, W, is

$$W = \frac{1}{2} \tau b^2 a_o = \frac{1}{4} \tau a_o^3 \quad . \tag{4}$$

Equating the work done to the energy increase yields

$$\tau = 1.66 \; z^2 e^2/Ka_o^4 \quad , \tag{5}$$

where z is the excess charge on the impurity ion. Four glide planes pass through the complex containing two trivalent ions so the effective concentration is twice the trivalent ion concentration. This introduces a factor of $\sqrt{2}$, so the effective shear strength is

$$\tau = 2.34 \; z^2 e^2/Ka_o^4 \quad , \tag{6}$$

which yields a hardening coefficient of:

$$B = 4.7 \; z^2 e^2/Ka_o^4 \quad . \tag{7}$$

The static dielectric constant and lattice parameter for MgO are $K = 10.1$ (25 °C and 100_oHz to 1 MHz independent of dopant concentration) and $a_o = 4.203$ A , respectively, yielding a numerical value for B of

$$B = 34 \times 10^9 \; \text{dynes/cm}^2 \quad . \tag{8}$$

This compares favorably with the experimental value

$$B_{exp} = 50 \times 10^9 \; \text{dynes/cm}^2 \quad . \tag{9}$$

This simple theory describes quantitatively the solution hardening of MgO by trivalent impurities. The solution hardening coefficient, B , given by the theory differs from the observed value by 32%. The generality of this theory may be examined by comparing the concentration dependence of σ_y, equation (2), normalized with respect to B , equation (7), for I - VII compounds doped with group II elements and II - VI compounds doped with group III elements. Figure 5 includes data for MgO strengthened

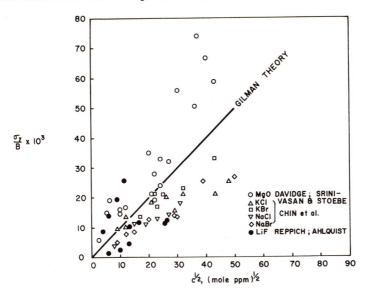

Fig. 5 The Concentration Dependence of Yield Strength normalized
with respect to B for I - VII Compounds (KCℓ, KBr, NaCℓ,
NaBr, LiF) doped with Group II Elements (Mg^{2+}, Ca^{2+}, Sr^{2+},
Ba^{2+}) and II - VI Compounds (MgO) doped with Group III
Elements (Aℓ^{3+}, Cr^{3+}, Fe^{3+}).

by Aℓ^{3+}, Cr^{3+} and Fe^{3+}, Figure 3, KCℓ, KBr, NaCℓ and NaBr
strengthened by Ca^{2+}, Sr^{2+} and Ba^{2+}, Chin et al. (12), LiF strengthened by Mg^{2+}, Reppich (14), and LiF strengthened by Ca^{2+}, Ahlquist
(15). The elastic constants of these crystals range from a low
of $c_{44} = 5.2 \times 10^{10}$ dynes/cm^2 for KBr to a high of $c_{44} = 148 \times 10^{10}$
dynes/cm^2 for MgO. Gilman's electrostatic theory, shown by the
solid line in Figure 5, predicts the yield strength to within a
factor of 2 using no adjustable parameters. This general agreement
lends credence to this model (13).

ACKNOWLEDGMENTS

The author wishes to thank Professor Seymour Geller for many
stimulating discussions and Dr. W. B. Westphal for measuring the
dielectric constants of the crystals studied. This research was
sponsored by the Air Force Office of Scientific Research under
Grant No. 74-2603.

REFERENCES

1.   G. W. Groves and M. E. Fine, J. Appl. Phys. 35, 3587 (1964).
2.   R. W. Davidge, J. Mat. Sci. 2, 339 (1967).
3.   M. Srinivasan and T. G. Stoebe, J. Appl. Phys. 41, 3726 (1970).
4.   E. W. Kruse III and M. E. Fine, J. Am. Ceram. Soc. 55, 32
      (1972).
5.   J. E. Wertz and P. Auzins, Phys. Rev. 106, 484 (1957).
6.   J. E. Wertz, J. W. Orton and P. Auzins, J. Appl. Phys. 33,
      12 (1962).
7.   R. L. Fleischer, Acta Met. 10, 835 (1962); J. Appl. Phys.
      33, 12 (1962).
8.   R. D. Shannon and C. T. Prewitt, Acta Cryst. B25, 925 (1969).
9.   Calculated from index of refraction and lattice constant of
      respective oxide using the Lorentz-Lorenz formula. See e.g.
      N. F. Mott and R. W. Gurney, Electronic Processes in Ionic
      Crystals, Dover Publications, Inc. New York (1964), p. 14.
10.   A. M. Alper, R. N. McNally, R. C. Doman and F. G. Keihn,
      J. Am. Ceram. Soc. 47, 30   (1964).
11.   A. M. Alper, R. N. McNally, P. G. Ribbe and R. C. Doman,
      J. Am. Ceram. Soc. 45, 264 (1962).
12.   G. Y. Chin, L. G. Van Uitert, M. L. Green, G. J. Zydzik and
      T. Y. Kometani, J. Am. Ceram. Soc. 56, 369 (1973).
13.   J. J. Gilman, J. Appl. Phys. 45, 508 (1974).
14.   B. Reppich, Acta Met. 20, 557 (1972).
15.   C. N. Ahlquist, accepted for publication Acta Met. 22 (1974).

COHERENT PRECIPITATE STRENGTHENING MECHANISMS IN MgO

B. Reppich and H. Knoch

Institut of Materials Science I, University
of Erlangen-Nürnberg, D-852 Erlangen, W.-Germ.

(1) INTRODUCTION

MgO containing magnesia ferrite particles is a model
system for the study of the interaction between dislocations and second phase particles in ceramic
materials. Therefore, several investigations have been
performed on two-phase MgO under this aspect [1-7].
Magnesia ferrite crystalizes as inverse spinel with
disorder [8]. The precipitate particles are coherent,
stress-free, octahedral-shaped and give rise to an increase of the critical resolved shear stress (CRSS)
depending on particle size. A broad maximum of the
CRSS at particle sizes of about 50 Å is followed by a
decrease of the CRSS with increasing particle size
[4,7]. Fine [4] and Wicks and Lewis [6] have given an
interesting interpretation of the dislocation-particle
interaction in MgO. According to these authors
strengthening is attributed to a "stacking fault" in
the spinel particles when a matrix dislocation cuts
particles. However, a rough calculation based on this
idea [4] yields a size-independent CRSS in contrast to
the observation. In a later study Kruse and Fine [7]
explain the CRSS decrease by chemical hardening, that
is creation of new matrix-particle interface during
cutting. As demonstrated later a marked contribution
of chemical hardening to the CRSS can be expected only
for particle sizes less than 100 Å even when relatively high values of surface energy are assumed. For
particles greater than 100 Å the chemical hardening

is negligible and cannot explain the decrease of the
CRSS alone.

(2) THEORY

MgO containing spinel particles can be considered as
a system consisting of non-ordered matrix with
"ordered", coherent particles. As predicted by Gleiter
and Hornbogen [9,10] the mechanical behavior of such
systems is characterized by pairwise dislocation movement. The leading dislocation of the pair creates disorder in the form of an antiphase domain boundary
(APB) on the slip plane within the sheared particle.
The second trailing dislocation removes the APB
created by the first one. The APB thus causes pair
coupling which increases with increasing particle size
so that the required cutting stress is reduced considerably and the Orowan process does not take place.
In this way particles of arbitrary size can be sheared
and the controlling cutting stress decreases with increasing particle size. For two-phase metallic systems
containing coherent, stress-free, ordered zones the
theory of Gleiter and Hornbogen has been verified experimentally and also the pairwise dislocation movement was observed often [10-14].

     Starting with the concept of Gleiter and Hornbogen
[9] and following the ideas of Fine [4,7] and Wicks
and Lewis [6] a hardening model for MgO containing
spinel particles is proposed and formulated quantitatively in [16]. The CRSS is calculated in dependence
on typical parameters such as particle size, volume
fraction, APB-energy *) and surface energy. The
estimation of the CRSS rests on the assumption that
the dislocation-particle interaction is controlled by
APB formation and pairwise dislocation motion as well
as by chemical hardening only. Other factors which can
influence the strengthening of materials with coherent
particles, e.g. coherency strains [17,18], differences
in shear modulus of matrix and of particle [14] or
differences in stacking fault energy of dissociated

*) Instead of the term "stacking fault" used by Fine
[4,7] and Wicks and Lewis [6] we propose the term
"APB" because the concept of "stacking fault strengthening" is occupied already by other mechanisms in the
literature [14,15].

dislocations [15] as well as changes in the glide
systems within the particle, will probably not give a
contribution to the CRSS. The glide system in the coherent, stress-free magnesia ferrite particles having
the spinel lattice may be the same as in the MgO
matrix, namely {110} ⟨110⟩. The Burgers vector in the
spinel lattice is then twice as large as in the MgO
matrix (fig. 1). The oxygen lattices of the two
structures are continuous. A matrix dislocation of one
of the four {110} ⟨110⟩ dodecahedral slip systems may
shear the spinel particle. As illustrated in fig. 1
the cations which occupy octahedral sites and tetrahedral sites come into antiphase position after the
{110} ⟨110⟩ shearing. In this way a planar fault is
generated [6], similiar to APB in long range ordered
crystals, attracting a second trailing dislocation.
The APB on the slip plane of the sheared spinel particle couples the cutting dislocations pairwise. The
APB energy γ and the surface energy γ' may be constant
locally as well as with respect to aging time and particle size.

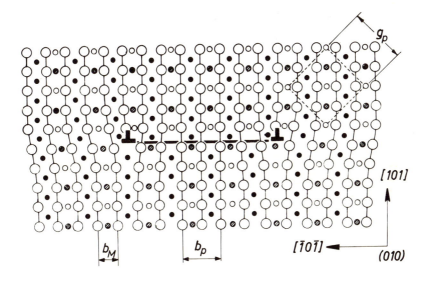

Fig. 1: Edge dislocation pair and antiphase boundary
         (APB) in (ideal) spinel lattice [6].

Starting from the balance of the forces acting on a coupled pair of dislocations, i.e.

the force due to the applied shear stress,
the repulsive force between the two paired dislocations,
the attracting force of the APB,
and the force producing new matrix-particle surface,

the forward stress on the first leading dislocation can be derived. As the determining parameters the obtained equation contains the effective particle spacings along the dislocations, l_I and l_{II} (fig. 2a). Hence, the dislocation configurations must be discussed. According to the calculations of Gleiter and Hornbogen [9], which are demonstrated experimentally [10] as well as supported by our own TEM observations, the following cases can be distinguished (fig. 2a):

(i) Weak Pair Coupling (Small Particles)

The dislocations of the pair are loose-coupled and lie in different particles. The dislocation II is nearly straight; we have $d_{II}/l_{II} = f$, where f is the precipitate volume fraction, d is the particle size. Bending and particle spacing along the leading dislocation I may be determined by the Friedel condition [19-21], l_F = Friedel spacing. For the present octahedral morphology the relation between planar particle spacing L, particle size d, and volume fraction is L = $d(\sqrt{\sqrt{2}/3f} - 1/\sqrt{2})$. The cutting stress for paired edge dislocations is then given by formula (1) in fig. 2a, where $\Delta\tau_Q = \tau_Q^o - \tau_Q^M$ is the increase of the CRSS caused by the particles; τ_Q^o is the measured total CRSS, τ_Q^M is the contribution of the matrix [16,22], b is the Burgers vector.

(ii) Medium Pair Coupling

With coarser dispersions at larger stresses, e.g. near the peak strength, the pair coupling becomes stronger so that the distance of the paired dislocations is closer. But, the dislocations still lie in different particles (fig. 2a, middle). Because the leading dislocation I bends forward more, it may be no longer appropriate to use the Friedel spacing. It seems to be better to take roughly $l_I = L$, the mean planar particle

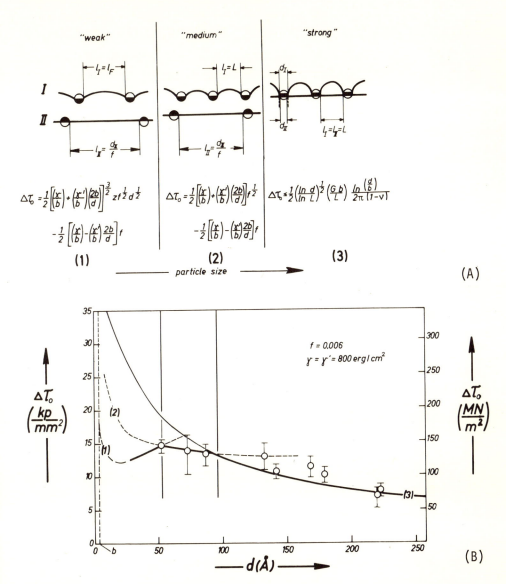

Fig. 2: (a) Idealized representation of the assumed
              dislocation configuration and of the
              symbols and formulas used. For simplifi-
              cation the particles are drawn as
              circles. Dark line: APB.

          (b) Comparison of calculated and measured in-
              crease of the CRSS, $\Delta \tau_o$, in dependence of
              particle size d, see text.

spacing. One obtains the formula (2) in fig. 2a. As in formula (1) the factor of pair coupling, 1/2, appears.

(iii) Strong Pair Coupling

The paired dislocations lie within the same particle (fig. 2a, right). The leading dislocation I forms nearly the critical Orowan bowing while the second dislocation may be a straight line. Using this dislocation configuration, Gleiter and Hornbogen have derived that the cutting stress for pairs can be approximated by half of the Orowan stress of single dislocations: $\Delta\tau_o = 1/2\ \tau_{OR}$ (single dislocations).

The classical Orowan stress is $\tau_{OR} = (2T/bL) = (Gb/L)$, G = shear modulus, $T = Gb^2/2 =$ constant line tension. However, in the last years the Orowan formula has been modified with respect to the particle spacing L and line tension T. As demonstrated in the newest computer calculations of Kocks and coworkers [23] the Orowan stress must be corrected by two factors (fig. 2a). The first factor, (lnd/lnL), is a statistical factor which accounts for the randomness of the particles in the glide plane. The second factor, $\ln(d/b)/2\pi\ (1-\nu)$, ν = Poissons ratio , is the dipole extension formula proposed by Ashby [24] which describes the effect of the mutual loop interaction. Both corrections lower the flow stress drastically from the classical value (Gb/L).

Using the modified Orowan formula for single dislocations the cutting stress for strong coupled screw dislocations is given by eq. (3) in fig. (2a).

(3)   EXPERIMENTS

Undoped MgO single crystals of sufficiently high purity (total impurity level 100 ppm) as well as doped with 7500 or 8500 wt ppm iron have been investigated.[*] All samples were solution treated for 1 day at 1773 K in air, and quenched to RT with a cooling rate of about $50°$/min. After isothermal ageing at 1073 K in air, magnesia ferrite particles precipitate.

---

[*] W. & C. Spicer, England

The samples for the deformation tests were $\langle 100 \rangle$ oriented and have {100} outside faces. The sample length varied between 5-10 mm; the ratio of length to thickness was about 2. The CRSS was derived from stress-strain curves in compression tests at temperatures between RT and 2000 K in air *) The deformation rate is $\dot{\epsilon} = 6 \cdot 10^{-4} s^{-1}$ unless otherwise stated. Testing was performed in a Instron machine (model 1114) equipped with a high-temperature compression jig [27] with Al_2O_3 rams and sapphire monocrystal spacers.

Particle size, particle distribution and precipitated volume fraction, as well as the dislocation structure of deformed crystals, were determined by TEM. {100} transmission foils can be obtained with the following procedure. Discs which were parallel or perpendicular to the sample axis were first sawed out of the middle of the deformed specimen (fig. 3a), then mechanically ground a n d chemically polished in hot orthophosphoric acid by a method developed by Hüther [28]. Here the dielectric properties of MgO are used to control the foil thickness and formation of holes during the thinning process. The TEM micrographs were taken in a Siemens Elmiskop Ia or Philips EM 300; acceleration voltage was 100 kV. For the determination of the size distributions, the TEM micrographs were evaluated with the particle counter Zeiss-TGZ-3. The foil thickness was ascertained by a simple optical method [26]. An examination of the obtained values for the foil thicknesses by the aid of the precise latex ball method recently published by Heimendahl [29] gave satisfactory agreement.

(4) RESULTS AND DISCUSSION

After the thermal treatment described above, regular, randomly distributed, coherent, octahedral-shaped particles with a spinel structure precipitate in the MgO matrix. Their position in the $\langle 100 \rangle$ oriented single crystal is shown in fig. 3a. The base axis of precipitate and matrix are parallel (fig. 3b), the habit planes are {111} oxygen planes. The precipitated octahedrons project in a $\langle 100 \rangle$ projection as squares

---

*) For the graphic determination of the CRSS see fig. 1 in [25]. Each of the RT values of the CRSS have been obtained by "wing size" measurements (cf. [26]).

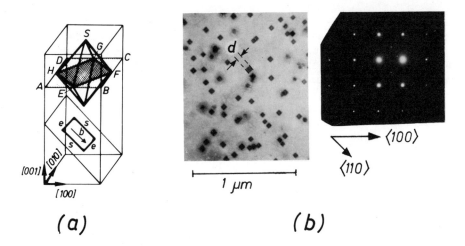

(a)   (b)

Fig. 3: (a) Glide geometry and position of a magnesia
            ferrite octahedron in a [001] oriented MgO
            single crystal with {100} outside faces.
            One of the four ⟨110⟩ {110} dodecahedral
            glide systems with a dislocation loop;
            e: edge, s: screw component.

        (b) TEM micrograph of spinel particles in MgO
            (left); electron diffraction pattern
            (right); [001] projection.

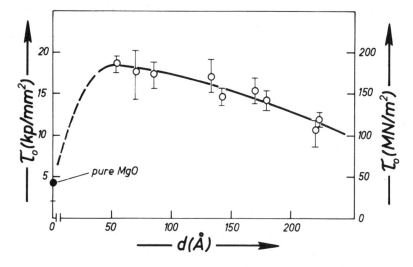

Fig. 4: The CRSS at RT as a function of particle size,
        d.

(hatched plane EFGH in fig. 3a). The particle size may
be characterized by the edge length, d, lying parallel
to $\langle 110 \rangle$, fig. 3b. The particle size distribution
curves are symmetrical. The precipitated volume fraction is constant for the applied ageing times (longer
than 20 min) and particle sizes ($\geqslant 50$ Å) [30] given in
the following, and amounts to f = 0,006 $\pm$ 0,0015.

(4.1) The CRSS

The influence of the spinel particles on the RT values
of the CRSS, τ_o, is illustrated in fig. 4. As observed
by other authors [4,7], particles less than 50 Å cannot
be resolved with the electron microscope. Starting
from the value of pure undoped MgO as the appropriate
reference value, a maximum is found at the particle
size of about 50 Å. With increasing particle size τ_o
decreases markedly.

In the following the measured increase of the
CRSS, $\Delta\tau_o = \tau_o - \tau_o^M$, shall be compared regionwise with
those values estimated theoretically in section (2).
From the τ_o values in fig. 4 the contribution of the
matrix, $\tau_o^{MO} = 4$ kpmm^{-2}, which is identical with that
for the undoped MgO [16,22], is substracted. The $\Delta\tau_o$
values so obtained are plotted in fig. 2b as a function of particle size d.

For large particle sizes, d $\geqslant 100$ Å, the curve (3)
in fig. 2b describes well the decrease of $\Delta\tau_o$ inversely to increasing d, indicating that the theoretically
assumed strong pair coupling is justified. Note that
no fitting parameter appears.

For particle sizes less than 100 Å, in the peak
strength region, the case of medium pair coupling is
realized, curve (2). In equation (2) the cutting
stress depends linearly on the interacting parameters
APB energy γ and surface energy γ' which appear as
free fitting parameters. According to crystallographic
aspects, as well as the TEM observations [16], both
energies should be of the same order of magnitude, as
also assumed by Kruse and Fine [7]. The fit of curve
(2) to the data points is achieved if we chose $\gamma =$
800 erg/cm^2 = γ'. These values are higher than those
exhibited for metallic systems [9-14]. The position
and shape of curve (2) is fixed mainly by the first
term in formula (2), and in this term dominates the
term with γ. On the other hand, the term containing γ',

which characterizes the influence of the chemical
hardening, is responsible for the significant dependence on particle size for d $<$ 50 $\overset{o}{A}$. But, it is seen
that for d $\geqslant$ 100 $\overset{o}{A}$ chemical hardening can be neglected,
curve (2) runs horizontally.

For small particle sizes, d $\leqslant$ 50 $\overset{o}{A}$, the case of
weak pair coupling should be realized, curve (1) in
fig. 2b. Typically, the 3/2 dependence on the interaction parameters γ and γ' appears in the dominating
first term of eq. (1). The contribution of the chemical hardening, $(\gamma'/b)(2b/d)$, is small compared with
the contribution of the APB energy γ, and leads to a
decrease of $\Delta\tau$ for d $<$ 20 $\overset{o}{A}$ only. Unfortunately, only
data points for d $\geqslant$ 50 $\overset{o}{A}$ were available; but, using the
same values for γ and γ' as above, resulting in the
range of medium pair coupling with formula (2), one obtains the curve (1) plotted in fig. 2b. The thickly
drawn full line in fig. 2b demonstrates reasonable
agreement between experimental values and the theoretically estimated cutting stress for pairs.

(4.2) The Dislocation Structure

The dislocation shapes and arrangements observed in
TEM are shown in fig. 5-7, supporting the assumed
configurations in the model. Fig. 5 demonstrates the
dislocation form for the case of weak pair coupling
at small particle sizes as illustrated in fig. 2a,
left. The leading dislocation I of the pair is
slightly bowed, the trailing dislocation II does not
lie in the same particle and is nearly straight. Fig.
6b showes strong pair coupling, compare fig. 2a,
right. Both dislocations lie in the same particles
(arrows). The framed dislocation shapes correspond
exactly with those calculated theoretically by Gleiter
and Hornbogen when the applied shear stress τ is small
or zero (fig. 6a, bottom). Because the TEM photo is
taken from a sample which was reloaded after deformation, this condition was fulfilled, as both dislocations bow out symmetrically in opposite directions
outside of the particles because of the mutually repulsive force between them.

The cutting process could be observed directly in
the electron microscope (see also [16]). Fig. 7 demonstrates that even very large particles are sheared.
From the comparison of the direction of the dislocation
lines and of the (dotted) trails of the observed sec-

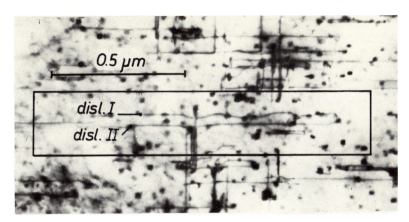

Fig. 5: Dislocation configuration at weak pair-coup-
        ling in MgO containing spinel particles.
        Deformation: T = 1073 K; ϵ = 0.5 %.

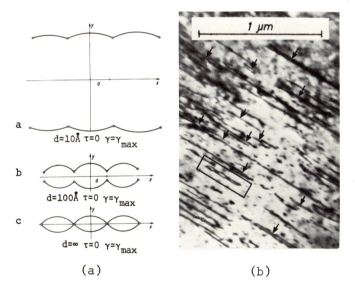

(a)   (b)

Fig. 6: (a) Theoretical calculated dislocation confi-
            guration and pair coupling in dependence
            on particle size, d, for the case that the
            applied shear stress τ is zero, according
            to Gleiter and Hornbogen, fig. 12 in [9].

        (b) TEM micrograph showing the dislocation
            configuration for the case of strong pair
            coupling in MgO containing spinel partic-
            les of particle size of about 150 Å.
            Deformation: T = 1073 K; ϵ = 0.5 %.

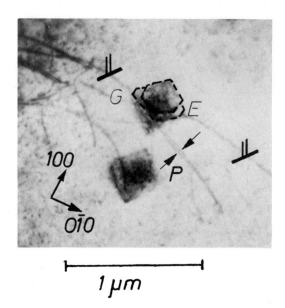

Fig. 7: TEM micrograph showing the cutting of very
        large spinel particles in MgO by strong coupled
        pairs of edge dislocations (P), compare fig.
        3a. Particle size 3000 Å.
        Deformation: T = 1073 K; ϵ = 0.5 %.

tions with fig. 3a, it follows that the cutting dislocations have edge character. At "P" it is marked
that this edge dislocation is a strong coupled pair.

ACKNOWLEDGEMENT

Financial support by the Deutsche Forschungsgemein-
schaft is gratefully acknowledged.

REFERENCES

1. G.W. Groves, M.E. Fine, J. Appl. Phys., 35, 3587
   (1964)
2. R.J. Stokes, J. Am. Ceram. Soc., 48, 60 (1965)
3. R.W. Davidge, J. Mater. Sci., 2, 339 (1967)
4. M.E. Fine, Trans. JIM, 9, Supplement, 527 (1968)
5. M. Srinivasan, T.G. Stoebe, J. Appl. Phys., 41,

3726 (1970)

6. B.J. Wicks, M.H. Lewis, phys. stat. sol. (a), 6, 281 (1971)

7. E.W. Kruse, M.E. Fine, J. Am. Ceram. Soc, 55, 32 (1972)

8. G.P. Wirtz, M.E. Fine, J. Appl. Phys., 38, 3729 (1967)

9. H. Gleiter, E. Hornbogen, phys. stat. sol., 12, 235 (1965)

10. H. Gleiter, E. Hornbogen, phys. stat. sol., 12, 251 (1965)

11. D. Raynor, J.M. Silcock, Metals Sci. J., 4, 121 (1970)

12. L.K. Singhal, J.W. Martin, Acta Met., 16, (1968)

13. S.M. Copley, B.H. Kear, Trans. AIME, 239, 977 and 984 (1967)

14. L.M. Brown, R.K. Ham, in: A. Kelly, R.B. Nicholson (ed.) "Strengthening Methods in Crystals", Appl. Sci. Publ. Ltd., London, S. 12 (1971).

15. V. Gerold, K. Hartmann, Trans JIM, 9, Supplement, 509 (1968)

16. B. Reppich, in preparation

17. V. Gerold, H. Haberkorn, phys. stat. sol, 16, 675 (1966)

18. H. Gleiter, Z. angew. Phys., 23, 108 (1967)

19. J. Friedel, "Dislocations", Addison-Wesley, (1964)

20. R.L. Fleischer, in: D. Peckner (ed.), "The Strength of Metals", Reinhold Publ. Co., New York, (1964)

21. U.F. Kocks, Can., J. Phys., 45, 737 (1967)

22. B. Reppich, in preparation, see: Frühjahrstagung der Deutschen Physikalischen Gesellschaft (DPG), Arbeitsgemeinschaft "Metallphysik", Freudenstadt, April 1974, Vortrag M 19

23. D.J. Bacon, U.F. Kocks, R.O. Scattergood, Phil. Mag., 28, 1241 (1973)

24. M.F. Ashby, Proc. Second Bolton Landing Conf. on Oxide Dispersion Strengthening, Gordon u. Breach, New York, S. 134, (1968)

25. B. Reppich, Acta Met., 20, 557 (1972)

26. H. Knoch, Diplomarbeit, Univ. Erlangen-Nürnberg (1972)

27. B. Reppich, W. Blum, B. Ilschner, Ber. DKG, 44, 41 (1967)

28. W. Hüther, Diplomarbeit, Univ. Erlangen-Nürnberg (1972)

29. M.v. Heimendahl, Micron, 4, 111 (1973)

30. G.P. Wirtz, M.E. Fine, J. Am. Ceram. Soc., 51, 402 (1968).

SLIP SYSTEMS IN STOICHIOMETRIC $MgAl_2O_4$ SPINEL

L. Hwang*, A. H. Heuer, and T. E. Mitchell

Dept. of Metallurgy & Materials Science
Case Western Reserve University
Cleveland, Ohio 44106

ABSTRACT

The mechanical properties of stoichiometric $MgAl_2O_4$ spinel single crystals have been investigated at high temperatures ($\sim1800°C$). The slip system has been determined by slip trace analysis and electron microscopy to be {111} <110>. This behavior is rationalized by analyzing the changes in atomic configuration around dissociated dislocations as they move through the lattice. The change to {110} slip planes in non-stoichiometric alumina-rich spinel is explained in terms of the influence of cation vacancies on the dislocation structure and movement.

1. INTRODUCTION

The crystal structure of $MgAl_2O_4$ spinel is based on an approximate fcc close packing of oxygen anions with half the octahedral interstices occupied by Al cations and an eighth of the tetrahedral interstices occupied by Mg cations. Hornstra[1] predicted that dislocation glide would occur parallel to the close packed {111} oxygen anion planes and in the close-packed <110> directions. He furthermore suggested that dissociation of dislocations with $\vec{b}$ of 1/2 <110> into quarter partials would occur in a manner similar to the synchro-shear mechanism suggested by Kronberg[2] for basal slip in sapphire. The dislocation reactions predicted for spinel are:

---
*Now at Truline Castings, Cleveland, Ohio

$$1/2 \; <110> \; \rightarrow \; 1/4 \; <110> \; + \; 1/4 \; <110> \qquad (1)$$

$$1/4 \; <110> \; \rightarrow \; 1/12 \; <121> \; + \; 1/12 \; <21\bar{1}> \qquad (2)$$

Equation (1) maintains the fcc anion packing but generates stacking errors in the cation sublattice while equation (2) generates a stacking fault in the anion sublattice as well, with a consequently higher energy. Lewis[3] confirmed that dissociation into half-partials (equation (1)) occurs in non-stoichiometric alumina-rich spinel, and that the separation of the partials is more than 200Å. More recently, Welsch et al.[4], as part of the present study, reported that the same dissociation occurs in stoichiometric spinel with a smaller separation (∿100Å). Thus, the cation stacking fault energy apparently decreases with increasing deviation from stoichiometry. However, in neither stoichiometric nor non-stoichiometric spinel has the dissociation in quarter partials (equation (2)) been observed.

Although Lewis[4] confirmed the predicted <110> slip direction, questions remain as to the operating slip plane. Lewis[4] and others[5,6] showed that {110} was the operative slip plane in various non-stoichiometric compositions except for one reported case of {111} slip in MgO:3.0 Al_2O_3 spinel[5]. Work on stoichiometric $MgAl_2O_4$ has been limited because of the difficulty of obtaining large crystals, although Radford and Newey[5] reported {111} slip planes in their investigation. The availability of large Czochralski-grown crystals of stoichiometric spinel made the present study possible. Specimens were deformed in compression at elevated temperatures (∿1800°C); the slip character has been studied by slip trace analysis, etch pit techniques and transmission electron microscopy.

2. EXPERIMENTAL PROCEDURES

Compression specimens, 2.5 x 2.5 x 5.9mm, were cut from a Czochralski-grown single crystal of stoichiometric composition purchased from the Crystal Products Division of Union Carbide Corporation. Specimens were mechanically polished, finishing with 0.3μm Al_2O_3 abrasives, to remove surface damage. The orientation of the compression axis was chosed to be 45° from (111) and ($\bar{1}$01) to give a Schmid factor of 0.5 for (111) [$\bar{1}$01]. Compression tests were carried out on an Instron testing machine at a strain rate of 1.43 x 10^{-4}/sec. Temperatures of 1790°C to 1895°C

were obtained with a Brew furnace; an argon environment
was used to prevent inconguent evaporation of MgO. Foils
for transmission electron microscopy were prepared by ion
bombardment and examined in a Hitachi HU-650B electron
microscope operating at 650kV.

3. RESULTS

3.1 Slip Plane

Two sets of slip lines of roughly equal density were
observed on each surface, as shown in Fig. 1. The corresponding slip planes were determined by two surface trace
analysis to be (111) and ($1\bar{1}1$). (111) is the expected
primary slip plane but ($1\bar{1}1$) is the cross-slip plane.
Taking [$\bar{1}01$] as the slip direction (confirmed in the next
section), the cross-slip system has a Schmidt factor of
0.14 and is therefore quite unexpected. It is suggested
that the high yield stresses (section 3.3) give rise to a
high normal stress which results in a "squeezing" of the
primary slip plane. Dislocations might then prefer to
use the steeply inclined cross-slip plane which has a
smaller normal stress acting on it.

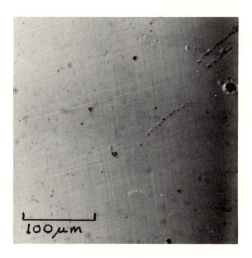

FIGURE 1

Slip lines on the surface of spinel deformed in compression
(axis vertical on the page). The slip traces correspond to
the primary slip plane (nearly horizontal lines) and the
cross-slip plane (nearly vertical lines).

Etch-pit techniques have also been used on lightly
deformed specimens and revealed the alignment of dislocations on {111} slip planes, confirming the slip trace
analysis[7].

3.2 Slip Direction

The slip direction was found by analyzing the burgers
vector of dislocations produced by deformation. This was
done by the usual $\vec{g}.\vec{b}$ analysis of transmission electron
microscopy. In this technique, micrographs are taken of
the same area with different operating $\vec{g}$ reflections;
dislocations are in minimum contrast when $\vec{g}.\vec{b}=0$. An example is shown in Fig. 2, where dislocations having $\vec{b}$ of
$1/2\ [\bar{1}01]$ are in strong contrast for $\vec{g}=\bar{4}04$ and weak contrast for $\vec{g}=040$.

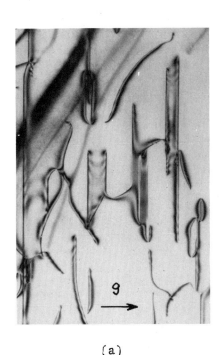 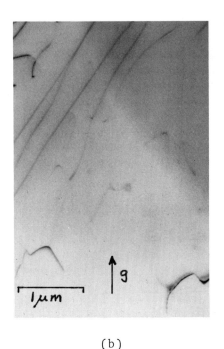

(a)   (b)

FIGURE 2

(101) foil of spinel deformed in compression. Disloca-
tions with $\vec{b}=1/2\ [\bar{1}01]$ are in strong contrast in (a) where
$\vec{g}=\bar{4}04$ and in weak contrast in (b) where $\vec{g}=040$. Dislocations with secondary burgers vectors are also visible.

The particular foil in Fig. 2 was cut parallel to the
(101) plane. Other foils were cut parallel to the (111)
and (1$\bar{1}$1) planes. The analysis of all these foils showed
that dislocations with the primary 1/2 [$\bar{1}$01] burgers vector
are most common, as expected, but many secondary dislocations are also present, mostly in the form of networks.
These secondary dislocations probably result from localized multiple slip and are not responsible for the slip
lines observed on the surface.

3.3 Deformation Substructures

     Transmission electron microscopy has been used to
observe dislocation substructures in foils cut parallel
to the (111), (1$\bar{1}$1) and (101) planes. Two types of
dislocation substructures were observed, often in close
proximity. The first is shown in Fig. 2, where long
primary dislocations are aligned in edge orientation to
form small-angle tilt boundaries. The second consists
of randomly oriented planar networks of primary and
secondary dislocations, as shown in Fig. 3. The dislocation density was measured to be $\sim$ 6 x 10^8/cm^2 in specimens deformed 0.25%.

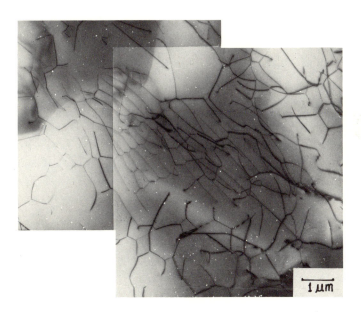

FIGURE 3

Planar dislocation networks in (1$\bar{1}$1) foil.

Dislocations were never observed to be in well-defined slip planes (neither primary nor cross-slip). Lewis[3] also reported that dislocations were not confined to particular slip planes in non-stoichiometric spinel. Presumably the deformation temperature is so high that dislocations are able to climb rapidly into lower energy configurations during deformation.

3.4 Dislocation Dissociation

As reported previously[4], we have used the weak-beam technique to study dislocation dissociations. Separation into half-partials according to equation (1) was observed, and the calculated stacking fault energy was $\sim$180 ergs/cm^2, considerably higher than that in non-stoichiometric spinel[3,4]. However, the further dissociation of equation (2) was not observed and, if it exists, the separation must therefore be less than $\sim$20Å, the resolution of weak-beam electron microscopy.

3.5 Stress-strain Curves

Fig. 4 shows shear stress-shear strain curves obtained in compression at strain-rates of 1.4×10^{-4}/sec for temperatures from 1790 to 1895°C. Specimens yielded quite sharply without yield drops. Yield stresses varied exponentially with reciprocal temperature, giving an apparent activation energy of 53 kcal/mole. The true activation energy is obtained by multiplying the apparent activation energy by n, the stress exponent in the strain-rate equation $\dot{\varepsilon} = A\sigma^n \exp(-\Delta H/RT)$, which we have not yet determined. The very high yield stresses, e.g. 22 kg/mm^2 at 1795°C, are unusual, and, to the authors' knowledge, are only exceeded by sapphire oriented to suppress both basal and prismatic slip.

Fig. 4 also shows that the initial work-hardening rates are high ($\sim\mu$/70, where μ is the shear modulus) except at the highest deformation temperature, 1895°C. Work softening sets in at higher strains which may be due to recovery by climb. The initially high work hardening rates are probably due to the localized secondary slip described in section 3.3, which is due in turn to the high stresses. This makes an interesting comparison with sapphire deformed by basal slip, where the work hardening rate is several

times smaller than in spinel. Pletka et al[8] found that
the work hardening mechanism in sapphire is the interaction of dislocations on parallel slip planes to form dipoles, which is less effective than the interaction with
dislocations on intersecting secondary planes, as is the
case in spinel.

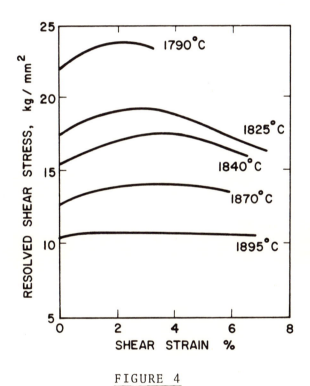

<u>FIGURE 4</u>

Shear stress-shear strain curves of stoichiometric spinel
deformed in compression at various temperatures
($\dot{\varepsilon}$ = 1.43 x 10^{-4}/sec).

4. DISCUSSION

From the results described on stoichiometric spinel, the following aspects of the plastic deformation of spinel require further discussion:

(a) Slip occurs on the {111} <110> system in stoichiometric spinel, as predicted by Hornstra; however, the predominant slip plane in non-stoichiometric spinel changes to {110}, although the slip direction remains the same.

(b) The yield stress is very high but decreases with increasing deviation from stoichiometry.

(c) The cation stacking fault energy (equation (1)) also decreases with increasing deviation from stoichiometry.

(d) The mixed stacking fault energy is so high that dissociation into quarter partials (equation (2)) is not observable.

All of these points can be addressed by focussing on the atomistics of slip on {111} compared with {110} planes. 1/2 <110> is the shortest lattice vector in spinel and so there is no question as to the burgers vector of a perfect dislocation. This is shown in Fig. 5(a) which is a photograph of a model of a spinel unit cell oriented to reveal the close-packed <110> directions. Fig. 5(b) is the same model oriented to reveal the close-packed oxygen anion layers parallel to {111} planes. Fig. 5(c) is again the same model, now oriented to call attention to the {110} planes. Comparison of Figs. 5(b) and (c) demonstrates that, although {111} is the close-packed plane from the point of view of the anions, {110} is the close-packed plane if both anions and cations are considered. Thus, if close-packing is of paramount importance, {110} should be the operative slip plane, as observed in non-stoichiometric spinel. Clearly other factors, involving the detailed interaction of the dislocation with the crystal lattice, must be considered. The necessary analysis can be performed by considering the atomic arrangement around a dislocation in the half-slipped position; this analysis is thus similar to that of Gilman[9] for NaCl crystals, where he was able to rationalize the preference for slip on {110} planes rather than the close packed {100} planes. In the present case, since half partials have been observed in spinel, it is necessary to consider 1/4 <110> dislocations in the half-slipped position.

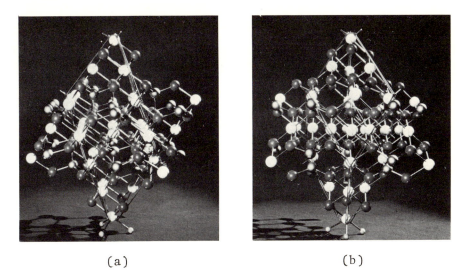

(a)   (b)

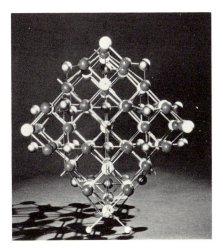

(c)

FIGURE 5

Photographs of spinel crystal model along various directions. Dark balls are oxygen, white balls are Al, and
intermediate are Mg.
  (a)Model viewed approximately along <110>, showing close-
packed lines of atoms.
  (b)Model viewed with <111> vertical, showing close-packed
layers of oxygen.
  (c)Model viewed along <100>, showing horizontal and vertical close-packed {110} planes of anions and cations.

TABLE I

Anion-cation distances during {110} slip. Distances
are expressed in terms of the anion-cation distance in
the perfect lattice.

		Oxygen Nearest Neighbors					
	Sites	1	2	3	4	5	6
Perfect Lattice	Octahedral	1	1	1	1	1	1
	Tetrahedral	0.866	0.866	0.866	0.866	-	-
Half -Slipped	Octahedral	1	1	1	1	1	-
	Tetrahedral	0.866	0.866	0.866	1.118	1.118	-

(a) Slip on {110} planes. Fig. 6(a) shows two {110}
layers of spinel. If a half partial slips half its burgers
vector (i.e. 1/8 <110>, the atomic arrangement shown in
Fig. 6(b) occurs. In the latter position, tetrahedral
cations gain and octahedral cations lose an oxygen nearest
neighbor. Also, as shown in Table I, some of the cation-
anion distances are appreciably different from those in
the perfect lattice and this severe distortion represents
essentially the Peierls energy.

(b) Slip on {111} planes. Fig. 7 shows a close-
packed layer of oxygen anions along with a kagomé layer of
Al cations which occupy 3/4 of the octahedral sites. As
seen in Fig. 5(b), mixed {111} layers of octahedral and
tetrahedral cations also occur between anion layers. However, it seems much more likely that slip will occur in
the simpler kagomé layers.[1] For slip to occur in Fig. 7
by a 1/4 <110> direction, a cation such as that denoted by
1 must move to the 2 position. All other ions above the
close-packed plane must also move by this distance.
(Since 1/4 <110> corresponds to the distance between
anions, the fcc stacking of anions is undisturbed by passage of this partial dislocation). The Al cation can move
from 1 to 2 either via 3 (corresponding to a 1/6 <211>
partial) or via 4 (corresponding to a 1/8 <110> displacement). Position 4 seems unlikely since the cation would
be forced to come very close to anion 5 in Fig. 7. On the

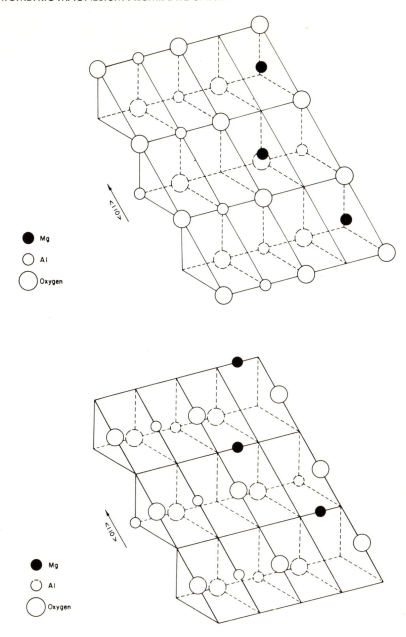

FIGURE 6

Slip on {110} planes. Two {110} layers of atoms are
shown with the dashed atoms underneath. (a) is the
perfect lattice. (b) is the atomic arrangement after the
top layer has been displaced by 1/8 <110>.

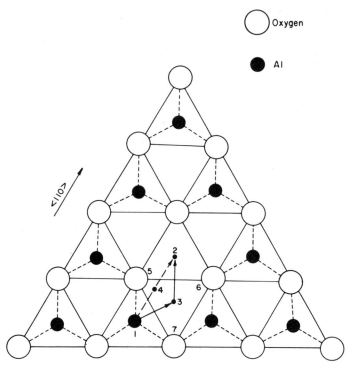

FIGURE 7

Close-packed {111} layer of oxygen ions and an adjacent kagomé layer of aluminum ions. The slip process is described in the text.

other hand, position 3 seems much more likely since it becomes a new octahedral site for the Al cation as the lattice is sheared. Admittedly position 3 creates the mixed stacking fault already described which was not detectable by transmission electron microscopy; however, this position does not disturb nearest neighbor distances. The important Peierls barrier must therefore occur between sites 1 and 3. Note that the movement via site 3 represents the synchro-shear of equation 2: the lack of evidence for the associated mixed fault still does not preclude its existence as part of the partial dislocation core structure.

The fact that stoichiometric spinel crystals slip on {111} planes is entirely consistent with the above discussion. The intermediate position for the half

partial is more distorted for {110} than for {111} slip
planes.

The fact that non-stoichiometric spinel crystals slip
on {110} planes needs to be considered in light of the way
in which the excess Al_2O_3 is accommodated. Jagodinski and
Saalfield[10] determined that the cation vacancies introduced in alumina-rich compositions occur primarily on the
octahedral cation sublattice. The existence of additional
cation vacancies on the kagome layer in Fig. 7 should make
little difference to the shear stress for slip on this
{111} plane since the difficult process is still movement
from 1 to 3. On the other hand, Fig. 6(b) shows that, for
{110} slip at the half-slipped position, a large fraction
of the Peierls barrier results from the repulsive interaction due to the close approach of octahedral and tetrahedral cations; additional octahedral vacancies will therefore
lower the shear stress for {110} slip. In addition, the
cation vacancies will be mobile at the high temperatures
of deformation especially near the core of a dislocation,
allowing the cations to adjust their position to facilitate
dislocation motion on {110} planes.

Additional experimental observations and discussions
will be given in a further paper.[11]

ACKNOWLEDGMENT

This work was supported by the Army Research Office
under Grant No. AROD 3112473G95.

REFERENCES

1. J. Hornstra, J. Phys. Chem. Solids 15, 311 (1960).

2. M.L. Kronberg, Acta. Met. 5, 507 (1957).

3. M.H. Lewis, Phil. Mag. 17, 481 (1968).

4. G. Welsch, L. Hwang, A.H. Heuer and T.E. Mitchell,
    Phil. Mag. 29, (1974).

5. K.C. Radford and C.W.A. Newey, Proc. Brit. Ceram.
    Soc. 9, 131 (1967).

6. N. Doukhan, R. Duclos and B. Escaig, J. Physique 34,
    C9-379 (1973).

7. L. Hwang, M.S. Thesis, Case Western Reserve University, (1974).

8. B.J. Pletka, T.E. Mitchell and A.H. Heuer, J. Am. Ceram. Soc., in press.

9. J.J. Gilman, Acta Met. $\underline{7}$, 608 (1959).

10. H. Jagodinski and H. Saafeld, Z. Krist. $\underline{110}$, 197, (1958).

11. L. Hwang, T. E. Mitchell and A.H. Heuer, J. Am. Ceram. Soc., to be published.

IMPURITY AND GRAIN SIZE EFFECTS ON THE CREEP OF POLYCRYSTALLINE MAGNESIA AND ALUMINA

Paul A. Lessing and Ronald S. Gordon

Division of Materials Science and Engineering

University of Utah, Salt Lake City, Utah 84112

ABSTRACT

Steady state, constant load, creep experiments have been conducted on pure and doped, polycrystalline magnesium and aluminum oxides to determine:
1) the effect of solid solution impurities (e.g. Fe, Cr, Ti) on deformation mechanisms, particularly as they relate to the concentration of lattice defects
2) the effects of impurities, atmosphere, and grain size on the relative contribution to viscous deformation of lattice and grain boundary diffusion of both cationic and anionic species
3) the relative roles of viscous and non-viscous deformation mechanisms.

I. INTRODUCTION

The creep deformation of polycrystalline ceramics can take place by one or more mechanisms which operate alone or simultaneously with others. These mechanisms can be classified in terms of inter- and intra-granular processes. Inter-granular mechanisms include grain boundary sliding which can be accomodated by intra- or inter-granular mass transport[1] (i.e. diffusional creep), localized plastic deformation (dislocation climb and glide), or by the formation of voids and cracks along grain boundaries and triple points. Typical examples of intra-granular processes include dislocation glide controlled by climb[2] and dislocation climb involving sub-grains or dislocation arrays[3].

In polycrystalline ceramics dislocation deformation modes are
inhibited to some degree because of a limited number of slip systems (e.g. Al_2O_3) and relatively large (when compared to metals)
energies required for dislocation motion. Consequently, it is
expected that in the limit of small grain sizes, low stresses, and
reasonably high temperatures ($T \geq 0.5 T_m$; T_m = melting temperature),
polycrystalline ceramics should deform viscously by one or more
diffusional mechanisms. This type of deformation, which is commonly
referred to as diffusional creep involves the transport of ions
through the grains (i.e. lattice or volume diffusion) and/or the
grain boundaries. Gordon[4] has invoked the theories of Nabarro-
Herring[5], Coble[6], and Raj-Ashby[1], which are only strictly
valid for polycrystalline metals, and the principles of coupled
mass transport to derive the following general expression for the
steady state diffusional creep rate ($\dot{\varepsilon}$) of the polycrystalline
compound, $A_\alpha B_\beta$ *

$$\dot{\varepsilon} = \frac{44 \ \Omega_v \ \sigma}{kT \ (GS)^3} \ D_{complex}$$

(1)

$$D_{complex} = \frac{\frac{1}{\alpha} \left[\frac{(GS)}{\pi} \ D_A^\ell + \delta_A D_A^b \right]}{1 + \frac{\beta}{\alpha} \frac{\left[\frac{(GS)}{\pi} \ D_A^\ell + \delta_A \ D_A^b \right]}{\left[\frac{(GS)}{\pi} \ D_B^\ell + \delta_B \ D_B^b \right]}}$$

Ω_v is the molecular volume of $A_\alpha B_\beta$, σ is the stress, (GS) is the
grain size,**k is Boltzmann's Constant,T is the absolute temperature,
D_A^ℓ and D_A^b are the $A^{\beta+}$ ion lattice and grain boundary diffusivities,
D_B^ℓ and D_B^b are the $B^{\alpha-}$ ion lattice and grain boundary diffusivities, and δ_A and δ_B are the effective widths of enhanced diffusion
near the grain boundaries for $A^{\beta+}$ and $B^{\alpha-}$ ions, respectively.
Gordon[4] has shown how Equation (1) can be simplified in limiting
conditions to the Nabarro-Herring (i.e. $\dot{\varepsilon} \ \alpha \ (GS)^{-2}$) and the Coble
(i.e. $\dot{\varepsilon} \ \alpha \ (GS)^{-3}$) creep equations. In these limiting conditions
creep rates can be controlled in principle by either the cation or
anion diffusing in the lattice or in the grain boundary. The
characteristic features of diffusional creep are a linear strain
rate - stress dependence and a strong inverse relation ($\dot{\varepsilon} \ \alpha \ (GS)^{-m}$;

---

*For compounds like MgO, BeO, NaCl, etc. set $1/\alpha$ and β/α to unity[4].

** In this paper, all grain sizes are linear intercepts multiplied
by the factor 1.5 unless otherwise indicated.

$2 \leq m \leq 3$) between the creep rate and grain size. Impurities and atmosphere will play a significant role in diffusional creep since they can alter the values of the various diffusion constants via the control of defect concentrations.

As the grain size and stress increase diffusional modes become less important and non-viscous (i.e. $\dot{\epsilon} \propto \sigma^N$; $N > 1$) mechanisms should begin to dominate creep deformation. Many models for non-viscous creep by dislocation motion abound in the literature. Only a few examples, which may be important in the creep of polycrystalline MgO and Al_2O_3 will be mentioned here.

Perhaps the most frequently quoted model is that by Weertman[2] in which the rate of creep is controlled by dislocation climb, while dislocation glide yields most of the strain. In this theory the creep rate is given by

$$\dot{\epsilon} = \frac{A\, D\, \sigma^{4.5}}{N_d^{\frac{1}{2}}\, kT} \tag{2}$$

A is basically a geometrical constant, D is the appropriate self-diffusion coefficient describing the climb process (presumably the $B^{\alpha-}$ ion in the compound $A_\alpha\, B_\beta$), and N_d is the density of Frank-Read dislocation sources. The stress exponent of 4.5 is due to the fact that not only the rate of climb is increased, but also the dislocation density when the stress is increased. In an earlier theory of Weertman[7] the dislocation density was assumed to be independent of stress and the stress exponent was 3 instead of 4.5.

If a crystal possesses fewer than five independent slip systems deformation cannot occur by glide alone as Weertman predicted[8]. Consequently, a model developed by Nabarro[3] is probably more useful for understanding the mechanism of dislocation creep in polycrystalline ceramics. In this theory grains are assumed to be large so that dislocations are the principal sources and sinks for vacancies. Vacancy fluxes within sub-grains formed by dislocations lead to climb and the resulting deformation. In this model, the creep rate is

$$\dot{\epsilon} = \frac{\pi\, \gamma^2 \Omega\, \sigma^3\, D}{10b^2\, G^2\, kT} \tag{3}$$

Again D is the self diffusion coefficient, γ is a constant (~ 1), Ω is the molecular volume, b is the Burger's vector, and G is the shear modulus. The presence of climb as a possible deformation mechanism allows for dislocation motion in directions other than those of slip. This model predicts a stress exponent of 3 as compared to 4.5 for Weertman's climb-glide model. Weertman[9] has also derived a creep equation in which $\dot{\varepsilon} \propto \sigma^3$. Instead of a climb mechanism the model is based on the assumptions that dislocation glide is Newtonian viscous and rate controlling. The Weertman and Nabarro theories are based on intra-granular deformation processes (glide and climb) and thus no explicit dependence of the creep rate on grain size is expected.

In this paper steadystate creep data on both pure and doped (Fe,Cr, Ti), polycrystalline MgO and Al₂O₃ will be reviewed in light of the theories described above. Particular attention will be focused on the roles of impurities, atmosphere, and grain size in dictating the relative contributions of viscous and non-viscous deformation mechanisms. Furthermore, the relative roles of cation and anion diffusion processes in the viscous creep regime will be analyzed in terms of these variables.

II. EXPERIMENTAL

The preparation of pure and iron-doped (0.05-5.3 cation %) magnesium oxide powders is described extensively elsewhere[10-11]. High purity (99.999%) α-Al₂O₃ powder was prepared by an organo-metallic technique. Aluminum metal was reacted with an excess of isopropanol to form aluminum isopropoxide. The resulting solution was then hydrolyzed with water leading to the precipitation of an aluminum hydroxide gel. The gel after filtration was calcined at 1150-1200°C to form α-Al₂O₃. Procedures for doping the powder with iron (0.2 and 1 cation %) and chromium (1 cation %) are described elsewhere[12].

Dense (> 98% of theoretical), polycrystalline billets of pure and doped Al₂O₃ and MgO were prepared by vacuum hot pressing at temperatures between 1400 and 1700°C in graphite dies at pressures up to 7000 psi [11,12]. In some cases pure and iron-doped MgO powders were pressed in TZM dies at pressures up to 40,000 psi and temperatures between 1000 and 1250°C[10]. In some situations hot pressed specimens were annealed in air or vacuum at temperatures up to 1900°C to produce microstructures with either large or stable grain sizes. Preannealing at higher temperatures frequently prevented grain growth from occurring during the creep test at a lower temperature. The grain-size ranges for pure MgO, iron-doped MgO, pure Al₂O₃, iron-doped Al₂O₃, and chromium-doped Al₂O₃ were 30-218 μm, 8-487 μm, 9-72 μm, 11-110 μm and 15-35 μm, respectively.

For creep tests rectangular beams ($\sim$ 2 x 5 x 30mm) were cut from hot pressed billets parallel to the pressing direction. All specimens were tested in four-point, dead-load experiments using single crystal sapphire rods for support and load members (11,12).

Detailed descriptions of the creep testing facilities are given
elsewhere [10-12]. Stresses ($10-550 kg/cm^2$) and strain rates (10^{-5}
$-10^{-3} h^{-1}$) were calculated from relations developed by Hollenberg,
et.al.[13] for viscous and non-viscous creep of a specimen in four-
point bending on which load-point deflections are measured. These
equations are valid for small deflections (1-2% outer fiber strain).

Oxygen partial pressures(P_{O_2}) were measured with a CaO-stabil-
ized ZrO_2 probe. A spectrum of P_{O_2} was obtained by mixing air,
pure O_2, N_2, CO and CO_2 in various combinations.

III. RESULTS

3.1 Diffusional Creep of Polycrystalline $MgO-FeO-Fe_2O_3$ Solid Solutions

Substantial data have been accumulated on the steady state
creep properties of polycrystalline $MgO-FeO-Fe_2O_3$ solid solutions.
Viscous or diffusional creep is the dominant deformation mechanism
at low stresses ($<300 Kg/cm^2$) (10,11,14) over a wide range of
experimental conditions (i.e. iron concentrations between 0.05 and
5.3 cation % , temperatures between 1100 and 1450°C, grain sizes
between 6 and 100μm, and oxygen partial pressures between 1 and
10^{-9} atmospheres). Under these conditions steady state creep
is always observed provided the grain size is stable[11].

After an extensive analysis of the data[4,15] it appears that
three distinct mechanisms of diffusional creep are operable in
this system: (1) Coble creep controlled by magnesium grain boundary diffusion(4,10,14,15), (2) Nabarro-Herring creep controlled by
magnesium lattice diffusion (4,10,11,14, 15) and Coble creep
controlled by oxygen grain boundary diffusion (4,11,15), All three
of these limits have been predicted in a theoretical analysis of
mass transport in this system[4].

Limit I: At relatively small grain sizes (6-23μ) and low
temperatures (1100-1300°C) Coble diffusional creep has been observed
in iron-doped MgO (0.05-0.27 cation %) [10,14] when creep tested
in air. This type of creep behavior is characterized by a reciprocal dependence between the creep rate and the cube of the grain
size which is predicted from equation (1) in the limit of small
grain sizes and low values for the lattice diffusion coefficients.
This limit will apply at low temperatures since it is expected that
the activation energy for volume diffusion is greater than that for
grain boundary diffusion.

Grain size effects in this regime were determined from a
combined analysis of creep data and simultaneous grain growth (10,

14,16). In this limit creep rates increased with increasing iron
dopant levels. Since it is expected that oxygen grain boundary
diffusion is rapid in iorn-doped MgO[4], it is believed that magnesium grain boundary diffusion is rate controlling in Limit I.

Limit II: At larger grain sizes (15-100μm) and higher temperatures (>1250°C) Nabarro-Herring creep has been identified[11].
Extensive experiments (Figure 1) at 1350°C in oxygen on the effects
of grain size at the 0.53% composition revealed a grain size
exponent m of 1.94 in good agreement with that predicted by the
Nabarro-Herring relation (i.e. m=2). Creep in limit II was characterized by a relatively high activation energy (117 ± 10 kcal/mole)
and creep rates which increased as the oxygen partial pressure was
raised[11]. Creep specimens doped between 0.05 and 0.53 cation %
Fe exhibited creep in this limit over a wide range of oxygen partial
pressures (1-10^{-9}atm). At higher dopant levels (2.65 and 5.3
cation %) the oxygen partial pressure had to be reduced to 10^{-2}
-10^{-5} atm before Limit II was operating[11]. The effect of oxygen
partial pressure indicated that creep rates increase with increasing
concentration of trivalent iron and, hence, the concentration of
magnesium vacancies.

The impurity and oxygen partial pressure effects along with
the results on activation energies and grain size effects suggest
that Nabarro-Herring creep of iron-doped MgO in Limit II is controlled by the diffusion of magnesium ions through the lattice.
Furthermore diffusion coefficients which were calculated from creep
data in Limit II were in good agreement with magnesium tracer values
[11](See Table 1). This conclusion implies that a rapid transport path (e.g. grain boundaries) must exist for the slower moving
(i.e. in the lattice) oxygen ion.

Limit III: A third limit has been identified in the viscous
creep of iron-doped MgO[11]. Steady state creep in this regime is
characterized by:
1) High iron concentrations (≥2.65 cation %)
2) Oxidizing atmospheres (high Fe^{3+}/Fe^{2+} ratios)
3) Creep rates which are nearly independent of oxygen
   partial pressure and iron dopant level (e.g. refer to
   Fig. 1 for creep specimens at the 2.65 and 5.3% dopant
   levels)
4) A creep activation energy (81 ± 5 kcal/mole) which is
   considerably lower than that encountered in Limit II(i.e.
   117 ± 10 Kcal/mole)
5) Grain size effects which are consistent with a significant
   contribution of Coble creep: 2 < m < 3.

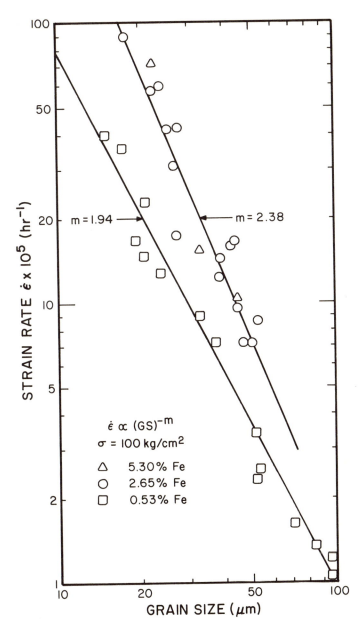

Fig. 1 Effect of Grain Size on the Diffusional Creep Rates of Iron-Doped, Polycrystalline MgO (After Reference 11)

The transition to Limit III is a natural consequence of the inequality: $(GS/\pi) D_{Mg}^{\ell} \gg \delta_0 D_0^b$. In the limit of large grain size and high cation lattice mobility (high dopant levels and oxygen partial pressures) cation lattice diffusion becomes too rapid to remain the rate-controlling step. At this stage oxygen grain boundary diffusion must become rate-limiting.

The transition oxygen partial pressure between limits II and III was very sensitive to the iron concentration, grain size and temperature[11]. In fact, at low dopant levels (≤ 0.53 cation %) the transition pressure was above atmospheric for the grain sizes which have been studied (15-100μm), since Limit III was not observed at these concentrations (11).

Extensive data were collected on grain size effects in Limit III at the 2.65% dopant level[11]. These are plotted in Figure 1 for grain sizes between 18 and 53μm. A grain size exponent of 2.38, intermediate to that of Nabarro-Herring and Coble creep, was found. This result implies that Limits II and III actually overlap at this concentration and in this range of grain sizes. In this case equation (1) reduces to

$$\dot{\varepsilon} = \frac{14 \, \Omega_v \, \sigma \, D_{Mg}^{\ell} \delta_0 \, D_0^b}{\left[kT \, (GS)^2 \, (\delta_0 \, D_0^b + (GS/\pi) \, D_{Mg}^{\ell}) \right]} \tag{4}$$

Using the $\dot{\varepsilon}$ -(GS) data in Figure 1, D_{Mg}^{ℓ} and $\delta_0 D_0^b$ were evaluated from Equation (4) at 1350°C[11,15]. They are 9.0 x 10⁻¹² cm²/ sec and 1.6 x 10⁻¹⁴ cm³/ sec, respectively, and are in good agreement with expected values of magnesium lattice and oxygen grain boundary diffusion in polycrystalline, iron doped $M_gO(4)$.

In summary two regimes of Coble creep have been identified in polycrystalline MgO-FeO-Fe_2O_3 solid solutions: Limit I controlled by magnesium grain boundary diffusion at small grain sizes and low temperatures and Limit III controlled by oxygen grain boundary diffusion at large grain sizes, high temperatures, and high concentrations of trivalent iron. Between these two regimes Nabarro-Herring creep (limit II), which is rate-limited by magnesium lattice diffusion) has been identified. In Table 1 representative diffusion coefficients which have been calculated for $\delta_{Mg} D_{Mg}^b, D_{Mg}^{\ell}$ and $\delta_0 D_0^b$, are summarized.

According to the data in Figure 1 for grain sizes over 100μm, the diffusional creep rate should be small and non-viscous modes should become important particularly at higher stress levels

Table 1

DIFFUSION COEFFICIENTS CALCULATED FROM CREEP DATA

LIMIT I - COBLE CREEP

Composition (cation % Fe)	Temperature (C°)	P_{O_2}(atm)	Grain Size (μm)	$\delta_{Mg_3} D^b_{Mg}$ (cm/sec)	Reference
0.05	1200	0.18	6-9	1.5×10^{-17}	14
0.47	1200	0.18	8-13	1.8×10^{-16}	14
0.05	1300	0.18	8-23	1.5×10^{-16}	10,14
0.27	1300	0.18	8-23	2.7×10^{-16}	10,14
0.47	1300	0.18	8-23	3.9×10^{-16}	10,14

LIMIT II - NABARRO-HERRING CREEP

				D^ℓ_{Mg} (cm^2/sec)	
0.05	1350	0.86	27-33	0.8×10^{-12}	11
0.53	1350	0.86	15-100	2.3×10^{-12}	11
2.65	1350	0.86	18-53	9.0×10^{-12}	11,15
2.65	1350	10^{-8}	18-53	1.0×10^{-12}	11
5.3	1350	10^{-8}	23-45	1.5×10^{-12}	11
Magnesium Tracer Diffusion (1350°C)				1.2×10^{-12}	17
Oxygen Tracer Diffusion (1350°C				1.1×10^{-14}	18

LIMIT III - COBLE CREEP

				$\delta_0 D^b_0$ (cm^3/sec)	
2.65	1350	0.86	18-53	1.6×10^{-14}	11,15

3.2 Non-Viscous Creep of Polycrystalline MgO-FeO-Fe$_2$O$_3$ Solid Solutions

(> 100 kg/cm^2). To this end a series of polycrystalline specimens were fabricated at the 0.53% dopant level with grain sizes ranging between 137 and 487μm. Initial creep experiments were attempted at 1350°C so that a direct comparison could be made with the viscous data in Figure 1. Non-viscous creep was observed ($\dot{\varepsilon} \alpha \sigma^N$, N>1); however, steady state was never achieved in that creep rates continually decreased with time. Since no grain growth occurred in these specimens, this time decay of the strain rate (i.e. primary creep) was attributed to a "dislocation hardening process". Subsequent tests, which were conducted at higher temperatures (1425-1500°C) resulted in steady state creep. Recovery processes resulting in steady state creep behavior should become predominant as the temperature is increased.

At 1425-1500°C several stress change creep experiments were conducted at stresses between 30 and 300 kg/cm². In general, non-viscous creep behavior was observed with stress exponents (N) varying between 1.6 and 3.4. A typical stress change experiment is shown in Figure 2 for a specimen with a grain size of 258μm. A summary

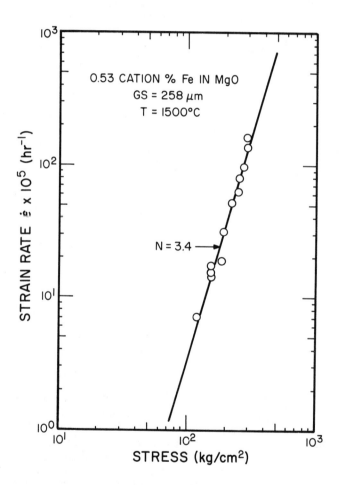

Fig. 2 Non-viscous creep of Iron Doped (0.53 cation %), Polycrystalline MgO with a Large Grain Size (258μm).

TABLE 2

STRAIN RATE - STRESS DATA FOR IRON-DOPED (0.53 cation %),
POLYCRYSTALLINE MgO AT 1500°C

Grain Size(μm)	Stress Range(kg/cm^2)	Apparent Stress Exponent(N)	Non-Viscous Stress Exponent(N')
487	90-350	3.1	3.6
258	100-300	3.4	4.1
179	120-270	2.9	3.2
180†	30-236	3.0	3.4
137	100-300	1.6	2.8
		Ave(2.8±0.6)	Ave(3.4±0.5)

†1425°C

of all the strain rate-stress experiments is given in Table 2. Inspection of these data reveals that the apparent stress exponents,
N, ($\dot{\varepsilon} \propto \sigma^N$) decrease with decreasing grain size. A particularly
interesting result of this trend is shown in Figure 3 where the
effect of grain size on the creep rate is shown at two stress levels.
The solid lines represent the viscous creep data (Figure 1) which
have been extrapolated to 1500°C from 1350°C using an activation
energy of 117 kcal/mole*. The data with the open symbols represent
non-viscous creep for grain sizes over 100μm. Several features are
worthy of note: At the low stress level a grain size effect is
present and the creep rates are only slightly higher than the extrapolated diffusional creep rates particularly as the grain size
approaches 300μm. However, at the higher stress of 300 kg/cm^2 the
non-viscous creep rates are apparently independent of grain size
and they are over an order of magnitude larger than those expected
from diffusional creep. In addition it appears that a sharp transition exists between the diffusional and non-viscous creep regimes
around 60μm. At the lower stress level the transition is $\sim$ 300μm.

The data in Figure 3 clearly demonstrate that viscous creep in
polycrystalline MgO-FeO-Fe$_2$O$_3$ solid solutions gives way to non-
viscous creep at lower grain sizes as the stress level is increased.
Secondly, the dependence between the grain size and the apparent
stress exponent (Table 2) is indicative of significant contributions
of diffusional creep to the overall creep rate particularly as the
grain size is decreased below 300μm at low stresses ($\leq$ 100 kg/cm^2).

---

*This extrapolation is reasonable since viscous creep with this
activation energy has been reported up to 1400°C[11]. It would
have been impossible to determine the diffusional creep rates experimentally at 1500°C because of the problem of the simultaneous
grain growth.

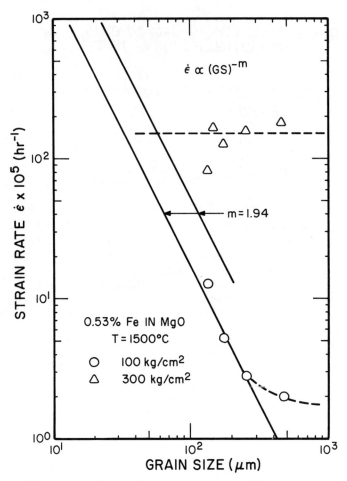

Fig. 3 Effects of Grain Size on the Non-Viscous Creep of Iron-Doped Polycrystalline MgO

To check if the reason for the dependence of the grain size on the stress exponent is due, in fact, to a contribution of viscous creep let us assume that the two processes (i.e. diffusional and non-viscous creep) are additive, i.e..

$$\dot{\varepsilon} = K\sigma^N = K'\sigma + K''\sigma^{N'} \tag{5}$$

K, K', and K" are constants independent of stress. N is the apparent stress exponent and N' is the actual non-viscous stress exponent. To determine the non-viscous stress exponents (N'), equation (5) must be arranged into the following form:

$$\dot{\epsilon} - K'\sigma = K''\sigma^{N'} \qquad (6)$$

A plot of log $[\dot{\epsilon} - K'\sigma]$ versus log σ should result in a line with slope equal to N'. Construction of this plot requires a knowledge of the viscous component ($K'\sigma$). Using a diffusion coefficient of 2.3×10^{-12} cm^2/sec(refer to Table 1) at 1350°C, an activation energy of 117 kcal/mole, a linear stress dependence, and a reciprocal square grain size dependence, values of $K'\sigma$ were evaluated at 1500°C for different grain sizes as a function of stress. These estimates of $K'\sigma$ were substracted from all measured non-viscous creep rates to obtain estimates of $[\dot{\epsilon} - K'\sigma]$. Typical log-log plots of $\dot{\epsilon}$ and $[\dot{\epsilon} - K'\sigma]$ versus stress are shown in Figure 4 for a stress change experiment on a creep specimen with a grain size of 179μm. It is clear that the diffusional contribution is significant particularly at the lowest stress level ($\sim$ 100 kg/cm^2). The corrected plot of only non-viscous creep data results in a stress exponent (N') which is higher (3.2 versus 2.9). In Table 2 a comparison is given between values of N and N'. The actual non-viscous stress exponents (3.4±0.5) are higher than the apparent values (2.8±0.6) particularly at smaller grain sizes. They are also independent of grain size.

The determination of the activation energy for non-viscous creep was complicated by the problem of time (or strain) hardening of the creep rate particularly at lower temperatures ($\sim$ 1350°C). In temperature change experiments on single specimens with grain sizes between 135 and 180μm, activation energies of 50-70 kcal/mole were determined in temperature rise (1350-1500°C) experiments. However for decreasing temperature experiments the apparent activation energies were much higher ($\sim$ 102-109 kcal/mole). Probably the most reliable activation energy ($\sim$ 71 kcal/mole) to date was obtained using the Dorn instantaneous temperature change method on a 74μm sample creep tested at a relatively high stress (315 kg/cm^2) and in a lower temperature range (1225-1300°C). Activation energies in the range of 50-70 kcal/mole are comparable to that predicted for extrinsic[18] oxygen lattice diffusion in MgO ($\sim$ 62 Kcal/mole).

In summary the creep of coarse-grained, polycrystalline MgO-FeO-Fe$_2$O$_3$ (0.53 cation % Fe) solid solutions consists of two processes:

1) a viscous (diffusional creep) process which is significant at small stresses ($\leq$ 100 Kg/cm^2) and grain sizes (< 300μm)
2) a non-viscous process with a stress exponent of $\sim$ 3 in which the creep rate is independent of grain size.

The non-viscous process is consistent with Nabarro's sub-grain model described earlier; however, microstructural data on the substructure need to be obtained prior to making any firm conclusions along these lines.

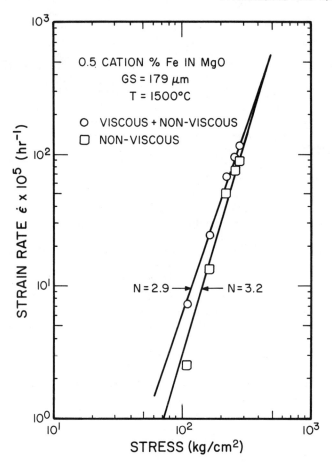

Fig. 4 Apparent and Actual Non-Viscous Creep Rates of Iron-Doped
Polycrystalline MgO (179μm).

3.3 Non-Viscous Creep of Pure, Polycrystalline MgO

Ample evidence exists in the literature that power law creep
($\dot{\epsilon} \propto \sigma^N$; $N \sim 3$) describes the deformation behavior of pure polycrystalline MgO over a range of grain sizes (18-102μm), stresses
(70-1410 kg/cm²), and temperatures (1200-1500°C). Hensler and
Cullen[19] reported a stress exponent of 2.6 ± 0.6 and an average
activation energy of 108±12 Kcal/mole for compression creep experiments at 1200-1500°C and grain sizes between 19 and 102μm*. On the

*Grain sizes reported in references 19 and 20 were assumed to be
   linear intercepts and have been multiplied by 1.5.

other hand, Langdon and Pask[20] who worked at a lower temperature (1200°C) and very high stresses (352-1410 Kg/cm^2) reported a stress exponent of 3.3 and an activation energy of 51 $\pm$ 5 Kcal/mole for tests in compression. Furthermore they reported no effect of grain size on the creep rate in the range of 18-78μm. It should be noted that at the high stresses and low temperature (1200°C) studied by Langdon and Pask, diffusion creep is not expected to be important. Terwilliger et.al[10] reported average stress exponents of 2.4 $\pm$ 0.7 and 3.6 $\pm$ 0.2 for bending creep tests at 1300 and 1400°C on specimens with grain sizes between 35 and 83μm. In general the highest stress exponents were found in samples with the largest grain size indicating the possibility of mixed diffusional creep and dislocation creep modes at grain sizes below 30μm in this temperature range. Furthermore, time (strain) hardening of the creep rate for specimens with stable grain sizes was observed.

In Figure 5 steady state creep data at 1500°C on a pure polycrystalline MgO specimen with a grain size of 218μm are compared with earlier measurements by Hensler and Cullen and by Langdon and Pask. The agreement between the three studies is remarkably good considering the range of grain sizes (19-218μm) and the magnitude of the temperature extrapolation required for the data of Langdon and Pask. These studies give strong support to a three power creep law which is independent of grain size [21].

Included in Figure 5 for comparison are the non-viscous data at two grain sizes (258 and 487) for specimens doped with 0.53% iron. Assuming again the absence of any grain size effects, the creep rates of iron-doped specimens are slightly slower than the pure MgO particularly at lower stress levels (∿ factor of 2-5 at 90 kg/cm^2). This decrease in rate might be expected if oxygen lattice diffusion is rate controlling. Doping with Fe^{3+} should enhance D_{Mg}^{ℓ} and depress D_{O}^{ℓ} [4].

For comparison the expected diffusional creep rates (Limit II) for the 0.53% composition are shown also in Figure 5. The viscous creep rate of pure MgO (yet to be measured) will be at least an order of magnitude lower at this same grain size. Thus, it is clear that no contribution of viscous creep is expected at stresses over 100 kg/cm^2 until the grain size gets much smaller (< 20-30μm). Experimentally it has been virtually impossible to obtain viscous creep data at temperatures over 1300°C on small grain size MgO because of rapid grain growth which takes place during the creep test[22].

The data in Figure 5 also indicate the role of iron doping. While a possible inhibition of non-viscous creep may exist, the biggest effect is probably the enhancement of the diffusional (viscous) creep rate through the creation of cation lattice defects. Increased concentration of dopant (i.e. cation vacancies) will

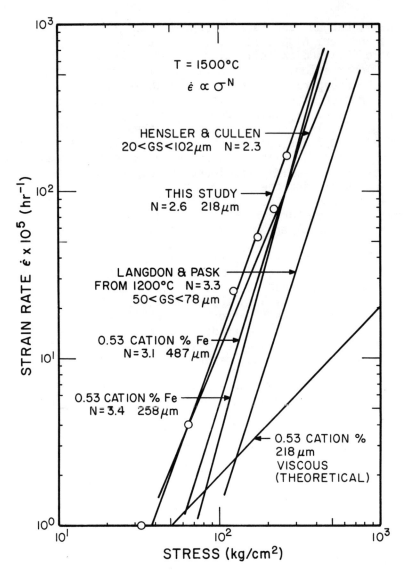

Fig. 5   Non-Viscous Creep of Pure, Polycrystalline MgO.

shift the transition to non-viscous creep to larger stresses at a given grain size and temperature.

In summary, non-viscous creep data on pure, polycrystalline MgO are in good agreement with a 3 power creep law in which the creep rate is independent of grain size. A similar creep law holds for the non-viscous creep of polycrystalline, iron-doped MgO. No reliable data exist for diffusional creep in pure, polycrystalline MgO. However, from the data available on iron-doped MgO, viscous creep is not expected to be important in pure MgO at temperatures between 1200 and 1500°C unless the grain size is well below 20μm and stress levels are low ($\sigma \leq 100$ kg/cm^2).

3.4 Creep of Polycrystalline Al_2O_3 - Cr_2O_3, Al_2O_3 - FeO - Fe_2O_3, Al_2O_3 - Ti_2O_3 - TiO_2 Solid Solutions

Substantial data have been accumulated on the creep in bending of pure and doped (Cr, Fe, Ti) polycrystalline Al_2O_3 at stresses between 10 and 550 kg/cm^2, at temperatures between 1375°C and 1525°C, and for grain sizes between 4.5 and 110μm. In general nearly all of the creep data are characterized by slightly non-viscous stress exponents which vary between 1.03 and 1.43. A summary of all data both in this study and that of Hollenberg and Gordon[12] is given in Table 3. No systematic variation of the stress exponents with temperature, grain size, or atmosphere has been observed; although occasionally a few examples of viscous (N = 1) or near viscous (1.03) behavior has appeared in small grain size (15μm) specimens.

Reports of non-viscous creep of polycrystalline Al_2O_3 abound in the literature. Heuer, et.al.[23] were the first investigators to document this behavior. They reported stress exponents between 1.08 and 1.67 for bending creep tests on pure and MgO-doped Al_2O_3 (1.2 < GS < 11μm;*1300 < T < 1700°C) at stresses between 70 and 2114 Kg/cm^2. At this time, Heuer, et.al. reanalyzed the earlier bending creep data of Folweiler[24] (7-34μm)* and Warshaw and Norton[25] (3-13μm)* and concluded that stress exponents ranging between 1.07 and 1.67 were more representative of their work than viscous creep. Later Cannon[26] (14-34μm)* and Sugita and Pask[27] (3-7μm)* reported slightly non-viscous stress exponents (1.2 and 1.1-1.3) for compression testing of MgO-doped Al_2O_3. Apparently stress exponents up to ∿ 1.7 are characteristic of the creep deformation in bending and compression of polycrystalline Al_2O_3, pure and doped, over a wide range of experimental conditions.

At high stresses (> 300 kg/cm^2), high temperatures (> 1600°C) and large grain sizes (>50-75μm) some evidence (21, 25, 28, 29) exists for high power creep (i.e. N = 3-4). Whether or not this behavior is due to a dislocation creep mechanism remains to be seen. In two of these studies (25, 28) inter-granular separations were observed, while in another[29], porous specimens (2-5% porosity) were tested.

---

*Linear Intercepts

TABLE 3

STRAIN RATE-STRESS EXPONENTS (N) IN POLYCRYSTALLINE Al_2O_3

Composition	N	Grain Size (μm)	Temperature (°C)	Stress (Kg/cm²)	P_{O_2} (atm)	Number Specimens	Reference
Pure Al_2O_3	1.30 ± 0.15	9-72	1450	40-500	0.86	5	This study
0.2 cation % Fe	1.14 ± 0.06	15	1400-1500	10-100	0.86	3	12
"	1.14	27-38	1450	40-500	0.86	2	This study
1 cation % Fe	1.0	15	1500	10-150	0.86	1	12
"	1.25 ± 0.06	26-107	1450	40-500	0.86	3	This study
"	1.25 ± 0.05	42	1400-1500	50-550	0.86	5	12
"	1.0	15	1500	10-100	$10^{-2.2}$	1	12
"	1.29	38	1450	40-500	10^{-8}	1	This study
1 cation % Cr	1.30	9	1400-1500	50-550	0.86	2	12
"	1.03	15	1450	40-500	0.86	1	This study
"	1.21	32	1450	40-500	0.86	1	This study
1/2 cation % Ti	1.30	63	1525	10-100	0.86	1	12
"	1.07	63	1525	30-300	10^{-9}	1	12

In Figure 6 creep rate - grain size data at 1400°C and a low stress of 50 kg/cm^2 are plotted over an extensive range of grain sizes (9-110μm). In the case of pure and chromium doped Al_2O_3 the data give an excellent fit to the Nabarro-Herring creep equation.* In the earlier work of Folweiler[24] and Warshaw and Norton[25] on MgO-doped aluminas, a reciprocal square dependence between the creep rate and grain size also was reported. For specimens (25-110μm) doped with 1 cation % iron the grain size effect on the creep rate appears to be more pronounced than in the case of undoped material (m=2.24 as compared to 2.0). The higher grain size exponent suggests a mixed diffusional creep mechanism in which both lattice (aluminum?) and grain boundary (oxygen?) diffusion processes are comparable in magnitude.

Also of interest in Figure 6 are the effects of impurities in solid solution on the steady state creep rate. An addition of 1 cation % Cr had no effect while additions of 0.2 and 1 cation % Fe enhanced the creep rates in air by factors of ∿ 2 and ∿ 7, respectively. Ti-doping (1/2 cation %) also has been shown to enhance the creep rate of polycrystalline Al_2O_3[12].

The steady state creep rates of Fe- and Ti-doped, polycrystalline Al_2O_3 are very sensitive to changes in the oxygen partial pressure[12]. Reducing the oxygen partial pressure from 1 to 10^{-6} atm. leads to nearly an order of magnitude increase in the creep rate of iron-doped (1 cation %) Al_2O_3. Conversely, a reduction in P_{O_2} from 1 to 10^{-10} atm. leads to nearly an order of magnitude decrease in the creep rate of Ti-doped (1/2 cation %) Al_2O_3. These effects suggest that both a divalent (Fe^{2+}) and a quadrivalent (Ti^{4+}) ion in solid solution enhance the creep rate over and above that of pure and chromium-doped, polycrystalline Al_2O_3.

Apparent creep activation energies (1350-1525°C) have been determined in temperature change experiments on pure, chromium (1 cation%), iron (0.2 cation %) and iron (1 cation %) - doped aluminas. They are 130±5, 117±10, 134±5, and 120±5 kcal/mole, respectively. In earlier work by Hollenberg and Gordon[12] creep activation energies of ∿ 137 and 148 kcal/mole were determined for Ti (1/2 cation %) and Fe (1 cation %) -doped aluminas, respectively. All of these values are greater than those expected for Al ion diffusion[4] in Al_2O_3 (∿ 114 kcal/mole).

In view of the grain size and P_{O_2} effects which are suggestive of a diffusion-controlled creep process, cation diffusion coefficients have been calculated from the Nabarro-Herring equation[4,12]. It has been assumed that grain boundary diffusion dominates the overall

---

*This fit extends up to stresses of 400 kg/cm^2 since no systematic variation in the stress exponents with grain size exists.

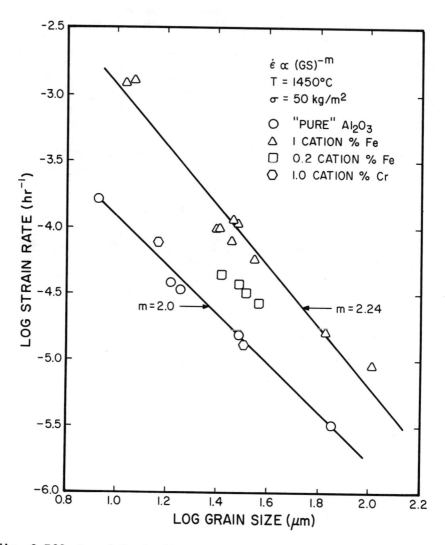

Fig. 6 Effects of Grain Size on the Creep of Pure and Transition
Metal-Doped, Polycrystalline Al_2O_3

mobility of the oxygen ion and that $\delta_0 D_0^b \gg 3/2 \ [(GS)/\pi]D_{Al}^\ell$ (4b, 15). These calculations, which are presented in Figure 7, support the hypothesis that Nabarro-Herring creep controlled by aluminum lattice diffusion is significant in the creep of pure and transition metal-doped, polycrystalline Al_2O_3. Calculated values of D_{Al}^ℓ were greater than measurements of aluminum tracer diffusion. Although these values might be overestimated because of slightly non-viscous creep behavior, this problem should be minimized since the stress level chosen for the calculations was low (50 kg/cm^2) and likely in the viscous limit. Values of $D_{Al}^\ell > D_{Al}$ (tracer) have been obtained in previous creep studies on MgO-doped Al_2O_3 at higher temperatures(24,25).

The most consistent explanation for the P_{O_2} and grain size effects, the apparent creep activation energies, and the correlations with the Nabarro-Herring diffusional creep theory is that the creep of pure and transition metal-doped, polycrystalline Al_2O_3 (4-110μm, 1375-1550°C) is a cation diffusion controlled process and that the intrinsic defects in Al_2O_3 are of the Frenkel type (cation vacancies and interstitials). Doping Al_2O_3 with Fe^{2+} creates extrinsic aluminum ion interstitials; whereas doping with Ti^{4+} leads to extrinsic aluminum ion vacancy formation. Both of these defects will enhance aluminum lattice diffusion and hence lead to increased creep rates over and above those of pure and chromium-doped Al_2O_3. These results in combination with the conductivity data of Brook, et.al.(30), in which divalent ions increased the ionic conductivity, strongly support the proposition that aluminum ions can migrate in the corrundum structure by either a vacancy or an interstitialcy mechanism.

Finally since slightly non-viscous creep behavior has characterized this and previous work on polycrystalline Al_2O_3, mechanisms other than diffusional which can accomodate grain-boundary sliding in Al_2O_3 should be mentioned. Heuer, et.al.(23) attributed slightly non-viscous stress exponents to (unaccomodated?) grain boundary sliding. This type of deformation is usually represented by void formation at triple points or grain boundary separation and, possibly, by a reciprocal grain size dependence ($\dot{\varepsilon}$ α (GS)$^{-1}$). No evidence, metallographic features or grain size effects, consistent with these features has been found in this study. It is possible that local climb or glide of dislocations near grain boundaries might be a partial contributor to the accomodation of grain boundary sliding in addition to the diffusional transport of ions. Further studies are needed on this aspect.

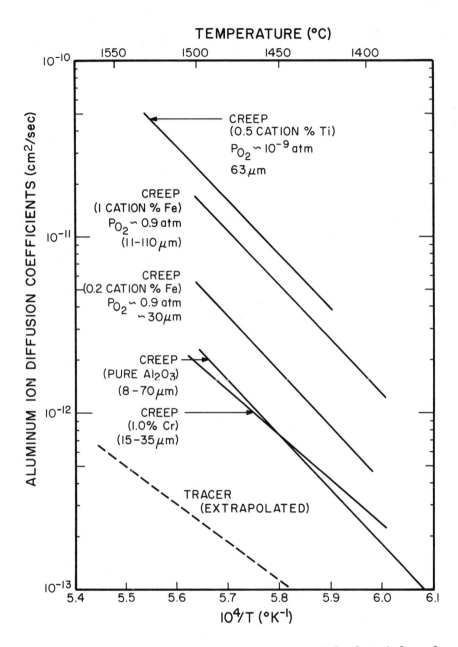

Fig. 7 Aluminum Ion Diffusion Coefficients Calculated from Creep
        Data

IV. SUMMARY

Diffusional creep is the dominant deformation mechanism in polycrystalline iron-doped MgO at low stresses ($\leq$100-300 kg/cm^2) and small grain sizes (8-100μm). Three diffusional mechanisms have been identified as rate controlling in certain regimes:

(1) Coble creep controlled by magnesium grain boundary diffusion
(2) Nabarro-Herring creep controlled by magnesium lattice diffusion which is enhanced with increasing amounts of trivalent iron in solid solution and
(3) Coble creep controlled by oxygen grain boundary diffusion

As the grain size is increased over 100μm and stress levels exceed 100 kg/cm^2, non-viscous creep is dominant (T > 1425°C). In this regime strain rate-stress exponents (N) of $\sim$ 3 and creep rates which are grain size independent have been observed.

Pure, polycrystalline MgO deforms by a power law (N = 3) mechanism in which the creep rate is independent of grain size and slightly faster than iron-doped material. No reliable data exist for viscous creep in undoped, polycrystalline MgO.

In polycrystalline Al$_2$O$_3$ it has been demonstrated that both a divalent (Fe^{2+}) and a quadrivalent (Ti^{4+}) impurity in solid solution in increasing amounts enhance the creep rate over and above the creep of pure and chromium doped, polycrystalline Al$_2$O$_3$. Creep is characterized, for the most part, by slightly non-viscous stress exponents (1.03-1.45) over a range of stresses (up to 550 kg/cm^2) and temperatures (1375-1525°C). Grain size and P$_{O_2}$ effects, activation energies, and calculated diffusion coefficients indicate that the Nabarro-Herring mechanism is a major contributor to creep deformation. Furthermore, it has been concluded that the steady state creep of pure and doped, polycrystalline Al$_2$O$_3$ is controlled primarily by the lattice diffusion of aluminum ions either by a vacancy or an interstitialcy mechanism.

V. ACKNOWLEDGEMENT

This work was supported by the Atomic Energy Commission under Contract AT (11-1) - 1591

VI. REFERENCES

1. R. Raj and M.F. Ashby, "Grain Boundary Sliding and Diffusional Creep", Trans AIME 2 1113 (1971).

2. J. Weertman, "Steady State Creep through Dislocation Climb", J. Appl. Phys. 28 196 (1957).

3. a. F.R.N. Nabarro, "Steady State Diffusional Creep", Phil Mag 16 231 (1967)
    b. J. Weertman, "Dislocation Climb Theory of Steady State Creep", ASM (Amer. Soc. Metals) Trans. Quart., 61 681 (1968).

4. a. Ronald S. Gordon, "Mass Transport in the Diffusional Creep of Ionic Solids", J. Amer. Ceram. Soc., 56 147 (1973).
    b. R.S. Gordon, "Ambipolar Diffusion and its Application to Diffusion Creep". Proceedings of Ninth University Conference on Ceramic Science-Mass Transport Phenomena in Ceramics, Case-Western Reserve University, June 1974 (to be published).

5. a. F.R.N. Nabarro, pp 75-90 in Report of a Conference on Strength of Solids, University of Bristol, H. Willis Physical Laboratory, Bristol, England 1947, Physical Society of London, 1948.
    b. Conyers Herring, "Diffusional Viscosity of a Polycrystalline Solid", J. Appl. Phys. 21 437 (1950).

6. R.L. Coble "Model for Boundary Diffusion Controlled Creep in Polycrystalline Materials", J. Appl. Phys. 34 1679 (1963).

7. J. Weertman, "Theory of Steady State Creep Based on Dislocation Climb", J. Appl. Phys., 26 1213 (1955).

8. R. Von Mises "Mechanism of Plastic Deformation of Crystals", Z. Angew Math Mech., 8 101 (1928).

9. J. Weertman, "Steady-State Creep of Crystals", J. Appl. Phys. 28 1185 (1957).

10. G.R. Terwilliger, H.K. Bowen, and R.S. Gordon, "Creep of Poly-crystalline MgO and $MgO-Fe_2O_3$ Solid Solutions at High Temperatures", J. Amer. Ceram. Soc. 53 241 (1970).

11. R.T. Tremper, R.A. Giddings, J.D. Hodge and R.S. Gordon, "The Creep of Polycrystalline MgO-FeO-Fe_2O_3 Solid Solutions", J. Amer. Ceram. Soc., 57 (Oct.) (1974).

12. Glenn W. Hollenberg and Ronald S. Gordon,"Effect of Oxygen Partial Pressure on Creep of Polycrystalline Al_2O_3 Doped with Cr, Fe, or Ti", J. Amer. Ceram. Soc., 56 140 (1973).

13. G.W. Hollenberg, G.R. Terwilliger and R.S. Gordon,"Calculation of Stresses and Strains in Four-Point Bending Creep Tests", J. Amer. Ceram. Soc., 54 196 (1971).

14. R.S. Gordon and G.R. Terwilliger, "Transient Creep in Fe-doped Polycrystalline MgO", J. Amer. Ceram. Soc. 55 450 (1972).

15. R.S. Gordon and J.D. Hodge, "Analysis of Mass Transport in the Diffusional Creep of Polycrystalline MgO-FeO-Fe_2O_3 Solid Solutions", J. Mater. Sci. (in Press) (1974).

16. G.R. Terwilliger and R.S. Gordon, "Correlations Between Models for Time-Dependent Creep with Concurrent Grain Growth", J. Amer. Ceram.Soc., 52 218 (1969).

17. B. J. Wuensch, W.C. Steele, and T. Vasilos, "Cation Self-diffusion in Single-Crystal MgO", J. Chem. Phys., 58 5258 (1973).

18. Y. Oishi and W.D. Kingery, "Oxygen Diffusion in Periclase Crystals, "J. Chem. Phys., 33 905 (1960).

19. J.H. Hensler and G.V. Cullen, "Stress, Temperature and Strain Rate in Creep of Magnesium Oxide", J. Amer. Ceram. Soc., 51 557 (1968).

20. T.G. Langdon and J.A. Pask, "Mechanism of Creep in Polycrystalline MgO," Acta Met 18 505 (1970).

21. W. Roger Cannon and Oleg D. Sherby, "Third-Power Stress Dependence in Creep of Polycrystalline Nonmetals", J. Amer. Ceram. Soc., 56 157 (1973).

22. R.S. Gordon, D. D. Marchant, and G.W. Hollenberg, "Effect of Small Amounts of Porosity on Grain Growth in Hot-Pressed Magnesium Oxide and Magnesiowüstite", J. Amer. Ceram. Soc., 53 399 (1970).

23. A.H. Heuer, R.M. Cannon, and N.J. Tighe, pp 339-65 in Ultra-fine-Grain Ceramics. Edited by J.J. Burke, N.L. Reed, and Volker Weiss. Syracuse University Press, Syracuse, NY, 1970.

24. R.C. Folweiler, "Creep Behavior of Pore-Free Polycrystalline Aluminum Oxide", J. Appl. Phys., 32 773 (1961).

25. S.I. Warshaw and F.H. Norton, "Deformation Behavior of Poly-crystalline Aluminum Oxide", J. Amer. Ceram. Soc., 45 479 (1962).

26. W.R. Cannon, "Mechanisms of High Temperature Creep in Poly-crystalline Aluminum Oxide", Ph.D. Thesis, Stanford University, Stanford, CA, 1971.

27. Tadaaki Sugita and J.A. Pask, "Creep of Doped Polycrystalline Al_2O_3", J. Amer. Ceram. Soc., 53 609 (1970).

28. R.L. Coble and Y.H. Guerard, "Creep of Polycrystalline Aluminum Oxide", J. Amer. Ceram. Soc., 46 353 (1963).

29. G.V. Engelhardt and F. Thümmler, "Kriechuntersuchungen unter 4 - Punkt - Beigebeamspruchung bei Holen Temperaturen", Ber Dt. Keram Ges 47 571 (1970).

30. R.J. Brook, J. Yee, and F.A. Kroger, "Electrochemical Cells and Electrical Conduction of Pure and Doped Al_2O_3", J. Amer. Ceram. Soc., 54 444 (1971).

CREEP OF LOW-DENSITY YTTRIA/RARE EARTH-STABILIZED ZIRCONIA

M. S. Seltzer* and P. K. Talty†

*Battelle-Columbus Laboratories, Columbus, Ohio 43201

†Aerospace Research Laboratories, Wright-Patterson AFB,
Ohio 45433

ABSTRACT

Constant-stress compression creep experiments have been performed on low-density yttria/rare earth-stabilized zirconia (YRESZ)
containing 12.5 and 14.5 weight percent Y_2O_3. Major microstructural modifications which occur in these materials at high
temperature serve to increase creep strength without increasing
the overall density. The constitutive equations describing
steady-state creep behavior of low-density YRESZ all contain a
stress exponent $n \approx 3.2$ and a creep activation energy $Q_c \approx 128$
kcal/mole. The pre-exponential structure factors vary with
thermal history.

INTRODUCTION

Zirconia, stabilized in the cubic fluorite structure with at
least ten weight percent of yttria and rare earth oxides, is the
leading candidate material for use in regenerative ceramic storage
heaters associated with "blowdown" wind tunnels under development
at several Air Force and NASA centers. Such ceramic heaters must
provide large flows of air heated to 2250°C at pressures to 3500
psi[1]. Under these conditions, a major concern is the creep
strength of the ceramic material, which is in the form of hexagonal
cored bricks. Any distortion of the bricks during operation can
lead to misalignment of the core structure and a decrease in
heater efficiency. While high-density bricks (greater than 90%TD)
appear to offer completely satisfactory creep strength, they are
prone to thermal shock cracking in operation; a problem which is

lessened by use of low-density material. The primary objective of
this program was, therefore, to establish the creep behavior of
low-density YRESZ at temperatures approaching the operating conditions of a storage heater facility. An extensive study has been
conducted on high-density YRESZ[2], and the results of the present
study could be evaluated by comparison with the earlier work.

EXPERIMENTAL PROCEDURES

Creep testing was performed under a vacuum of 10^{-5} torr in a
tungsten-mesh, constant-stress compression creep unit. Strain
over the specimen gage length alone was measured by determining
the difference in deflection between the tungsten platens above
and below the specimen. Tungsten push rods transmitted the deflection to a linear variable differential transducer located
above the leading frame. Deflections of 5×10^{-5} inches were
measured, and the strain was recorded continuously. A constant-
stress system ensured that the initial stress did not vary by more
than 0.5 percent as the specimen was strained over several percent,
without the necessity of adding load at regular intervals. Temperature was recorded continuously with a control of less than
± 5 °C over the specimen gage length.

Tests were performed on annular cylindrical specimens ultra-
sonically trepanned from cored bricks provided by Arnold Engineering
Development Center (AEDC). The specimens were 0.34 inch high with
an outside diameter of 0.31 inch and an inside diameter of 0.19
inch. The bricks, produced by Zircoa Division of Corhart Refractories Company, had densities of 70-75 percent of theoretical
and contained 12.5 or 14.5 weight percent Y_2O_3 and various rare
earths and 1000-5000 ppm of Mg, Al, Si, K, Ca, Fe, and Hf. Some
bricks were received in the as-fabricated condition, which included a final anneal at 1800°C, while others had been heated to
2240°C for various periods of time in the AEDC storage heater.

EXPERIMENTAL RESULTS

The microstructures of the different YRESZ samples tested in
this study were examined by optical metallography and scanning
electron microscopy (SEM) before and after each creep experiment.
Optical micrographs were obtained from samples which were etched
for several minutes in a boiling solution of 100 grams NH_4F in
100 cc H_2O. Scanning electron micrographs were obtained from the
surface of as-received and crept samples fractured at room
temperature.

An optical micrograph of an as-fabricated specimen containing
14.5 weight percent Y_2O_3 is presented in Figure 1. This micrograph

reveals a nonuniform grain structure consisting of colonies of fine-grained material, isolated large grains, and large areas containing voids. By contrast, the microstructure of a specimen containing 14.5 weight percent Y_2O_3, which had been heated to 2240°C in the AEDC heater is shown in Figure 2. In this case recrystallization and grain growth has obviously occurred during the high-temperature heating cycles. The grain structures of bricks containing 12.5 weight percent Y_2O_3 were essentially similar to the ones shown in Figures 1 and 2.

The striking differences in structure of YRESZ bricks before and after being heated to 2240°C are further illustrated in Figures 3 and 4. These show, respectively, the SEM micrographs of the very porous structure of as-fabricated material, and the re-crystallized structure of the annealed bricks. In spite of the apparent densification of the original as-fabricated brick as shown in Figure 4, a bulk-density measurement based on weight and volume of a specimen taken from a fired brick gave a value of 73 percent of theoretical based on an X-ray density of 6.01 gm/cm^3.

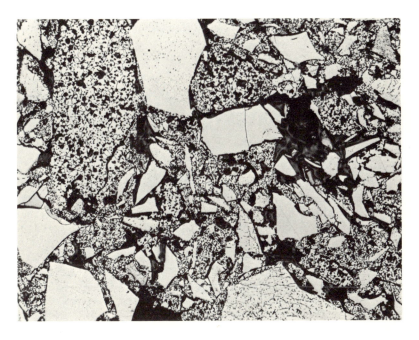

FIGURE 1.   OPTICAL MICROGRAPH OF ZN14.5L, AS-FABRICATED
            YRESZ CONTAINING 14.5 w/o Y_2O_3.   100X

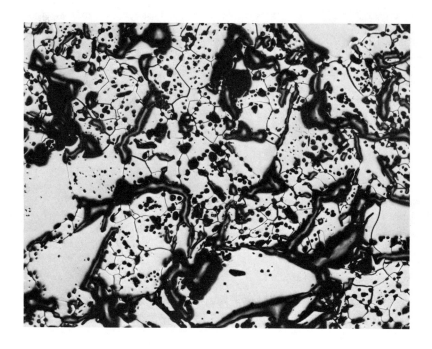

FIGURE 2.   OPTICAL MICROGRAPH OF ZF14.5L, YRESZ CONTAINING
            14.5 w/o Y_2O_3, FIRED IN AEDC HEATER TO 2240°C.
            100X

Steady-state creep rates obtained from specimens containing
14.5 weight percent Y_2O_3 are presented in Figure 5 versus the
applied stress. Specimen ZN14.5L-1 was machined from an as-
fabricated brick, while specimen ZF14.5L-1 was taken from brick
11-2, whose thermal history in the AEDC storage heater was reported by Tinsley[1]. This brick remained in the heater for forty
runs during which time it was heated to temperatures between 1650
and 2240°C. The numbers indicate the order in which data were
obtained. All of the data shown in Figure 5 can be fit by a power
law stress dependence with a stress exponent of 3.2. No evidence
was found for a viscous creep mechanism operating in these
specimens in the stress range of 300 to 3000 psi. The creep
strength of the brick previously fired in the AEDC heater is more
than an order of magnitude greater than that of the as-received
brick which was heated to 1882°C and tested after several days at
that temperature. The as-fabricated specimens were extremely weak
and crept very rapidly on initial heat-up beyond 1700°C.

FIGURE 3.   SCANNING ELECTRON MICROGRAPH OF ZN14.5L, AS-FABRICATED
YRESZ CONTAINING 14.5 w/o Y_2O_3.   500X

Results for duplicate specimens containing 12.5 weight percent
Y_2O_3 and heated in the AEDC heater (brick 6-2) are given in Figure
6 as steady-state creep rate versus applied stress. The creep
rates for ZF12.5L-1 and -2 are somewhat higher than those measured
for ZF14.5L-1, but the data can be fit by the same 3.2 power
stress dependence.

Optical micrographs of as-fabricated and of previously
annealed specimens after being creep tested at 1647 and 1882°C
are presented in Figures 7 and 8. As shown in Figure 7, even
after being crept at 1882°C and 1647°C, to 9.2 percent strain
(specimen ZN14.5L-1), the nonuniform grain distribution remains
and relatively little evidence of recrystallization can be found.
The grain structure of specimens fired in the AEDC heater and
crept at the above temperatures to 7.1 percent strain (see Figure
8) appears to be unchanged from that of the uncrept specimens.

These results are verified by the SEM micrographs of crept
specimens presented in Figures 9 and 10. The structure of the

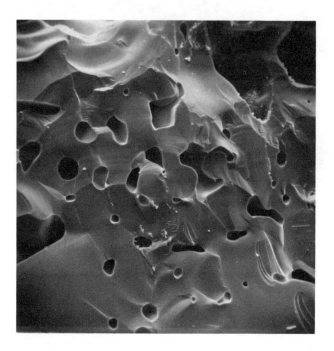

FIGURE 4.   SCANNING ELECTRON MICROGRAPH OF ZF14.5L,
             YRESZ CONTAINING 14.5 w/o Y_2O_3, FIRED IN
             AEDC HEATER TO 2240°C.   500X

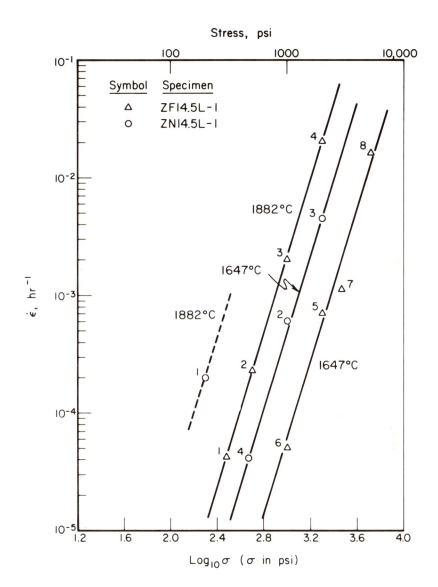

FIGURE 5. STEADY-STATE CREEP RATE VERSUS APPLIED STRESS FOR
YRESZ CONTAINING 14.5 w/o Y_2O_3, AS-FABRICATED, AND
AFTER BEING FIRED IN AEDC HEATER TO 2240°C

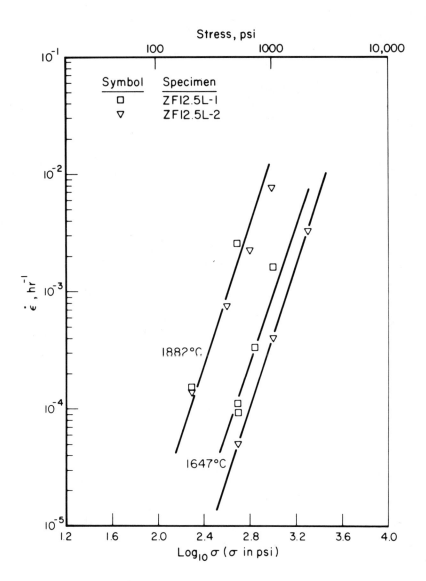

FIGURE 6.   STEADY-STATE CREEP RATE VERSUS APPLIED STRESS FOR
            YRESZ CONTAINING 12.5 w/o Y_2O_3, FIRED IN AEDC HEATER
            TO 2240°C

FIGURE 7.   OPTICAL MICROGRAPH OF ZN14.5L-1, AS-FABRICATED YRESZ
             CREPT AT 1882°C AND 1647°C TO 9.2 PERCENT STRAIN. 100X

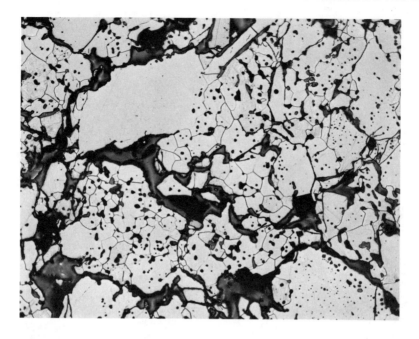

FIGURE 8.   OPTICAL MICROGRAPH OF ZF14.5L-1 YRESZ FIRED IN
            AEDC HEATER TO 2240°C AND CREPT AT 1882°C AND
            1647°C TO 7.1 PERCENT STRAIN

as-fabricated specimen after creep, shown in Figure 9, reveals
that some recrystallization and grain growth has occurred during
creep or equilibration at high temperature, while the structure
of the fired specimen after creep, shown in Figure 10, is similar
to that of the uncrept fired specimen presented in Figure 4. The
lack of viscous creep behavior, normally associated with fine-
grained crystalline material, is not surprising in view of the
structure of these samples.

                            DISCUSSION

     It is clear that low-density YRESZ, in the as-fabricated
condition, undergoes significant microstructural modifications as
a storage heater material; probably during initial heat-up to
2240°C. Most of the creep deformation observed in practice[1]
could be expected to have occurred while the structure was being
transformed from the one containing open-porosity and bimodal

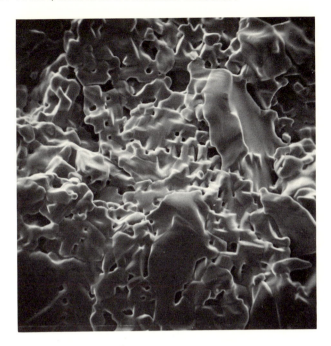

FIGURE 9.   SCANNING ELECTRON MICROGRAPH OF ZN14.5L-1,
             AS-FABRICATED YRESZ CREPT AT 1882°C AND
             1647°C TO 9.2 PERCENT STRAIN.   500X

grain distribution shown in Figures 1 and 3 to the more uniform
pore and grain size distribution shown in Figures 2 and 4.

    The evidence that steady-state creep rates for as-fabricated
YRESZ are an order of magnitude greater than those for material
fired to 2240°C (see Figure 5) is only a partial indication of
the relative initial strengths of these bricks, since the as-
fabricated brick undergoes some structural improvements during
creep testing, as shown in Figure 9.

    A comparison of the relative creep strengths of high- and
low-density YRESZ after similar heating cycles in the AEDC heater
is presented in Figure 11. At an applied stress of 1630 psi
the high-density brick is about one order of magnitude stronger
than low-density material. Extrapolation to the operating conditions of 2240°C and 20 psi (for the top-most bricks in the
heater) reveals that both kinds of as-fired bricks have sufficient
creep strength to meet the criteria of less than three percent

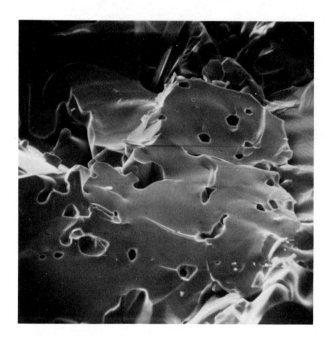

FIGURE 10.   SCANNING ELECTRON MICROGRAPH OF ZF14.5L-1, YRESZ
              FIRED IN AEDC HEATER TO 2240°C AND CREPT AT 1882°C
              AND 1647°C TO 7.1 PERCENT STRAIN.   500X

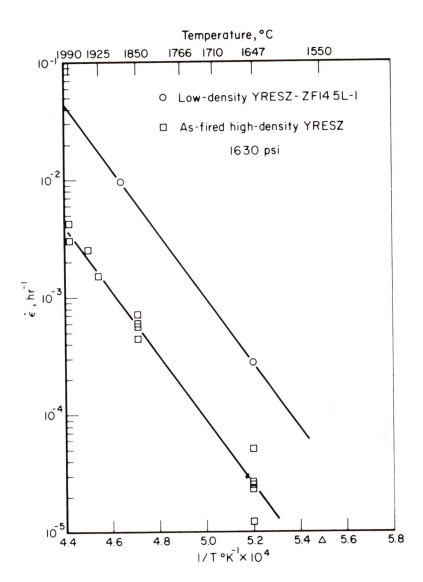

FIGURE 11.   COMPARISON OF STEADY-STATE CREEP RATES FOR LOW- AND
HIGH-DENSITY YRESZ FIRED IN AEDC HEATER

strain during the heater service life[3]. The extrapolated creep
rates for low-density fired bricks are on the order of $10^{-6} hr^{-1}$
or 1 percent per year for stress levels of 20 psi. Of course the
storage heaters are not maintained at 2240°C for long periods of
time, so that very little steady-state creep deformation should
be expected during the heater lifetime. The 3 percent creep
strain criteria was established under the assumption that tertiary
creep occurs at strains of this magnitude. While this is true for
compression creep at temperatures below 1600°C results of this
study suggest that much higher total strains (approaching 10 percent), can be accommodated at service temperatures before failure
occurs.

It has been shown that yttria content above 10 weight percent
has little effect on the creep strength of high-density YRESZ[2].
The moderate increase in creep strength with increasing yttria
content observed for low-density material is probably associated
with differences in grain size and density rather than with yttria
concentration, but the data are not sufficient to establish this
hypothesis.

A third power stress dependence has been found for both as-
fabricated and fired specimens. Several theories have been
developed to rationalize a σ^3 law[4,5]. Nabarro[4] showed that
for the case where the grain size is large and the stress and
dislocation density are high, dislocations may act as sources or
sinks for vacancies, leading to a dislocation climb diffusional
creep mechanism in which creep rates are proportional to the
third power of the stress.

The creep rate for the Nabarro climb model is given by[6]

$$\dot{\epsilon} = \pi D A \sigma^3 \Omega / (10 b^2 \mu^2 kT) \quad , \tag{1}$$

where D is the self-diffusion coefficient of the slower moving
species, $A \approx 1$, Ω is the vacancy volume, b is the Burgers vector,
μ is the shear modulus, and k is Boltzmann's constant. For the
case of YRESZ the value of D is unknown, precluding an accurate
test of this equation.

For calcia-stabilized zirconia, zirconium diffusivities are
approximately 10^6 slower than oxygen diffusion rates in the range
of 1750-2150°C[7-9], so the cation is clearly the slower diffusing
species in these materials. The creep activation of 128 ± 10
kcal/mole is not, however, comparable to either the value for
zirconium diffusion (92.5 kcal/mole) or oxygen diffusion (30
kcal/mole).

Cannon and Sherby[5] have shown that many nonmetals exhibit a third-power stress dependence for creep, including Al_2O_3, BeO, MgO, and UC. They demonstrated that the σ^3 dependence is not merely a transition dependence which might be observed by performing tests over a narrow stress range between regions in which $n = 1$ and $n = 5$. The lack of a grain-size dependence for creep rates in the $n = 3$ regime was taken as convincing evidence for no contribution from a viscous creep mechanism. Cannon and Sherby suggested that the third power stress dependence has been found in general when the ratio of ionic radii, r_{anion}/r_{cation} is greater than 2, although it has also been found for UC where this ratio is 1.85. For the case of YRESZ $r_{Oxygen}/r_{Zirconium} = 1.75$. These authors pointed out that in addition to the dislocation climb model, it is possible to obtain a σ^3 dependence from a model involving dislocation glide and climb, if the glide is Newtonion viscous and rate-controlling[10]. Cannon and Sherby suggested that nonmetals may exhibit a σ^3 dependence when creep occurs by the glide of charged dislocations dragging along a cloud of charged point defects, such as vacancies, interstitials, or impurity atoms. Either of the proposed dislocation mechanisms may be operative for the case of YRESZ.

ACKNOWLEDGEMENTS

This research was supported by the Aerospace Research Laboratories, Wright-Patterson Air Force Base, Ohio, under Contract No. F33615-73-C-4111. The low-density YRESZ bricks were kindly provided by Mr. G. Arnold of the Arnold Engineering Development Center.

REFERENCES

(1)   C. R. Tinsley, "Research and Development Testing of Yttria/ Rare Earth Stabilized Zirconia Matrix Bricks in the Pilot Test Unit (PTU) at AEDC", AEDC-TR-72-161, November, 1972.

(2)   M. S. Seltzer, A. H. Clauer, and B. A. Wilcox, "High-Temperature Creep of Ceramics", First Annual Report, to Aerospace Research Laboratories on Contract No. F33615-73-C-4111, February 28, 1974.

(3)   L. L. Fehrenbacher, F. P. Bailey, and N. A. McKinnon, SAMPE Quarterly, 2 (3) 48 (1971).

(4)   F.R.N. Nabarro, Phil. Mag., 16, 231 (1967).

(5)   W. R. Cannon and D. D. Sherby, J. Amer. Ceram. Soc., $\underline{56}$, 157 (1973).

(6)   J. Weertman, ASM Trans. Quart., $\underline{61}$, 681 (1968).

(7)   W. H. Rhodes and R. E. Carter, J. Amer. Ceram. Soc., $\underline{49}$, 244 (1966).

(8)   L. A. Simpson and R. E. Carter, J. Amer. Ceram. Soc., $\underline{49}$, 139 (1966).

(9)   W. D. Kingery, J. Pappis, M. E. Doty, and D. C. Hill, J. Amer. Ceram. Soc., $\underline{42}$, 393 (1959).

(10) J. Weertman, J. Appl. Phys., $\underline{28}$, 1185 (1957).

RELATIONSHIP BETWEEN PRIMARY AND STEADY-STATE CREEP OF UO_2 AT ELEVATED TEMPERATURE AND UNDER NEUTRON IRRADIATION[*]

A. A. Solomon

Materials Science Division, Argonne National Laboratory

Argonne, Illinois 60439

I. INTRODUCTION

Primary creep has typically been observed in polycrystalline ceramic materials, but it has received little quantitative study because of experimental difficulties and the absence of theoretical models. The experimental difficulties derive from strain resolution, microstructural stability, and, in compressive creep experiments, from specimen seating transients or, in bending experiments, from contact-point indentations. Passmore and Vasilos[1] measured the apparent activation energy Q for primary creep of fine-grained alumina, and Passmore et al.[2] obtained Q for MgO. In both investigations, it was noted that Q was strain independent and steady-state creep was not established even after ∿2% strain. Primary creep has also been observed in UO_2[3-7] and PuO_2.[8,9] Solomon[10] studied primary creep of stoichiometric UO_2 helices at low stresses (≵5000 psi) and elevated temperatures (1000–1600°C) and obtained the following expression for the primary creep of as-sintered helical-spring specimens:

$$\dot{\gamma}_p = \left[A \frac{\sigma}{d^{1.5}} \exp\left(\frac{-Q}{RT}\right) \right] t^{m-1} , \tag{1}$$

where $\dot{\gamma}_p$ is primary strain rate, σ is stress, d is grain size, t is time after the application of stress, and A and m are constants.[**]

---

[*]Work performed under the auspices of the U.S. Atomic Energy Commission.

[**]Equation (1) applies under all measurable conditions, but it leads to an infinite initial strain rate and must be applied with caution.

No change in σ, d, or density occurred in these experiments. For
grain sizes $\sim$10-30 μm, m $\approx$ 0.7 $\pm$ 0.1; A was found to be a function
of purity.[11] Again, steady-state creep was not observed after $\sim$2%
outer fiber shear strain.

Similar results were obtained by Solomon[12,13] for the fission-
induced creep of UO_2 helices at low temperatures and stresses. The
time dependence of the fission-induced creep rate $\dot{\gamma}_f$ was found to
be proportional to t^{m-1}, where m $\approx$ 0.6 $\pm$ 0.1; however, in the case
of fission-induced creep, $\dot{\gamma}_f$ was shown to be temperature independent.[12]

The same "power-law" time dependence is commonly observed for
dislocation-controlled creep of metals at elevated temperatures[14]
and would be expected for the same mechanism in ceramics. No
satisfactory theoretical model has been proposed to explain the
"power-law" time dependence, but it serves as a useful empirical
relation. Thus, justification exists for consideration of the
significance of this form of primary creep transient in terms of
the steady-state creep rates.

II.   ANALYSIS OF PRIMARY AND STEADY CREEP

Creep of ceramics and metals has, for the most part, been
examined in terms of strain vs. time curves or strain rate vs. time
curves. Figure 1 shows typical strain-time curves for UO_2 at
elevated temperatures (a) and under irradiation (b). These curves
show the typical decelerating creep rates for as-sintered polycrystalline UO_2. Casual inspection would suggest that the creep
rates approach or are in true steady state. When these data are
plotted as strain rate vs. time, Fig. 2, again it would appear that
steady state has been approximated. However, when the thermal creep
data are plotted in terms of log γ vs. log t, it is clear from Fig.
3 that no deviation from the initial m value of $\sim$0.77 occurs, indicating steady-state is not approached. If the distinct assumption
is made that the total strain rate $\dot{\gamma}_t$ is the algebraic sum of $\dot{\gamma}_p$ and
the steady state strain rate, $\dot{\gamma}_s$, the initial m value should show a
gradual transition from m = 0.77 to m = 1 as steady-state is obtained.

The absence of any deviation from m = 0.77 is significant because
it imposes an upper bound on the steady-state creep rate. This may be
seen by examining the log $\dot{\gamma}$ vs. log t curves. <u>Independent</u> measurements of the strain rates, as a function of time, are also plotted
in Fig. 3. Since Eq. (1) is obeyed, $\dot{\gamma}_p$ should be proportional to
t^{m-1} so the slope of log $\dot{\gamma}_p$ vs. log t should equal m-1. Since the
measured slope is -0.225, m = 0.77 by this method, which agrees with
the log γ vs. log t result. Now, in terms of strain rates, $\dot{\gamma}_s$
appears as a horizontal line in Fig. 3, but since steady state was
not obtained, all that is known is the maximum possible value or

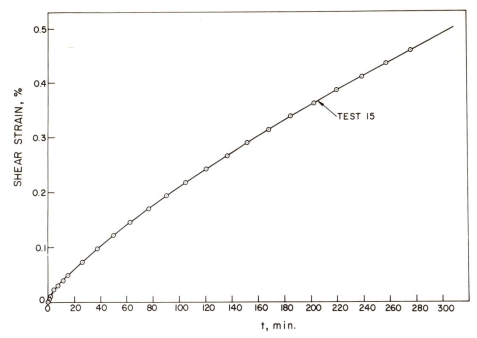

Figure 1(a). Creep curve for UO$_2$ helical-spring specimen at
1500°C and τ = 760 psi.

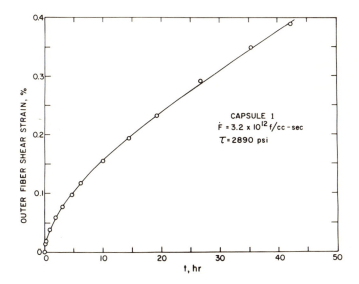

Figure 1(b). Radiation-induced creep curve for UO$_2$ helical-spring
specimen at a fission rate of 3.2 x 10^{12} fissions/
cc-sec, τ = 2890 psi and T = 100°C.

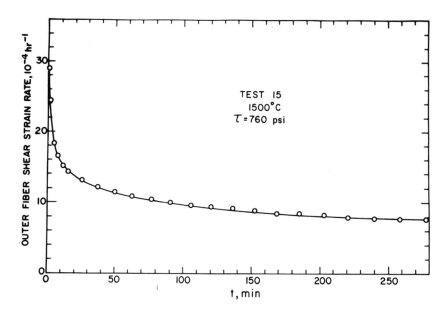

Figure 2(a).   Strain rate vs. time for UO_2 helical spring at
elevated temperature [see Fig. 1(a)].

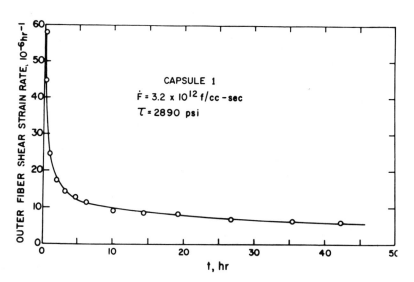

Figure 2(b).   Strain rate vs. time for UO_2 helical spring under
neutron irradiation.

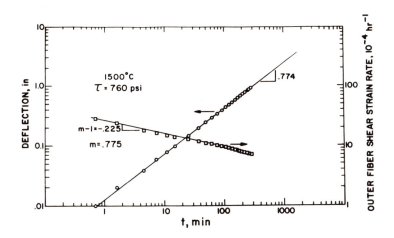

Figure 3.   Primary creep of UO$_2$ at elevated temperature.

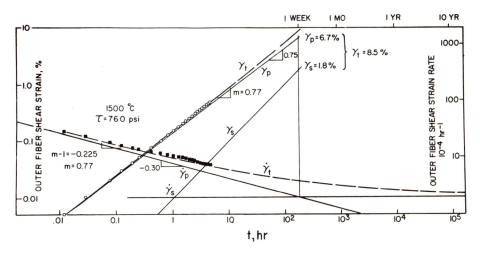

Figure 4.   Relation between primary creep and steady-state creep
of UO$_2$ at elevated temperatures.

<u>upper bound</u>. An upper bound of $\dot{\gamma}_s = 2 \times 10^{-4} hr^{-1}$ was selected such that $\dot{\gamma}_t = \dot{\gamma}_p + \dot{\gamma}_s$ is approximately linear over the region of the measured data (Fig. 4). This value is probably an overestimation, since lower values of 1.47×10^{-4} and $0.9 \times 10^{-4} hr^{-1}$ have been reported under similar conditions by Bohaboy et al.[14] and Poteat and Yust,[15] respectively (normalizing their data assuming Q=90 kcal and $\dot{\gamma} \propto d^{-2}$). Their data are, in turn, probably overestimates of the true steady-state strain rates, as shown below. In addition, the results shown in Fig. 3 were for a high-purity specimen, which would <u>minimize</u> the primary creep component.

The steady-state <u>strain</u> contribution γ_s is also plotted in Fig. 4 to evaluate the relative magnitude of γ_p and γ_s. Again, the sum of $\gamma_p + \gamma_s$ must be linear over the available data.

The striking result shown in Fig. 4 is that $\dot{\gamma}_t$ exhibits a broad transition from primary to steady-state creep. At the crossover point, when $\dot{\gamma}_s = \dot{\gamma}_p$, $\gamma_t = 8.5\%$ and t = 1 week (effective tensile strain = 4.9%). Thus, even at an effective tensile strain of ∿5% the steady-state strain rate is half the measured rate, and $\gamma_p = 6.7\%$ and $\gamma_s = 1.8\%$ so the strain is dominated by the primary component. If the steady-state rate shown in Fig. 4 were lowered even slightly, the crossover point would occur at large strains and times. Therefore, true steady-state strain rates are indeed difficult to obtain under these conditions.

As seen in Fig. 4, the m values obtained from log $\dot{\gamma}$ vs. log t and log γ vs. log t curves agree. However, if the upper bound steady-state creep rate is added to the primary creep rate, then $\dot{\gamma}_t$ must fit the data. Therefore, the true m value for γ_p is reduced to 0.75 (see Fig. 4) when determined from log γ_p vs. log t and 0.70 when determined from log $\dot{\gamma}_p$ vs. log t. This disagreement implies that the true steady-state creep rate is lower than the upper bound value.

A similar analysis was performed on the deformation-induced[*] transient for UO_2 under irradiation, Fig. 5, where only $\dot{\gamma}_p$ is shown. $\dot{\gamma}_s$ was obtained from the reported[12,16-18] steady-state strain rates normalized to the fission rate and stress shown, assuming $\dot{\gamma}_s$ is linearly proportional to stress and fission rate. In this case, the crossover occurs at $\gamma_t = 11.9\%$ and t = 10 years! Therefore, the attainment of true steady-state seems questionable under these conditions.

---

[*]The irradiation-induced substructure was well developed prior to loading.

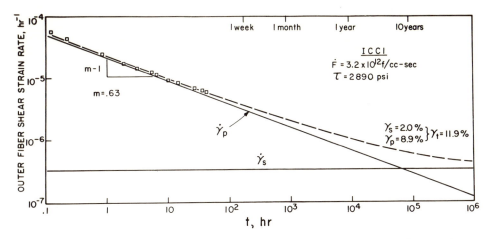

Figure 5.   Relation between radiation-induced primary and steady
            state creep of UO$_2$

III. DISCUSSION

     The analysis in the previous section demonstrates the importance
of primary creep and the difficulty encountered in obtaining the true
steady-state condition. The generality of this result, however, may
be questioned. Therefore, all published and some unpublished creep
curves for radiation-induced and thermal creep of UO$_2$ were examined
to see if steady state had been achieved as claimed. These data
must be viewed with some caution because microstructural stability
during the creep test was not always assured. Nevertheless, Fig. 6
shows the same "power-law" time dependence with no deviation from
linearity for all available data. Therefore, steady state creep
was not obtained in any of the investigations examined.*

     Often, investigators have conducted several creep tests at
different stresses on the same specimen. Because of the strain
recovery observed in UO$_2$,[10] a load <u>reduction</u> would superimpose this
recovery on the creep curve and could yield a pseudo "steady state."
However, the time dependence of this recovery is logarithmic and the
magnitude is small so the recovered strain component should become
negligible if plastic strains ⅓1% are accumulated after a load
reduction.

---

*The results of Burton and Reynolds[7] show a low-temperature creep
 curve in which a saturation strain was claimed.

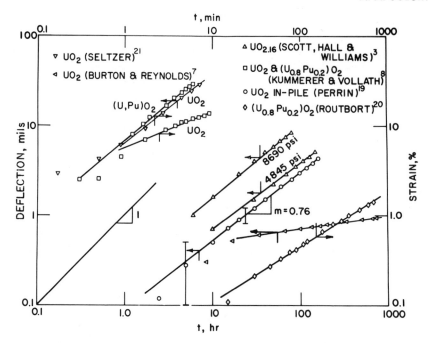

Figure 6.   Primary creep transients from previous investigations.
Data of Burton and Reynolds has been shifted vertically
for convenience.

No attempt was made to convert the results in Fig. 6 to a
uniform strain vs. t basis. Prior prestrain tends to change the
value of A in Eq. (1), and the degree of prestrain was usually not
reported.

Since the available evidence indicates that steady-state creep
of UO_2 has not been obtained, the question arises, what errors have
been introduced in the analysis of stress, temperature, or microstructural dependencies of steady-state creep using primary creep
results? All we know is the upper bound for $\dot{\gamma}_s$ from the available
primary creep curves, and no quantitative calculations concerning
steady-state creep can be made on that basis. However, since the
primary creep results are available, the errors in the determination
of the stress, temperature, and microstructural dependencies for
primary creep can be examined. Solomon[10] has proposed that the
separable time dependence indicated in Eq. (1) allows a determination
of the other dependencies shown, if they are determined at constant
time. Normally, creep tests are conducted until (a) a maximum time
has elapsed (dictated by the experimenter's patience), (b) a
maximum strain is obtained (to prevent barreling, etc), or (c) a
minimum change in strain rate with time is attained. The third
criterion is usually employed prior to a change in test conditions.

The maximum strain condition may be easily derived by first taking a shortened form of Eq. (1),

$$\gamma_p = At^m , \tag{2}$$

and differentiating with respect to time, which yields

$$\dot{\gamma}_p = A\, m\, t^{m-1} . \tag{3}$$

Substituting Eq. (2) into Eq. (3),

$$\dot{\gamma}_p = m\gamma_p / t .$$

We know m, and γ_{max} is selected arbitrarily for a given experimental setup so that

$$\dot{\gamma}_p t \geq m\, \gamma_{max} , \tag{4}$$

or

$$\log \dot{\gamma}_p \geq - \log t + \log m\, \gamma_{max} \tag{5}$$

for all creep tests. The "greater than" sign in Eq. (4) has the significance that tests are conducted until the $\dot{\gamma}_p t$ product is reduced to $m\, \gamma_{max}$. This condition is shown in Fig. 5 as the γ_{max} limit.

Similarly, the condition for maximum change in strain rate may be derived by taking the time derivative of Eq. (3) and substituting for $\dot{\gamma}_p$

$$\frac{d\dot{\gamma}_p}{dt} = Am(m-1)t^{m-2} = \frac{\dot{\gamma}_p}{t}(m-1) . \tag{6}$$

Therefore, tests are conducted such that

$$\frac{d\dot{\gamma}_p}{dt} = \left| \frac{\dot{\gamma}_p(m-1)}{t} \right| \geq \left| -C \right| , \tag{7}$$

where C is an arbitrarily selected positive constant for all experiments. Taking logarithms of both sides of Eq. (7) yields

$$\log \dot{\gamma}_p \geq \log t + \log \frac{C}{1-m} . \tag{8}$$

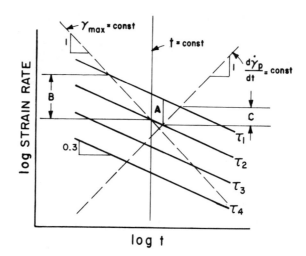

Figure 7.   Errors introduced in the analysis of primary creep.

Figure 7 shows a series of creep curves at different stress and an m value of 0.7.   The three methods of conducting tests are shown in Fig. 7 as (a) t = const, (b) γ_{max} = const, and (c) d $\dot{\gamma}_p$/dt = const.   Tests are run until the appropriate boundary is approached from the left.   Now, if tests are conducted following method (a) the correct stress dependence determined from n = d ln $\dot{\gamma}_p$/d ln τ is obtained.   However, if method (b) were used, instead of observing the difference in strain rate given by A in Fig. 7 between τ_1 and τ_2 the difference B is obtained.   Simple geometry reveals the ratio of B to A of 1.48 or 48% error.   Similarly, if method (c) were used, a 13% error is obtained.   Of course, these errors depend sensitively on the m value.   Lower m values, which might occur for dislocation-induced primary creep, would yield even greater errors.

If two or more of the above test criteria are used, e.g., maximum time on low stress, low strain-rate tests, and maximum strain on higher stress and strain-rate tests, a changing stress dependence would be observed.   This might be taken erroneously as an indication of a change in deformation mechanism.

The temperature dependence for primary creep can be determined by the temperature-cycling experiment, if the substructure does not change with an "instantaneous" change in temperature.   With this assumption, a strain-independent activation energy has been found for the primary creep of metals[22] and ceramics.[1,2,10]   A second method of obtaining Q relies on the assumption that the substructure is constant at a fixed stress level in steady-state creep.   Thus, for this second method, the establishemnt of steady state is essential.   Since we can only determine the upper bound for $\dot{\gamma}_s$ we cannot assess the errors, but Q determined from temperature cycling and "steady state" appear to be in fair agreement. If, as above, we examine Q for primary creep, the same relative errors in strain rate for the maximum strain or maximum d $\dot{\gamma}_p$/dt criteria would be obtained.

SUMMARY

Elevated temperature and radiation-induced primary creep have been shown to be important and general phenomena in UO$_2$ and (U,Pu)O$_2$. Indeed, the existence of steady-state creep has not been conclusively established. The analysis of primary and secondary creep reveals a general difficulty in obtaining steady-state creep and suggests that the time dependence of primary creep should be analyzed before assuming steady state has been obtained.

Generally, creep tests on ceramics are conducted with several possible experimental constraints or limits, e.g., time, strain, or maximum strain-rate change. The errors introduced in the analysis of primary creep were shown to be significant. Since steady-state creep was not generally attained, no assessment of error was possible.

The importance of primary creep in oxide fuels and possibly other ceramics introduces complexity into many related phenomena and analyses. Hot-pressing, fracture, and thermal-shock behavior are affected. The engineering use of ceramics generally implies small strains in which primary creep is important. The prediction of behavior via deformation maps becomes much more complex because the time dependencies of the mechanisms may be different. These time dependencies must be determined accurately to predict the behavior of metals and ceramics.

REFERENCES

1. E. M. Passmore and T. Vasilos, J. Am. Ceram. Soc. 49:3, 166-168 (1966).
2. E. M. Passmore, R. H. Duff and T. Vasilos, J. Am. Ceram. Soc. 49:3, 594-600 (1966).
3. R. Scott, A. R. Hall and J. Williams, J. Nucl. Mater. 1, 39 (1959).
4. W. M. Armstrong and W. R. Irvine, J. Nucl. Mater. 9:2, 121-127 (1963).
5. W. M. Armstrong and W. R. Irvine, J. Nucl. Mater. 9, 121 (1963).
6. W. M. Armstrong and W. R. Irvine, J. Nucl. Materl 12:3, 261-470 (1964).
7. B. Burton and G. L. Reynolds, Acta Metall. v21, 1073-78 (1973).

8.   K. Kummerer and D. Vollath, Fast Reactor Fuel and Fuel Elements, M. Dalle Donne, Ed., Gesellschaft fuer Kernforschung mbH, 1970.

9.   J. L. Routbort, N. A. Javed and J. C. Voglewede, J. Nucl. Mater. 44, 247-259 (1972).

10. A. A. Solomon, to be published.

11. A. A. Solomon, Bull. Am. Ceram. Soc. 52:4, 398 (1973).

12. A. A. Solomon, J. Am. Ceram. Soc. 56:3, 164-171 (1973).

13. A. A. Solomon and R. H. Gebner, Nucl. Tech. 13, 177-184 (1972).

14. P. E. Bohaboy, R. R. Asamoto and A. E. Conti, Rept. No. GEAP 10054 (May 1969).

15. L. E. Poteat and C. S. Yust, Ceramic Microstructures, R. M. Fulrath and J. A. Pask, Eds., pp. 646-56, John Wiley and Sons, New York (1968).

16. D. Brucklacher and W. Dienst, J. Nucl. Mater. 36:2, 244-47 (1970).

17. D. J. Clough, Fast Reactor Fuel and Fuel Elements, Ed. by M. Dalle Donne, Gesellschaft fuer Kernforschung mbH, 1970.

18. E. C. Sykes and P. T. Sawbridge, CEGB Tech. Rept. RD/B/M-1489 (Nov 1969).

19. J. S. Perrin, J. Nucl. Mater. 39:2, 175-82 (1971).

20. J. L. Routbort, Argonne National Laboratory, private communication (1973).*

21. M. S. Seltzer, Battelle Columbus Laboratory, private communication (1974).**

22. F. Garofalo, Fundamentals of Creep and Creep-Rupture in Metals, p. 88, The Macmillan Company (1965).

---

*The creep curves analyzed were those used in Ref. 9.

**The creep curves analyzed were those used in the paper by M. S. Seltzer, A. H. Clauer and B. A. Wilcox, J. Nucl. Mater. 34, 351-53 (1970).

HIGH-TEMPERATURE PLASTICITY OF OXIDE NUCLEAR FUEL[*]

J. T. A. Roberts

Materials Science Division, Argonne National Laboratory

Argonne, Illinois 60439

INTRODUCTION

The oxide nuclear fuels based on UO_2 are the subject of detailed studies because of their importance to both thermal and fast-reactor systems. Of particular technological interest is the plasticity of the fuel, inasmuch as it determines the magnitude of fuel swelling and, hence, the stresses induced in the metal cladding (outer fuel container). The amount of permanent deformation allowed in the cladding before the fuel element must be removed from the reactor is limited by both operational and safety considerations. Thus, if less force could be exerted by the fuel on the cladding, the useful lifetime of the fuel element could be increased. This realization has been the impetus for several research programs in both the U.S. and Europe designed to understand the plastic nature of nuclear fuels in a high-temperature, fissioning environment. The principal objective is to develop a plastic fuel that would fill the internal space provided without unduly stressing the cladding.

Since the technological justification for studying the mechanical behavior of oxide nuclear fuels has been established, the purpose of the present paper is to describe some aspects of the deformation process at high temperatures ($\gtrsim 0.5\ T_m$), which also have important implications with respect to the plasticity of nonfissile ceramic oxides. Phenomena associated with low-temperature ($< 0.5\ T_m$) fission-induced plasticity will be the subject of another paper in this symposium.[1]

---

[*] Work performed under the auspices of the U.S. Atomic Energy Commission.

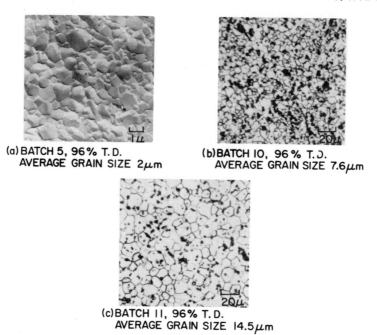

(a) BATCH 5, 96% T.D.
    AVERAGE GRAIN SIZE 2μm

(b) BATCH 10, 96% T.D.
    AVERAGE GRAIN SIZE 7.6μm

(c) BATCH 11, 96% T.D.
    AVERAGE GRAIN SIZE 14.5μm

Figure 1.   Microstructures of the three mixed-oxide materials.

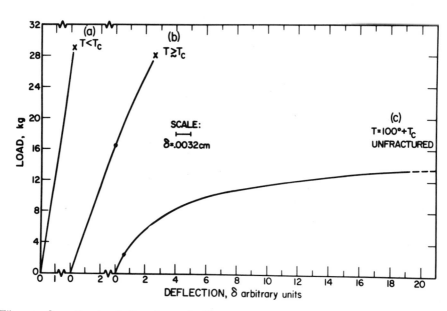

Figure 2.   Typical load vs deflection curves in four-point bending
            as a function of temperature.

It is well known that under certain conditions polycrystalline ceramics deform plastically by mechanisms which involve grain boundaries.[2-4] In fact, it appears that grain-boundary sliding (GBS) may enhance the ductility of fine grain ceramics which possess an insufficient number of independent slip systems.[4] Nevertheless, fracture occurs after relatively low strains, since the stress that builds up at discontinuities in the grain boundary (e.g., triple points) cannot be released by slip in the adjoining grains. It follows, therefore, that, if sufficient slip systems were available (at modest stresses) in a fine grain ceramic to accommodate the incompatibility due to sliding, extensive plasticity (superplasticity?) might be achieved. During the course of our oxide-deformation program, certain observations have led us to believe that this so-called superplasticity was achieved in relatively pure, fine grain UO_2-20 wt % PuO_2 materials and larger grain size UO_2-20 wt % PuO_2 containing impurities. These new observations will be described in the present paper, and the differences between these results and the normally observed high-temperature deformation behavior of UO_2[5,6] and UO_2-20 wt % PuO_2[6-9] will be discussed.

EXPERIMENTAL PROCEDURE

The UO_2-20 wt % PuO_2 specimens were prepared from mechanically blended powders. The fabrication procedure has been reported in detail elsewhere.[10] Tables I and II summarize the principal fabrication variables and sample characteristics, and Fig. 1 shows the grain and pore structures of the materials studied.

The specimens were deformed in four-point bending in a high-temperature, inert-atmosphere furnace described in Refs. 10 and 11. Conventional load versus deflection, strain-rate change and stress-relaxation tests were conducted in the strain-rate range of 0.1 to 0.4 h^{-1} and in the temperature range of 1500 to 1800°C.

RESULTS

All of the mixed-oxide specimens followed the same general deformation pattern, that is a sharp brittle-to-ductile transition was observed above 0.5 T_m (the transition temperature increased with an increase in strain rate[10,11]), and extensive ductility without fracture was recorded only ∿100°C above the brittle-to-ductile transition temperature (T_c) (Fig. 2). We are concerned here with the plastic regime, and the principal deformation parameters are steady-state flow stress σ_f and strain-rate sensitivity m.

TABLE I. Fabrication Variables for UO_2-20 wt % PuO_2 Specimens

Batch Number	Binder	Method of Binder Addition	Pressing Pressure (psi)	Sintering Time and Temp. (Atmosphere always He-6% H_2)
5	Carbowax 4000	1% binder solution, added to ball-milled slurry before drying	20,000	2 h at 1625°C
10	Carbowax 4000	1% binder solution, ball milled with powders for 8 h	20,000	3 h at 1700°C
11	Carbowax 4000	2% binder solution, ball milled with powders for 8 h	10,000	Same run as batch 10 (postsinter anneal, 3 h at 1800°C in He)

TABLE II. Characteristics of UO_2–20 wt % PuO_2 Batches

Batch Number	Immersion Density (% TD)	O/M Ratio	Grain Size (μm)	Impurity content (ppm)	Fe (ppm)	Si (ppm)	Pu Homogeneity (microprobe)
5	96.6±0.46	1.97±0.005	~2	400	250	42	Expected[a] (single phase)
10	96	1.94±0.005	7.6±0.4	~1340	970	–	Expected[a] (single phase)
11	96	1.94±0.006	14.5±2.3	~1420	1140	–	Expected[a] (single phase)

[a]Approximately the expected inhomogeneity for mechanically blended material.

The analysis of plastic stress and strains in a four-point bending test is complex. Heuer et al.[4] present a treatment of this problem in which m in bending is given by

$$\left(\frac{\partial 1nM}{\partial 1n\dot{\varepsilon}} \right)_{\varepsilon} , \tag{1}$$

where M is the bending moment equal to $\frac{Wa}{2}$ (W = load, and a = distance between inner and outer loading points), and the outer fiber strain rate is given by

$$\dot{\varepsilon} = \frac{12h\dot{\delta}}{3L^2 - 4a^2} , \tag{2}$$

where h is the specimen height, $\dot{\delta}$ is the deflection rate of the specimen at its midpoint, and L is the distance between the outer loading points.

The maximum stress in the outer fibers is given by[4]

$$\sigma_{max} = \frac{Wa}{bh^2} (2 + n + m) , \tag{3}$$

where b is the specimen width, and n is the strain-hardening coefficient, equal to zero when $\sigma_{max} = \sigma_f$,

$$\text{i.e., } \sigma_f = \frac{Wa}{bh^2} (2 + m) . \tag{4}$$

Thus, the calculated value of σ_f is less than or equal to the value of stress σ_R measured from elastic considerations[12]

$$\sigma_R = \frac{3Wa}{bh^2} . \tag{5}$$

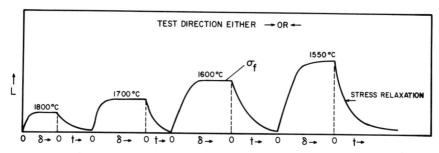

Figure 3. Schematic procedure for stress-relaxation tests.

In the following, the flow stress was calculated from load-deflection curves using Eq. (4), and the strain-rate sensitivity was measured from strain-rate change and stress-relaxation tests using Eq. (1). Inasmuch as the σ_f calculation requires knowledge of m, we will first describe these results for ~ 2, 7.6, and 14.5 μ UO_2-20 wt % PuO_2. The stress-relaxation experiments gave the most reproducible data and were therefore used to determine m as a function of temperature. Spot checks were made using the strain-rate-change test to compute m.

The experimental technique for the stress-relaxation experiments is shown in Fig. 3. The stress was allowed to relax from the steady-state flow stress at successively lower (or higher) temperatures, using the same specimen throughout one test direction. As shown in Fig. 4, stress-relaxation data were plotted as ln $\partial W/\partial t$ (i.e., load-relaxation rate $\equiv \dot{\epsilon}$) versus ln W ($\equiv M$) and, from Eq. (1), the inverse slope of the straight line gave m. The strain-rate-sensitivity parameter m is plotted as a function of temperature in Fig. 5. The two lines drawn through the data indicate the variation in data points at each temperature and the strong temperature dependence of m. By way of comparison, similar data for UO_2[13] are also plotted in Fig. 5.

The flow-strength data for UO_2-20 wt % PuO_2, from Eqs. (4) and (5), are plotted in Fig. 6. The stress calculations using purely elastic considerations, or the modified equation due to Heuer et al.,[4] give the same trend. The most significant observations are the strong temperature dependence of the flow strength and the increase in flow strength with an increase in grain size. The latter effect is illustrated in Fig. 7 for two temperatures.

The form of the temperature variation of flow stress appears to be exponential. However, a plot of log σ_f versus 1000/T in Fig. 8 indicates an apparent temperature dependence of the activation energy, presumably because all temperature-dependent parameters have not been isolated. According to Nicholson,[14] this is a common characteristic of superplastic behavior, and the correct method of treating the data is not clear.

In the majority of tests, the specimens were strained to the limits of the extensometer, which correspond to $\sim 3\%$ outer fiber strain, and then unloaded. However, to demonstrate the unusually high plasticity in these materials, plastic strain was accumulated in two tests by reversing the specimen and bending it back through zero deflection to 3% strain in the opposite direction. In this manner, $\sim 9\%$ strain was accumulated without fracture in a 2-μm grain-size specimen, and 15% strain was accumulated in a 14.5-μm grain-size specimen before slow cracking was observed.

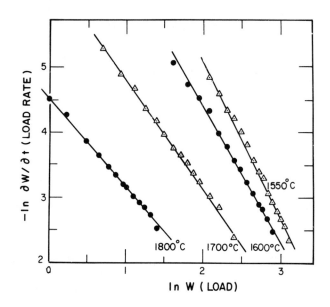

Figure 4.   Plots of ln ∂W/∂t vs ln W for 14.5-μm grain size mixed
oxide.

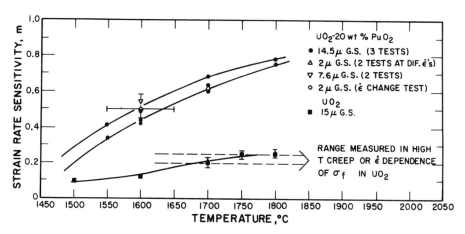

Figure 5.   The temperature dependence of the strain-rate sensitivity
in UO$_2$-20 wt % PuO$_2$ and UO$_2$ materials.

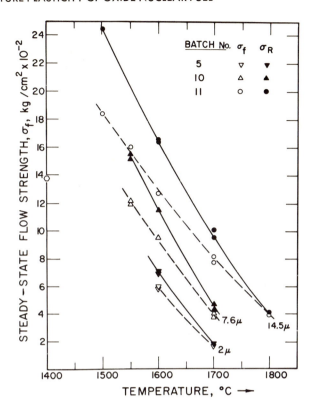

Figure 6.   The temperature dependence of flow strength for three
grain sizes of UO_2-20 wt % PuO_2 materials.

DISCUSSION

The deformation characteristics of these mixed-oxide materials
compare favorably with those of superplastic alloys.[14]   The major
points of comparison are (1) strain-rate sensitivities that increase with an increase in temperature and lie in the range of 0.4
to 0.8, (2) flow stress is a strong function of temperature, and
at high temperatures, decreases to almost vanishingly small values,
(3) flow stress increases with an increase in grain size over a wide
temperature range, and (4) a "temperature-dependent" activation

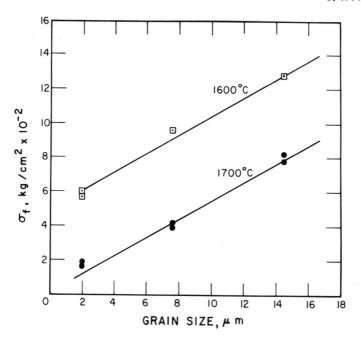

Figure 7.   Grain-size dependence of flow strength at 1600 and 1700°C.

energy.   In the following, we will discuss these characteristics in relation to observed creep and flow behavior in UO_2 and other UO_2-PuO_2 materials.

Figure 5 shows that the strain-rate sensitivities for 15-μm UO_2, in the same stress-strain-rate regime, are generally much lower than were observed for the mixed-oxide materials and are less sensitive to temperature.   The characteristic creep value of m = 0.22 (creep exponent = 4.5)[5,6] appears to be the limiting value at high temperatures.   Using these values of m and Eq. (4), it is of interest to compare the temperature dependences of flow stresses, at a strain rate of ∿0.1 h^{-1}, calculated from compressive creep data obtained elsewhere[5] and four-point bending data obtained in this research.

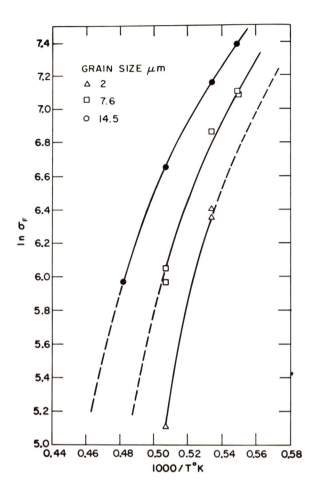

Figure 8.   Temperature dependence of flow strength for three grain sizes.

Figure 9 for UO_2 shows quite good comparison between bending and creep flow data, indicating that all UO_2 materials tested behave in a conventional manner. The difference in slopes of the two curves merely reflects slightly different activation energies, which can be explained by small differences in either oxygen-to-metal (O/M) or impurity levels in UO_2.[6] However, Fig. 10 shows a marked difference between mixed-oxide compressive creep data[7-9] and the present bending results. The creep data exhibit neither the grain-size effect nor the strong temperature dependence of σ_f. The mixed oxides prepared for this research obviously behave in an anomalous fashion, and, at this stage, the cause is speculative.

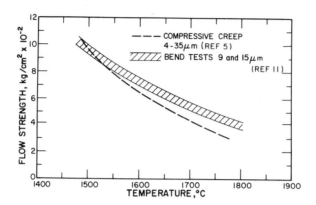

Figure 9.   Temperature dependence of flow strength for UO_2, comparison
of four-point bending and compressive-creep data.

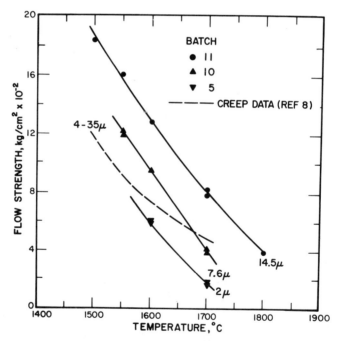

Figure 10.   Temperature dependence of flow strength for UO_2-20 wt %
PuO_2, comparison of four-point bending and compressive-
creep data.

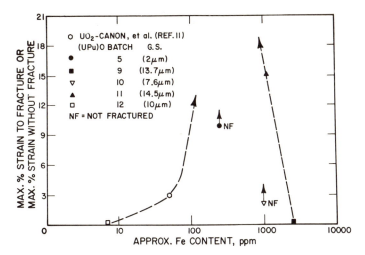

Figure 11.   Effect of iron content on ductility of mixed-oxide
             fuel materials.

     Superplasticity is a characteristic of fine-grain materials,[14]
and yet in these experiments superplastic characteristics persisted
in the 7.6- and 14.5-µm material. In fact, the temperature dependence of m and σ_f for the three grain sizes is remarkably similar,
and the grain-size dependence of σ_f shows no obvious anomaly. If
GBS is the dominant mechanism of superplasticity, then it follows
that some grain-boundary property in the mixed oxides is responsible
for retaining the superplastic-like properties beyond the expected
grain-size range. A preliminary thought is that an iron impurity
at the grain boundaries is responsible for this anomalous behavior
for the following reasons. As Table II shows, the iron content of
the two larger-grain mixed-oxide batches was high, ∿1000 ppm. In
another part of our mixed-oxide deformation and fracture research,
we had observed a possible trend between ductility and iron content,
reproduced in Fig. 11, and, in the most impure material that
fractures prematurely, we observed evidence of a once-liquid phase
at the grain boundaries. A tentative explanation for the observed
enhanced ductility in these materials could be that a liquid (or
near liquid) grain-boundary phase facilitates the GBS deformation
process. Figure 11 indicates that maximum ductility might be
achieved with materials containing between 100 and 1000 ppm of iron
and possessing between ∿2- and 14.5-µm grain size. Other impurities
may well contribute the same effect, and this would be an interesting
area for further study inasmuch as the fissioning process in-reactor
builds up a significant inventory of insoluble metals in the fuel.
Thus, the increased concentration of impurities could perhaps
compensate for the increase in grain size caused by the high fuel
temperatures, and a high level of plasticity could be maintained
throughout burnup.

Summary

Certain UO_2-20 wt % PuO_2 fuel materials fabricated at Argonne National Laboratory have demonstrated the possibility of super-plastic-like deformation at only 100°C above the brittle-ductile transition temperature. The enhanced ductility, which in one test attained 15% strain in bending, appeared to be facilitated by a grain-boundary-impurity liquid phase in larger grain-size materials.

REFERENCES

1.   A. A. Solomon, "Relation Between Radiation-induced and Thermal Primary Creep to Steady-state Creep in UO_2," This Symposium, p.

2.   J. H. Hensler and G. V. Cullen, "Grain Shape Change During Creep in MgO," J. Am. Ceram. Soc. 50, 584 (1967).

3.   C. S. Yust and J. T. A. Roberts, "On the Observation of Lattice and Grain Boundary Dislocations in UO_2 Deformed at High Temperatures," J. Nucl. Mater. 48, 317-329 (1973).

4.   A. H. Heuer, R. M. Cannon and N. J. Tighe, "Plastic Deformation in Fine-Grain Ceramics," Published in Ultrafine-Grain Ceramics, 1970 (Syracuse Univ. Press, Syracuse, N.Y.   USA).

5.   P. E. Bohaboy and R. R. Asamoto, "Compressive Creep Characteristics of Stoichiometric Uranium Dioxide," GEAP-10054 (May 1969).

6.   M. S. Seltzer, et al., "Review of Out-of-Pile and In-Pile Creep of Ceramic Nuclear Fuels," BMI-1906 (July 1971).

7.   P. E. Bohaboy and S. K. Evans, "Compressive Creep Properties of UO_2-PuO_2 Compounds," in Plutonium 1970 and Other Actinides, Part I, Ed., W. N. Miner, TMS-AIME, N.Y. (1970) p. 478.

8.   J. L. Routbort, N. A. Javed and J. C. Voglewede, "Compressive Creep of Mixed-oxide Fuel Pellets," J. Nucl. Mater. 44(3), 247-259 (Sept 1972).

9.   J. L. Routbort and J. C. Voglewede, "Creep of Mixed-oxide Fuel Pellets at High Stress," J. Am. Ceram. Soc. 56(6), 330-333 (June 1973).

10. J. T. A. Roberts and B. J. Wrona, "Deformation and Fracture of UO_2-20 wt % PuO_2," Argonne National Laboratory, ANL-7945 (June 1972).

11. R. F. Canon, J. T. A. Roberts and R. J. Beals, "Deformation of UO_2 at High Temperatures," J. Am. Ceram. Soc. 54(2), 105 (1971).

12. S. Timoshenko and D. H. Young, "Elements of Strength of Materials," 5th ed., D. VanNostrand Co., Inc., Princeton, N.J. (1968).

13. J. T. A. Roberts, "Mechanical Equation of State and High-temperature Deformation ($\geq 0.5\ T_m$) of Uranium Dioxide," Acta Met., in press.

14. R. B. Nicholson, "The Role of Metallographic Techniques in the Understanding and Use of Superplasticity," Intl. Mater. Symp., University of California, Berkeley (1971).

DEFORMATION OF OXIDE REFRACTORIES

Dr. Louis J. Trostel, Jr.

Norton Company

Worcester, Massachusetts 01606

INTRODUCTION

The majority of the published research work on the deformation of refractory oxides has been concerned with work on single crystal or uniformly fine grain bodies. Many have been on structures of near theoretical density.

However, the deformation of polycrystalline multiphase refractories is important in the performance at high temperature of structures such as roofs, walls and checkers, and when used as supporting members such as batts and saggers in kiln furniture.

Resistance to deformation of these refractories has been evaluated in the U.S. in hot load tests such as the ASTM test C16-70 "Load Test for Refractory Brick at High Temperatures" where the refractory is subjected to a load of 25 psi for 90 minutes at an elevated temperature. The percent deformation from before to after test is measured. In Great Britain and Europe a more common test is the "Refractoriness Under Load" test where the refractory is heated under load at a specified rate until 0.5% contraction occurs. This yeilds an RUL temperature.

Neither of these relatively fast tests duplicate use conditions nor do they allow time for the typical slow deformations to occur in the refractory bodies. For these reasons, other research work has been carried out on these commercial refractories using longer times under load. This work has proved useful in indicating mechanisms of slow deformation at high temperatures, types of refractories with good and bad resistance to hot deformation, and in

339

establishing engineering design parameters for the construction of refractory structures.

It is to these long-time under load investigations of the deformation of refractories that this paper is addressed. Fireclay, magnesia, and high alumina refractories are discussed in particular.

FIRECLAY REFRACTORIES

As a group, the fireclay refractories represent the largest tonnage of any sold. Chemically they are essentially Al_2O_3/SiO_2 based.

Based on naturally occurring minerals, fireclay refractories have quite complex structures. After firing they include major quantities of quartz, cristobalite, mullite and glass. The amounts of each of these phases change significantly in use. Softening under load may occur as low as $1100^{\circ}C$; however, they may still carry their own weight as high as $1650^{\circ}C$.

Eusner & Schaefer[1] attacked the problem of determining long time deformation properties of fireclay refractories evaluating 35 brands of Superduty and 52 brands of High Duty fireclay refractories. They adopted a 50 hour, 25 psi, $1380^{\circ}C$ test which better indicated the hot load deformation resistance of these refractories than did the short 90 minute hot load test. The longer time gave a chance for significant viscous flow to occur in the glassy phases and for crystallographic changes to occur which would better indicate service performance.

Results were presented as linear plots of deformation with time as shown in Figure 1. The curves with continuously decreasing slopes of widely differing values make analyses and comparisons difficult.

However, presenting the data on a log plot indicates the curves fall into two families as shown in Figure 2. The data can be fitted to the formula

$$E = AT^B \tag{1}$$

E = creep
T = time
B = slope, time exponent
A = indicates the relative creep between refractories
    with similar slopes (B).

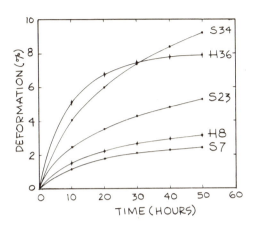

Figure 1. Hot load deformation of 5 fireclay refractories at 1380°C. "S" indicates Super Duty, "H" indicates High duty, the number indicates brand identification (1).

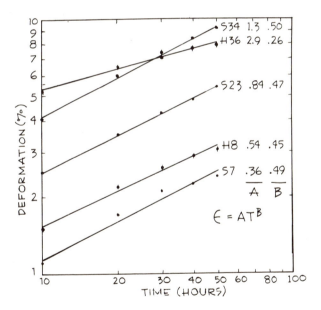

Figure 2. Logarithmic plot of creep of same 5 refractories in Figure 1 showing curves of two families of different slopes, B, of approximately 0.5 and 0.25.

One large family of these refractories has a typical slope (B) approaching 0.5 or the creep varies as the square root of time. The other family has a (B) time exponent of about 0.25 or the fourth root of time.

This indicates basic differences in mineralogical processes occurring in the two families.

Konopicky[2] used a similar approach for analyzing a number of fireclay refractories at temperatures from 1100° to 1400°C. From this type of data, characteristics such as activation energies can be calculated. These in turn can help indicate the type of process occurring in the refractories deformation.

Both McGee[3] & Astbury[4] report activation energies of about 100 Kcal/mole for the creep of fireclay refractories. This is close to the activation energy for viscous flow of fused silica.

Hulse & Pask[5] show decreasing strain rates with time for fireclay refractories of the 64% SiO_2 20% Al_2O_3 type in tests from 1000° to 1230°C. They suggest this represents changes in phase structure. In this work the refractories had been preheated to develop the mullite and glass contents. An activation energy of about 170 Kcal/mole was measured in the creep test, similar to that found for the creep of mullite rather than glass.

Similar tests on specimens not heat treated and thus with poor mullite development, gave an activation energy of only about 134 Kcal/mole.

MAGNESIA REFRACTORIES

Studies of high purity (+90% MgO) periclase refractories have shown the research on single crystal and dense fine grain very high purity magnesia specimens cannot explain all the deformation characteristics of these multi-grainsize, polyphase refractories.

The relatively small amount of second phase material surrounding the periclase grains in these refractories has been found to be of paramount importance in controlling the deformation properties of the refractories.

Fireclay refractories exhibit reasonable load bearing properties with glass contents as high as 50% because of the highly viscous character of silica-alumina glasses. In contrast, glass contents as low as 15% with magnesia refractories will produce failure under hot load conditions because of the much lower viscosity of the magnesium and calcium silicate glasses[3].

If the second phase is crystalline rather than glassy, the hot deformation characteristics of the refractory will reflect the properties such as melting point of the secondary phase. CaO (lime) and SiO_2 (silica) impurities in the periclase refractories occur almost exclusively in the secondary surrounding phases and not in the periclase grains. When present in a 2:1 ratio, they combine to form dicalcium silicate with a high melting point of $2150^\circ C$ making a very refractory second phase with good hot deformation resistance. In achieving good creep resistance at high temperature, the lime to silica ratio of approximately 2:1 has been found to be most important.

A variety of magnesia refractories were studied by Busby & Carter[6]. These ranged from 90 to 97% MgO with lime to silica ratios from 1:3 to 3:1, fired to a low and a high temperature (1550 & $1680^\circ C$). Creep testing was done in tension in this case at a load of only 7 psi at temperatures from 1300° to $1450^\circ C$.

The variation in the lime to silica ratio was tied to the variations found in the creep rates. The refractories with 2:1 lime to silica ratio gave lowest creep rates and were found to have dicalcium silicate present. Those with lower lime to silica ratios had monticellite and merwinite present. These minerals have melting points of $1490^\circ C$ and $1580^\circ C$ respectively, much lower than dicalcium silicate ($2150^\circ C$). Therefore, the more rapid deformation as test temperature approached their melting points is to be expected.

The highest fired refractories in all cases in this work, resulted in improved creep resistance due to increased solid to solid bonding between periclase grains.

Lower creep rates were obtained with the refractories with higher MgO contents and hence lower quantities of secondary phases present.

The same ordering of deformation results was obtained with these refractories in torsion creep and modulus of rupture tests.

Compressive creep was measured by Van Dreser[7] on various periclase refractories between 90 and 99+% MgO at $1600^\circ C$, under 25 psi load, for up to 72 hours. Some failed in shear during the test.

A sharp difference in creep character is seen in Figure 3 in those refractories whose lime to silica ratio was over 1.7:1 and those whose ratio was less than 1.7:1. Those over 1.7 all had lower creep deformation than similar MgO content refractories with lime to silica ratios of less than 1.7:1. In fact, the failures during the 72 hour test were all low lime to silica ratio refractories. Failure was probably due to excessive monticellite to sustain the load at $1600^\circ C$.

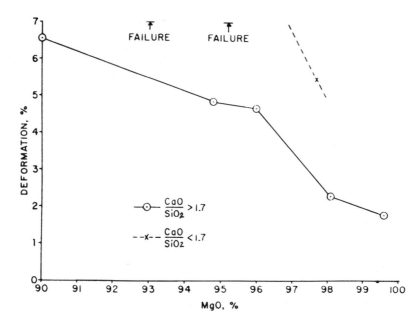

Figure 3.   Creep of periclase refractories at 1600°C showing
            effect of different lime to silica ratios (7).

Kroglo & Smothers[8] reported results on the compressive creep
at 1200° to 1500°C of magnesia refractories of 98% MgO with a lime
to silica ratio of 2:1. In the constant stress study at 25 and
50 psi they found two types of deformation.

From 1200° to 1430°C the strain rate varied as the fourth
power of the stress indicating plastic flow by dislocation movement
with an activation energy of 62 Kcal/mole. After creep testing the
specimens showed an increase in bulk density and modulus of
elasticity.

Creep testing at 1500°C; however, resulted in a 10% decrease
in density and shearing failure of the specimens. Failure occurred
through the fine grain matrix.

In a recent study, Shaffer[9] investigated 10 magnesia refractories subjecting them to constant stress compressive creep tests
at 25 to 200 psi from 1200° to 1500°C. These refractories ranged
from 92 to 97% MgO with lime to silica ratios of from 0.7 to 3.7:1.

In addition to the lime to silica ratio, he also considered
the amount of B_2O_3 present in total and in relation to the amount
of lime plus silica present.

The results of these creep tests indicate the stress exponent (n) in the strain to stress relationship (2) ranged from 1 to 2 for nine of the refractories and was almost five for the tenth.

$$\dot{\epsilon} = a\sigma^n \qquad (2)$$

The activation energies ranged from 85 to 135 Kcal/mole for the same nine refractories and was 170 Kcal/mole for the different tenth refractory.

The tenth refractory, with the markedly higher creep rates, was the one with a higher B_2O_3 content than allowed by the lime-silica factor (Eq. 3) reported by Davis & Havrank[10]. The excess B_2O_3 will flux the silicates increasing the quantity of rapidly deforming glass.

All creep failures were similar, failing in shear at 45° through the fines matrix as intergranular failure rather than transgranular fracture.

Shaffer proposed a pseudoplastic, viscous flow process which would account for the variations in stress exponents and activation energies and maintain the single mode of deformation.

$$B_2O_3 \text{ Max} = \frac{(\text{Lime + Silica})^2}{100} \qquad (3)$$

HIGH ALUMINA REFRACTORIES

In examining the deformation properties of high alumina re-fractories (over 50% Al_2O_3), the alumina-silica phase diagram is useful in considering the deformation characteristics of these refractories.

In refractories of lower Al_2O_3 content than mullite (72% by weight), the stable phases are cristobalite and mullite below about 1600^oC. Above 1600^oC a certain amount of glass will form and be in equilibrium with mullite. At 72% Al_2O_3, the composition of mullite, and higher, the situation abruptly changes and mullite or mullite plus corundum are equilibrium phases present at tempera-tures up to 1840^oC, 240^oC higher than Al_2O_3 compositions below mullite. This 240^oC change in glass forming temperatures in the alumina-silica system at 72% Al_2O_3 results in some quite abrupt changes between deformation characteristics of the various dif-ferent refractories spanning this composition range.

Clements & Vyse[11] studied a range of commercial high alumina refractories ranging from 55 to 96% Al_2O_3 under a compressive load of 28 psi at 1400° to 1600^oC for times of 24 to 96 hours. Deforma-

tions ranged from practically 0 up to 10%. Specimens were 2" D.
2" high cylinders. Analyzing the results of this work proved quite
complex but the major factor controlling creep was the structure
of these refractories. The structures were both complex and di-
verse. The refractories' structures included fused or sintered
alumina, bauxite, synthetic mullite, calcined kyanite, sillimanite,
andalusite, calcined clay grog, silica and a variety of bonding
materials.

One frequently used method of classifying this group of refractories is by their Al_2O_3 content. Clements & Vyse found no
significant correlation between creep rate and Al_2O_3 content as is
shown in Figure 4. Rather, the creep rate depended upon the form
in which the alumina was present, that is, the mineralogical
structure of the refractories.

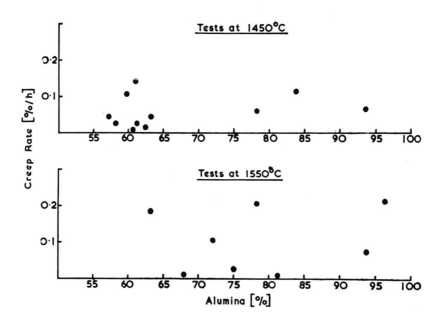

Figure 4.   Creep rate of refractories of several alumina contents
            indicating lack of correlation between creep rate and
            percent Al_2O_3 (11).

Refractories with the best resistance to creep had large grains consisting predominantly of fused mullite or of sillimanite with corundum grains that had reacted with the silicious bond to from mullite. The bond developed between the grains was mullite.

Poorest creep resistance was found with refractories whose large grain portion consisted of porous granules such as softer calcined alumina or bauxite and those refractories with open cracks in the fine grain matrix.

Kreglo & Smothers[12] investigated 60, 70 and 80% Al_2O_3 refractories under compressive loading of 25 psi at 1320° to $1430^\circ C$ for times up to 400 hours. Deformation ranged from 1% to 9%. Specimens were 4-1/2" x 2-1/4" x 1-1/4".

Although they did not characterize the structures of the refractories as completely as Clements & Vyse, X-ray phase identifications before and after the creep testing showed identifiable phases of cristobalite, mullite and alpha alumina present in all three refractories before testing. The cristobalite had disappeared by the end of the creep test; the amount of alpha alumina had been reduced, and the amount of mullite increased. These changes indicate lack of phase equilibrium in these refractories during the test.

The deformation versus time curves, Figure 5, show a decreasing rate of creep approaching a constant creep rate with extended time. The strain was directly related to stress indicating a viscous flow type of behavior.

Activation energy calculations emphasized the complex and changing nature of these three structures. The 60% Al_2O_3 refractory had a single activation energy of about 80 Kcal/mole over the entire temperature range of these tests, 1320° to $1430^\circ C$. The 70% and 80% Al_2O_3 refractories each showed two characteristic activation energies of 150 and 160 Kcal/mole at the lower temperatures and 85 and 100 Kcal/mole at the higher test temperature. This indicates two mechanisms of creep are present operating at different temperatures in each of these two refractories.

The differences these structure-temperature-time characteristics make are shown in Figure 5. At $1320^\circ C$, the 60% Al_2O_3 refractory had the most deformation of the three refractories but in tests run at $1430^\circ C$, the 60% Al_2O_3 refractory showed the least deformation.

In a study of a number of commercial mullite refractories (75% Al_2O_3) by Smith[13] specimens containing coarse grains of fused mullite with about 25% porosity in the refractory shape were heated at $1460^\circ C$ under 100 psi loads for times up to 30 hours. This fused

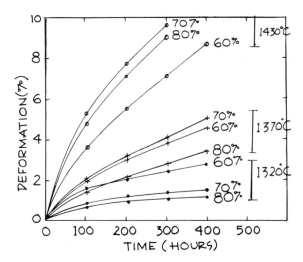

Figure 5. Hot deformation of 60, 70 and 80% Al_2O_3 refractories at
           increasing test temperatures from 1320° to 1430°C. Note
           the change in relative performance of the 60% Al_2O_3 re-
           fractory at various temperatures (12).

grain mullite refractory also decreased in creep rate with time,
Figure 6. The mullite refractory fired higher during manufacture
had a lower creep rate. This is due to better development of
mullite in the alumina-silica fine grain matrix.

        Work by the author characterized the deformation of high
alumina refractories of from 75 to 99.5% Al_2O_3 with coarse grain
portions of fused and highly calcined alumina and fused mullite
with alumina or alumina-silica fine grain matrices.

        These were dense refractories of 75 to 80% of theoretical
density with a gradation of grains of sizes from 1 μm to 1500 μm
(12 mesh).

        In one study[14] three refractories were examined of 75%, 90%
and 99% Al_2O_3. The 75% Al_2O_3 refractory consisted of coarse fused
mullite with an aluminum silicate fine matrix. The 90% Al_2O_3 refractory was about 1/3 mullite and 2/3 corundum with fused mullite,
calcined alumina and a fine grain alumina matrix. The 99% Al_2O_3
body was coarse fused alumina with a fine alumina matrix. Deformation tests were run at temperatures of from 1200° to 1400°C for

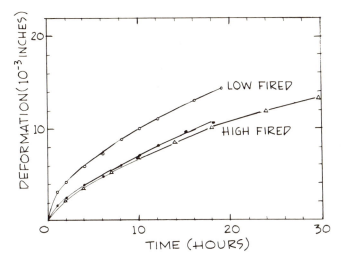

Figure 6.   Effect of high and low firing of fused mullite refrac-
            tories on hot deformation (13).

times up to 140 hours. Bar specimens were 2" wide by 3/4" thick.
Using an 8" span specimens were loaded as high as 100 psi. This
testing gave measures of the crossbending creep resistance.

     Creep rates for these 75 to 99% Al_2O_3 refractories were found
to decrease with time in all cases. The creep rate was a similar
function of time for the 90% Al_2O_3 body for all three loads and
quite similar for the 75% and 99% bodies as is shown by the similar
slopes in Figure 7.

     The importance of the mineralogical structure of the refrac-
tories is reemphasized when the total deformation after 140 hours
at temperature are summarized in Table I. At the lowest temperature, 1200°C, little or no deformation took place. At the intermediate temperature, 1300°C, little deformation took place at the
lower loads with some increase in deformation with higher loads.
At 1400°C, major differences were seen in deformation properties.

     At this highest test temperature, the 99% Al_2O_3 refractory
failed at progressively shorter times with increasing loads. The
90% Al_2O_3 refractory showed good creep resistance to loads as high
as 100 psi without failure. The 75% Al_2O_3 refractory showed very
marked differences in creep resistance depending upon whether it

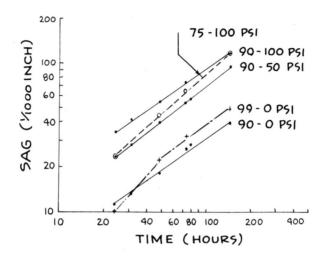

Figure 7. Logarithmic plot of deformation with time for 75% to
           99% Al$_2$O$_3$ refractories showing similar time dependency
           slopes for all cases (14).

was high or low fired in manufacture. The low-fired body showed
increasing creep deformation up to 100 psi loading with actual amounts
being the highest of the refractories studied. The high fired body
showed the least deformation of any of the bodies studied and sur-
vived the most severe test, 140 hours with an applied load of 200 psi.

   The dense coarse grain body with the well developed mullite
matrix shows excellent deformation resistance again, this time in a
crossbending test putting one outer surface of the specimen in
tension.

   Compressive creep tests have been run on 99.5% Al$_2$O$_3$ refractories at temperatures from 1200° to 1450°C at loads of from 10 to
400 psi for times up to 2300 hours[15]. Specimens were 1" D. cylinders, 4" high cut and ground from larger refractory blocks. These
refractories structures consisted of coarse fused alumina grains as
large as 1500 μm (12 mesh) with a fine alumina matrix as small as
1 μm. Porosity was about 24%.

   At the lowest test temperature, 1200°C, little or no creep
occurs at loads as high as 400 psi for times up to 2300 hours.
At 1300° to 1450°C, continuing creep occurs, Figure 8, with decreasing creep rates with time until just before failure where
enhanced creep occurs.

TABLE I

Deformation in Crossbend Test in 140 Hours

Refractory (% Al_2O_3)	Load (psi)	Deformation (10^{-3} inch) in Test at 1200°C	1300°C	1400°C
99	0*	-	0	50
	50	-	-	Fail 35 hours
	100	-	10	Fail 10 hours
	200	-	-	Fail 1 hour
90	0	0	0	25
	50	0	0	40
	100	4	20	90
	200	5	100	Fail 25 hours
75	0	-	0, 0**	30, 30
	50	-	-	30, 150
	100	-	0, 10	90, 170
	200	-	-	100, Fail 10 hours

*0 load indicates no load was added to the specimen. Any deformation is due to the bar specimen's own weight suspended across the 8" span.

**The 75% Al_2O_3 refractories were fired to two temperatures in manufacture. The first number is the deformation of the high fired refractory. The second is for the low fired refractory.

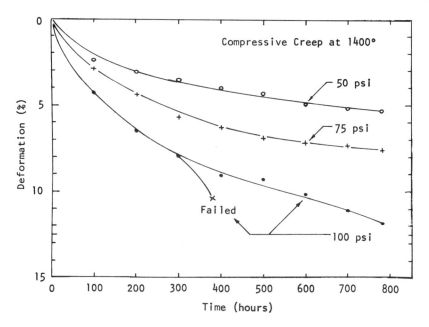

Figure 8.   Continuing compressive creep of 99.5% Al$_2$O$_3$ at 1400°C.
Creep rate decreases with time until just before
failure (15).

It was found at all temperature-load combinations yielding
continuous creep, the strain was directly proportional to the
stress.

The log plot of the deformation vs. time, Figures 9A and B,
indicates straight lines of similar scope for the temperature and
load conditions giving continuous creep indicating a general square
root time dependency for the creep rates.

By comparing the creep rates at common amounts of strain at
different temperatures, an activation energy of 184 Kcal/mole was
calculated.

Structural changes in the refractories were observed. A bowing
of the compressed specimens and shearing of the failed specimens
were found, Figure 10. Bulk densities of the specimens compressed
as much as 8% were unchanged or decreased indicating the barreling
out of the specimens compensated for the lengthwise shortening.
Examining the ground surfaces after testing, Figure 11, shows a
protruding of the coarse grains for the high deformation specimens.
Specimens cut in half axially show intergranular cracks parallel
to the direction of applied load in the highly deformed specimens.

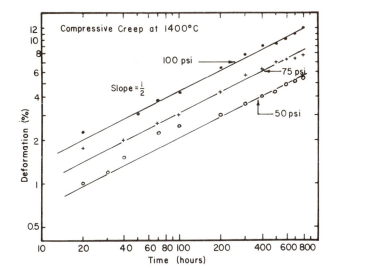

A

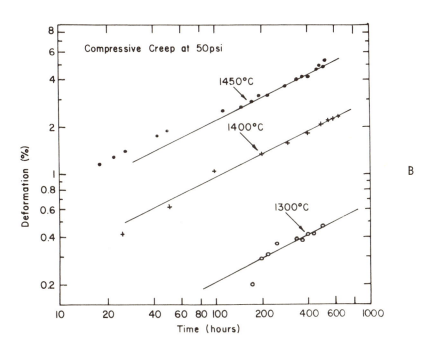

B

Figures 9A & B.   Similar dependency of deformation with time
                  (slope = 0.5) for dense 99.5% Al_2O_3 refractories
                  for loads from 50 to 100 psi (9A) and test temper-
                  atures from 1300° to 1450°C (9B) (15).

Figure 10. High alumina (99.5% Al_2O_3) specimens after test show
            differences from specimen on left which underwent
            little deformation; next one that crept 5% with a 1%
            decrease in bulk density, one that failed after 2%
            deformation and on the right one that failed after 8%
            deformation (15).

     The changes indicate the creep deformation in these high
alumina bodies is occurring as grain boundry sliding, yielding
through the fine grain matrix.

     SEM photographs were taken of a recent test's 99.5% Al_2O_3
creep specimens after 500 hours at 1400°C. Figures 12 and 13 are
of the ground surface of a specimen after testing which had deformed 3% without failing. Figure 12 (50X) shows a rough coarse
grain structure with fines sintering them together. Little coarse
grain pushout and no cracking is evident. Figure 13 (1000X) of
the fine grain matrix shows no cracking.

     The structure of the ground surface of a specimen that failed
after about 3% deformation is shown in Figures 14, 15 and 16.

     The coarse grains appear somewhat offset and cracks along
their boundries with the fine grain matrix appear in Figure 14
(50X). The presence of cracks is clearly shown in Figure 15
(1000X) taken in the obviously cracked central area of Figure 14.

Figure 11. Appearance of ground cylindrical surfaces after test.
On the left, a specimen that underwent 1/2% deformation
2000 hours. On the right, a specimen that underwent
8% deformation in 800 hours. Note protruding coarse
grains (15).

Further cracks in the fine grain matrix are suggested in Figure 16
(1000X) in what was apprently an uncracked fine grain area at the
bottom of Figure 14 viewed at lower magnification.

These photomicrographs help confirm the creep deformation and
failure in these 99.5% Al_2O_3 refractories is occurring through the
fine grain matrix.

SUMMARY

The mineralogical structure of multiphase polycrystalline refractories is extremely critical in determining their resistance to
deformation at elevated temperature under load for extended times.

In alumina-silica refractories, a dense alumina or mullite
coarse grain portion with a well bonded mullite fine grain matrix
has best creep resistance up to 1600°C.

The lime to silica ratio and the amount of B_2O_3 in the fine
grain or glassy phase surrounding the dense periclase grains are

Figure 12.   Ground surface of 99.5% Al$_2$O$_3$ specimen after 3% defor-
            mation in 500 hour 1400°C test. Little coarse grain
            pushout and no cracking evident in fine grains.
            Original magnification of 4'' x 5'' photo 50X.

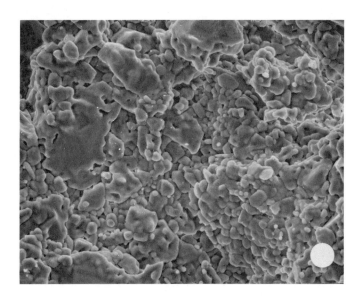

Figure 13.   Fine grain matrix of same surface as Figure 12 showing
            no evidence of cracking. White circle = approximately
            10μm. Original magnification 1000X.

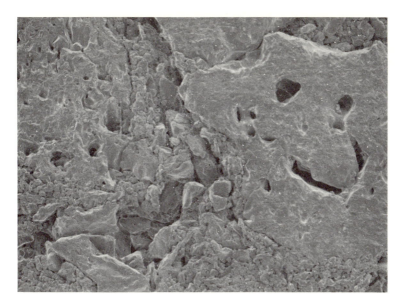

Figure 14. Ground surface of 99.5% Al_2O_3 specimen that failed
            after 3% deformation. Coarse grains are offset and
            cracks appear at boundries with fine grains. Original
            magnification 50X.

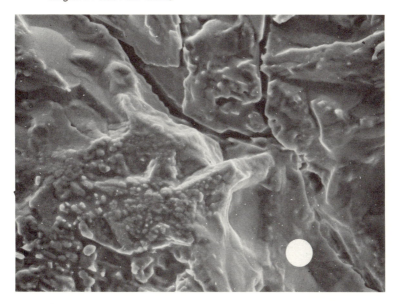

Figure 15. Cracking clearly in evidence in structure in enlarge-
            ment of cracked central section of Figure 14. White
            circle = approximately 10 μm. Original magnification
            1000X.

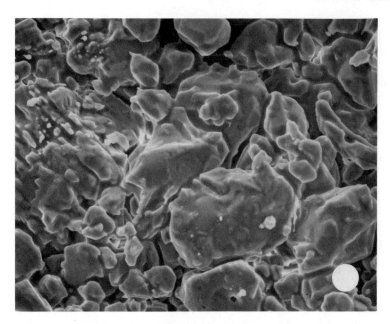

Figure 16. Enlargement of fine grain structure at bottom of
            Figure 14. Suggestion of cracks appear at this
          magnification in structure which looked uncracked at
          lower magnification. White circle = approximately
          10μm. Original magnification 1000X.

major controlling influences in the deformation characteristics of
magnesia refractories. A 2:1 ratio giving dicalcium silicate
appears optimum to reduce creep.

     With coarse portion of dense grains, creep deformation occurs
through the fine grain or glassy portion of all these refractories.

     A decreasing creep rate with time is characteristic of these
refractories due to the changing mineralogical structure of the
fine grain matrix.

     Over extended periods, the creep rate may approach a constant
figure as the structure stabilizes.

     Activation energies and stress to strain relationships can be
derived from creep tests on these refractories and help define the
processes occurring.

REFERENCES

1. G. R. Eusner & W. H. Schaefer, Jr., "Fifty Hour Load Test
    for Measuring the Refractoriness of Super-Duty and High-Duty
    Fireclay Brick", Am. Ceram. Soc. Bul., 35, 265-270, 1956.

2. K. Konopicky, Forschungsinstitut der Feuerfest-Industrie, 1965,
    private communication.

3. T. D. McGee, "Load Bearing Characteristics of Ceramic Refrac-
    tories", J. of Metals, 22-26, Dec. 1967.

4. N. F. Astbury, "Deformation and Fracture", Trans. British
    Ceram. Soc., 68, 1-7, 1969.

5. C. O. Hulse & J. A. Pask, "Analysis of Deformation of a Fire-
    clay Refractory", J. Am. Ceram. Soc., 49, 312-318, 1966.

6. T. S. Busby & M. Carter, "The Effect of Firing Temperatures
    and Compositions on the Creep Properties of Magnesite Bricks",
    Trans. Britich Ceram. Soc., 68, 205-210, 1969.

7. M. L. Van Dreser, "Developments with High Purity Periclase",
    Am. Ceram. Soc. Bul., 46, 196-201, 1967.

8. J. R. Kreglo, Jr. & W. J. Smothers, "Mechanism of Creep in
    MgO Brick", J. Am. Ceram. Soc., 50, 457-460, 1967.

9. G. E. Shaffer, "Compressive Creep of Magnesia Refractories",
    Thesis, Pennsylvania State University, 1973.

10. B. Davies & P. H. Havranck, Harbison-Walker Refractories
    Company, "Refractory", U.S. Patent 3,275,461, 1966.

11. J. F. Clements & J. Vyse, "Creep Measurements on Some High-
    Alumina Refractories", Trans. British Ceram. Soc., 65, 59-85,
    1966.

12. J. R. Kreglo, Jr. & W. J. Smothers, "Creep Characteristics of
    Selected High-Alumina Brick", J. of Metals, 20-22, July 1967.

13. P. C. Smith, "Creep of Mullite Refractories", Thesis, University
    of Missouri-Rolla, 1971.

14. L. J. Trostel, Jr., "High Alumina (Over 75%) Kiln Furniture
    Composition, Properties, Applications", presented 73rd meeting
    Am. Ceram. Soc., Chicago, Illinois, April, 1971.

15. L. J. Trostel, Jr., "Compressive Creep of High Purity Aluminum
    Oxide Refractories", Am. Ceram. Soc. Bul., 48, 601-605, 1969.

NON-ELASTIC DEFORMATION OF POLYCRYSTALS WITH A LIQUID BOUNDARY PHASE

F. F. Lange

Westinghouse Research Laboratories
Pittsburgh, Pennsylvania   15235

ABSTRACT

A conceptual view is taken to understand the deformation
behavior of a polycrystal with a liquid (or a quasi-elastic)
boundary phase. The analysis is based on the theories of liquid
adhesives, the fracture of liquids and the concepts of fracture
mechanics. It is shown that boundary separation rather than
boundary sliding is the step that controls the deformation rate.
Using this principal result, it is shown that the deformation
behavior of a polycrystalline material with a viscous boundary
phase is controlled by the flow characteristics, the volume content
of the boundary phase and the microstructure features of the polycrystal (viz., voids, solid inclusions and cracks). Polycrystalline
materials with a viscous boundary phase will exhibit a much greater
rate of deformation in tension than in compression.

INTRODUCTION

Grain boundary sliding first comes to mind in relating a
viscous boundary phase with deformation. Because individual grain
pairs are constrained by surrounding grains, grain boundary sliding
is restricted unless either the grains themselves can accommodate
the deformation produced by the sliding or the boundaries can
develop cavities (viz., voids or cracks). Present concepts relating
boundary sliding to the deformation behavior of polycrystalline
materials allow some accommodation to take place by either elastic
or plastic deformation within the grains.[1,2] As pointed out by
Raj and Ashby,[3] elastic deformation does not result in sufficient
accommodation to allow boundary sliding. Thus, if the formation of

cavities are to be prevented, the grains must exhibit plastic deformation by dislocation and/or diffusional processes.

Grain boundary cavities are frequently found in materials that have been forced to exhibit large deformations prior to fracture. Mechanisms and theories have been proposed to account for the formation and the growth of these cavities,[1,2] but no account has been taken for either the presence or the properties of a grain boundary phase. This is understandable since the models and the theories have been proposed to explain boundary cavitation in metals which were presumed not to contain a grain boundary phase.

In the present theory, the expected deformation behavior of a polycrystalline material containing a viscous grain boundary phase will be examined with respect to the volume fraction and the properties of the boundary phase. <u>It is assumed that the grains do not exhibit any plastic deformation</u>. Most of the analytical treatment will be concerned with the behavior of grains separated with a Newtonian liquid, but the general case of quasi-elastic boundary phases will also be discussed. The analysis of this material model is based on the theories of liquid adhesives, the fracture of liquids and the concepts of fracture mechanics. It will be shown that boundary separation rather than boundary sliding is the step that controls the deformation rate. Based on this principal result, it will be shown that the deformation behavior of a poly-crystalline material with a viscous boundary phase is controlled by the flow characteristic and the volume content of the boundary phase, microstructural features of the polycrystal (viz., voids, solid inclusions and cracks), and the mode in which the stress is applied (e.g., tension vs compression).

BOUNDARY SLIDING vs BOUNDARY SEPARATION

The model used to analyze the deformation behavior of a poly-crystalline material with a viscous boundary phase is shown in Fig. 1. It consists of cube-shaped grains, which are assumed to exhibit neither plastic deformation nor rotation, separated by a Newtonian liquid of thickness, s_0. Other features as, for example, the presence of vapor bubbles within the boundary phase, will be brought into the model at the appropriate point.

As discussed in the introduction, deformation by grain boundary sliding cannot be accommodated with elastic grains unless boundary separation occurs. For a polycrystal with a liquid boundary phase, boundary separation requires either the flow of liquid to occupy the increased volume between the separating grains or the growth of a vapor bubble. In either case, boundary separation requires a resolved tensile stress across the grains. Boundary sliding requires resolved shear stresses.

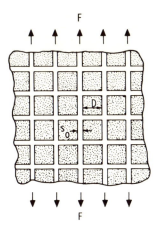

Fig. 1

The resistance of grain pairs to sliding and separation can be
examined separately as shown in Fig. 2 using two grains with cube
edges D separated by a Newtonian liquid with a viscosity η acted
upon by a force F. Figure 2a represents the case for boundary
sliding, i.e., the sliding of each grain with respect to the
surrounding grains. For this case, the liquid between the grains
is missing so that the volume increase between the separating
grains is not restricted by either the flow of liquid or the growth
of a vapor bubble. Figure 2b represents the case for boundary
separation, where the only liquid present is between the boundaries
to be separated.* In both cases, the strain in the direction of
force (F) is $\varepsilon \equiv (s - s_0)/(D + s_0)$ where s_0 is the initial separation
distance and s is the separation between the grain pair after a
period, t.

The classical equation that defines the rate of sliding (ds/dt)
of one surface of area A separated by a distance so from another
by a Newtonian liquid is [4]

$$\frac{ds}{dt} = \frac{s_0 F}{\eta A} .$$ (1)

*The condition of the limited amount of liquid between grain pairs
 is more representative of a polycrystalline material as discussed
 in the next section. The grain pair in Fig. 2a can also be
 surrounded by a large reservoir of liquid. Expressions [9]
 developed for this condition only differ from that given in Eq. (4)
 by a factor of 2.

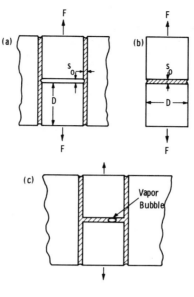

Fig. 2

Substituting $ds = (D + s_0)d\epsilon$ and the shear stress $\tau = F/A$, the strain rate $(\dot{\epsilon})$ for boundary sliding (Fig. 2a) is given by

$$\dot{\epsilon} = \frac{s_0}{(D + s_0)} \frac{\tau}{\eta} \qquad (2)$$

By integrating this expression and substituting[*] $s_0 = xD$, the time required to produce a certain strain in the direction of applied force is given by

$$t = \frac{(x + 1)\eta}{x\tau} \epsilon. \qquad (3)$$

   Solutions for the separation of boundaries with a sandwiched liquid (Fig. 2b) are found in the literature on adhesion. [5,6,7] Neglecting the capillary forces by assuming the surface tension of the liquid is zero,[**] Healey's [8] solution is

$$\frac{ds}{dt} = \frac{2\pi s^5}{3\eta V^2} F. \qquad (4)$$

---

[*] The relation between s_0 and D depends on the volume fraction of the liquid phase, V_ℓ. As shown in Appendix A, $x \approx V_\ell/(3(1-V_\ell))$ for the case of cube-shaped grains.

[**] Surface tension can be brought into this relation, [5] but its effect is not necessary for the argument.

By substituting the tensile stress across the boundary $\sigma = F/D^2$, the expression for the liquid's volume, $V = D^2 s_0$, $ds = (D + s_0)d\varepsilon$, and $s_0 = xD$, an expression for the strain rate in the direction of force is obtained for an arbitrary separation distance (s):

$$\dot\varepsilon = \frac{2\pi\, x^3\, s^5}{3(x + 1)\eta\, s_0^{\,5}}\, \sigma \qquad (5)$$

By integrating the above expression, the time required to produce a certain strain in the direction of the applied force is given by

$$t = \frac{3\eta}{8\pi\, x^2\, \sigma}\, [1 - \frac{1}{((\frac{1}{x} + 1)\,\varepsilon + 1)^4}] \qquad (6)$$

which can be approximated by

$$t = \frac{3\eta\,(x + 1)}{8\pi\, x^3\, \sigma}\, [\frac{\varepsilon}{(\frac{1}{x} + 1)\,\varepsilon + .25}] \qquad (7)$$

when $\varepsilon < 0.1$.

By comparing Eqs. (3) and (7), it can be shown that when $x \ll 1$ (the case where the volume fraction of the liquid phase is small) and when $\varepsilon \leq .1$, a much longer period is required to produce the same strain by boundary separation than by boundary sliding. Similar strains are only produced by both phenomenon in the same period when either the volume fraction of the liquid phase is large or after a long period of time when the grains become separated by large distances. Thus, when the two phenomenons are allowed to operate concurrently as shown in Fig. 2(c), it can be concluded that the deformation rate of the grain pair will be limited to the rate in which the boundaries can separate due to the applied tensile stress across the boundary. That is, boundary sliding is an incidental phenomenon whereas boundary separation is the rate limiting step.*

The deformation rate of the grain pair shown in Fig. 2c is therefore defined by Eq. (5). Similar equations are reported in Appendix B for the case where the boundary phase is either a non-Newtonian liquid or a Bingham solid.

---

* This conclusion is quickly reached by the casual experiment of separating two glass plates which are sandwiched together with either water or grease. It is easier to slide the plates than to pull them apart.

NEED FOR VAPOR-LIQUID SURFACES

The free flow of liquid between the boundaries is necessary for separation. The model used to examine boundary separation (Fig. 2b) assumed that the volume of liquid between the boundaries remained constant. As the boundaries separated, the vapor-liquid interface moved inward to account for the increased volume between the boundaries. A free flow of liquid would also occur if the grain pair were immersed in a large reservoir of liquid. In this case, the liquid in the reservoir moves to occupy the increased volume between the separating boundaries. The strain rate equation for this case differs from Eq. (5) by a factor of 2. [9]

Boundaries that are located on either external or internal surfaces of a polycrystalline body would resemble the case shown in Fig. 2b, i.e., the flow of liquid between the boundaries is unrestricted due to the movement of the vapor-liquid interface. Separating boundaries that are remote from a free surface would have to either contain a small vapor bubble (which would grow during separation) or borrow liquid from the surrounding boundaries.

The borrowing of liquid or the free flow of liquid from one boundary to another requires that some boundaries approach one another while others separate. Boundary approach requires compressive stresses. Although compressive stresses can arise across some boundaries in a polycrystalline material placed in tension, Eq. (5) illustrates that the rate of boundary approach ($s/s_0 < 1$) due to a compressive stress ($-\sigma$) is much smaller than the rate of boundary separation ($s/s_0 > 1$) due to a tensile stress (σ). Thus, for the ideal case where all boundaries are initially separated by a distance s_0 (Fig. 1), liquid is unlikely to flow from boundary to boundary. This restrictive flow of liquid requires that boundaries remotely located from free surfaces have a vapor bubble which can grow and allow the free flow of liquid between the separating boundaries. As discussed in the next section, the criterion for the growth of vapor bubbles depends on the applied tensile stress, the size of the bubble and the surface energy of the liquid.

CRITICAL STRESS FOR THE GROWTH OF VAPOR BUBBLES

A confined liquid subjected to a negative pressure (i.e., a tensile stress) is metastable; it will change to a two phase, liquid plus vapor, system. The vapor state takes the form of bubbles that grow until the liquid fractures. The growth of vapor bubbles was analyzed by Fisher [10] in a manner similar to that introduced by Griffith [11] for analyzing the fracture of solids. Fisher showed that for a given tensile stress, only vapor bubbles greater

than a critical size will grow. The relation between the critical
bubble diameter (d_c) and the applied tensile stress (σ) is given by

$$d_c = \frac{4\gamma}{\sigma} , \tag{8}$$

where γ is the vapor-liquid surface energy. Bubbles with diameters
less than d_c require free energy for further growth, whereas those
with larger diameters grow spontaneously with a decreasing free
energy. By rearranging this expression, it can be seen that a
bubble of diameter d requires a stress $\geq$ than a critical stress
(σ_c):

$$\sigma_c \geq \frac{4\gamma}{d} . \tag{9}$$

The growth of smaller bubbles requires larger stresses. When
the criterion for bubble growth is applied to the separation of
grain boundaries remote to a free surface, two conditions must be
satisfied. First, a vapor bubble must either pre-exist or nucleate
at the boundary. As pointed out by Fisher, [10] unless the liquid
has a zero contact angle with the solid, the pre-existence of a
vapor bubble is inevitable. Even for perfect wetting, a small
region of positive contact impurity is sufficient for the existence
of a vapor bubble. Lacking this, the vapor bubble must nucleate
due to a thermally activated process.

The second condition is that the tensile stress across the
boundary is $\geq \sigma_c$. The magnitude of the critical stress can be
estimated using Eq. (9) and a few assumptions. Assuming that
$\gamma = 0.35$ J/m^2 (a value for many silicate glasses [12] at 1200°C),
$d = s_0 \approx 0.02$ μm (for a grain size of 1 μm, and $V_\ell = 5\%$ -- see
Eq. (3a) in Appendix I), the smallest critical stress is
$\sigma_c = 60$ MN/m^2 (8700 psi). For this case, grain pairs remote to
a free surface will not begin to separate unless the applied
tensile stress is ≥ 60 MN/m^2. It should be noted that since $d \leq s_0$,
and s_0 depends on grain size for a given liquid content, a lower
critical stress is required to separate larger grain size materials.

EFFECT OF STRESS CONCENTRATORS

As discussed in a previous section, grain pairs that are
located on a surface are free to separate at any tensile stress.
When the surface is flat, the tensile stress across the properly
oriented boundary is equal to the applied tensile stress (σ_a). When
the surface has a curvature as shown in Fig. 3a, the tensile stress
across the boundary will be larger than the applied tensile stress
by a factor which depends on the position of the boundary along the
surface and the geometry of the surface. [13,14] This stress

concentration factor (k) is largest at the position where $\theta = 0$ as shown in Fig. 3a. At this position the tensile stress is

$$\sigma = k \, \sigma_a \qquad (10)$$

The factor k depends on the geometry of the surface, its position (i.e., external or internal surface) and the mode of loading. [14] This stress decreases with increasing distance from the surface into the body. When the dimension of the boundaries are much smaller than the dimensions that define the surface curvature, the tensile stress across the boundary can be approximated by Eq. (10). Thus, the strain rate exhibited by the grain pair at the location shown in Fig. 3a is

$$\dot{\varepsilon} = \frac{2\pi \, x^3 s^5}{3(x + 1)\eta \, s_0^5} \, k \, \sigma_a. \qquad (11)$$

A crack is also a free surface and a stress concentrator. The tensile stress distribution in the near vicinity of the crack front, along the symmetry plane for the mode of loading shown in Fig. 3b is [15]

$$\sigma = \frac{Y \, \sigma_a \, c^{1/2}}{(2r)^{1/2}}, \qquad (12)$$

where c is the crack length and r is the distance from the crack front into the material; Y is a dimensionless factor. Since this

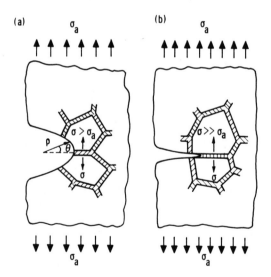

Fig. 3

expression exhibits a singularity* at $r = 0$, it presents a problem
in defining the average stress acting across the grain boundary
shown in Fig. 3b. For the purposes of this argument, it will be
assumed that the average stress is $\sigma = \sigma_a c^{1/2}/D^{1/2}$. Substituting
this expression into Eq. (5), the strain rate for a favorably
oriented grain pair at the crack front is

$$\dot{\epsilon} = \frac{2\pi x^3 s^5 Y}{3(x + 1)\eta s_o^5} \left(\frac{c}{D}\right)^{1/2} \sigma_a \qquad (13)$$

In summary, it can be seen that the deformation rate of a
grain pair depends on its location. The position of greatest tensile
stress is at a crack front; thus grain pairs located at a crack
front (or, as discussed in the next section, in the vicinity of
crack fronts) will exhibit the greatest deformation rate. Grain
pairs located along either surface notches or internal cavities**
will exhibit the next largest deformation rates. Next in order are
grain pairs on planar surfaces and those remote to any surface
(assuming $\sigma_a > \sigma_c$).

DEFORMATION ZONES

In the last section, the effect of stress concentrators was
only discussed in relation to grain pairs located on a free surface.
The effect of the stress distribution in the vicinity of these
concentrators, away from the free surface was neglected. In this
section, the expected zone of deformation associated with the
stress distribution around cracks and the expected change in geometry
of pre-existing, large cavities will be discussed.

Assume that all boundaries within the ideal polycrystal contain
vapor bubbles of size d_o. Whenever the tensile stress across these
boundaries is $\geq \sigma_c$ (see Section 4), boundary separation can take
place and these grain pairs can contribute to the total deformation
of the material. If the applied stress $\sigma_a < \sigma_c$, only those boundaries
around stress concentrators will separate. For the case of a crack,
only a certain volume of material in the vicinity of the crack front
will satisfy this condition. This volume can be defined by a limiting

---

* Equation (12) was derived assuming that the two surfaces meeting
  at the crack front to define a cusp. [15] As discussed in the
  next section, the radius of curvature at the crack front is finite
  due to the vapor-liquid surface at the grain boundaries. Thus,
  the stress distribution at the crack front is somewhat different
  than given by Eq. (12).
** Stress concentrators not associated with vapor-liquid surfaces,
   viz., solid, second phase inhomogeneous, will have a similar
   effect on grain pairs.

radius vector r_{lim} as shown in Fig. 4. An approximate equation for r_{lim} can be obtained from Eq. (12) but setting $\sigma = \sigma_c$; thus,

$$r_{lim} \overset{\sim}{-} (\frac{\sigma_a}{\sigma_c})^2 \quad c. \tag{14}$$

This equation shows that when $\sigma_a < \sigma_c$, only those boundaries within a cylindrical volume defined by r_{lim}, with an origin at the crack front, will separate. Boundaries closest to the crack front will exhibit the greatest separation rate. This is also true when $\sigma_a > \sigma_c$.

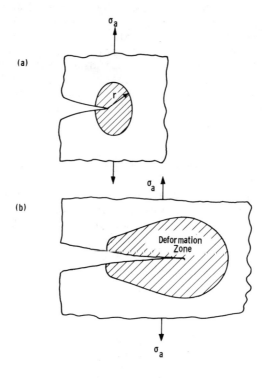

Fig. 4

It should be noted here that in a polycrystal with a liquid boundary phase, the crack front is not cusp-shaped as usually assumed, but it has a finite radius of curvature defined by the vapor liquid interface between the grain pairs at the crack front (see Fig. 3b). Thus, the stress concentration will be somewhat smaller than that usually reported for a sharp crack. As the grains at the crack front separate, the radius of curvature will increase as the vapor-liquid surface propagates to extend the crack's length. It is important to recognize that the curvature

cannot be increased by the deformation of adjacent grains as for
the case of dislocation motion in most metals, but that it remains
small due to limited amount of liquid between the separating grains.
Thus, as boundary separation occurs at the crack front, the crack
front extends without a significant change in the stress concentration despite the non-elastic deformation in the surrounding
material.

As the crack front grows due to successive boundary separation
at the crack front, the radius of the deformation zone will increase
linearly with c as shown in Fig. 4b. For the case where $\sigma_a > \sigma_c$,
all boundaries within the material will separate, but those within
the zone associated with the crack front will exhibit the greatest
rate of separation.

The initial deformation zone in the vicinity of other types
of stress concentrators will depend on their stress distribution.
The stress distributions for spherical and ellipsoidal cavities and
surface notches have been analyzed by Neuber. [14] The initial
deformation zone associated with these stress concentrators can
be defined in a similar manner used above for the case of a crack.
This will not be done here. The important observation concerning
these zones is that those boundaries at the position of highest
stress (viz., at the surface, where $\theta = 0$ in Fig. 3a) will exhibit
the greatest separation rate. Thus, after sufficient boundary
separation and sliding, a crack will form at the cavity and the
stress concentration will become greater.

DEFORMATION IN TENSION AND COMPRESSION

Boundaries that are perpendicular to applied compressive
forces will move closer together. The liquid between these
approaching boundaries must flow to other boundaries which must
separate by local tensile stresses. As pointed out in Section 3,
the rate of boundary approach ($s/s_0 < 1$) is much smaller than the
rate of boundary separation ($s/s_0 > 1$, see Eq. (5)). Thus, one
would expect a polycrystal with a liquid boundary phase to be more
resistant to deformation during compressive loading than during
tensile loading.

If boundary approach was the only phenomenon associated with
deformation during compressive loading, one might expect that once
all of the liquid between the grains was squeezed out, deformation
would stop. This is an unlikely occurrence because boundaries which
were separating to accommodate the 'squeezed-out' liquid would
continue to separate after the compressed boundaries stopped
approaching one another. Also, since the rate of boundary separation is much greater than boundary approach, the mechanics of

deformation, due to applied axial compression, can be assumed to be governed by the local tensile stresses which cause boundary separation.

As succinctly reviewed by Babel and Sines, [16] tensile stresses can arise within a body placed in axial (and biaxial) compression. These tensile stresses are located at inhomogeneties, viz., cavities, cracks and second phases. A properly oriented liquid boundary would also qualify as a location for tensile stresses during compressive loading. The tensile stress is largest at specific positions along the surface (or phase boundary) of these inhomogeneties. In general, much larger applied compressive stresses are required to produce the same local tensile stress than required by an applied tensile stress. Since tensile stresses are required for boundary separation, it is important to know the ratio of the applied compressional to the applied tensile stresses to produce the same local stress distribution at an inhomogeneity and thus, the same rate of boundary separation.

The ratio of applied compressional to tensile stresses to produce the same localized tensile stress distribution can be simply illustrated with a derivation borrowed from Babel and Sines. [16] The inhomogeneity used for this derivation is an elliptical hole shown in Fig. 5, which is either placed in compression $(\sigma_a)_{comp}$ or in tension $(\sigma_a)_{tens}$ by remote forces. The major axis of the ellipse is oriented such that the largest local tensile stress (σ_t) arises at the same position (shown in Fig. 5) on its surface for both cases of applied stress. The local tensile stress at this position is

$$
\sigma_t = \begin{cases}
(1 + 2\,\frac{a}{b})\,(\sigma_a)_{tens}, \text{ when } (\sigma_a)_{comp} = 0 & (15a) \\
\qquad\qquad\qquad \text{and} \\
\dfrac{(\frac{a}{b} + 1)^2}{4\,\frac{a}{b}}\,(\sigma_a)_{comp}, \text{ when } (\sigma_a)_{tens} = 0 & (15b)
\end{cases}
$$

The ratio, $R = (\sigma_a)_{comp}/(\sigma_a)_{tens}$ required to produce the same local tensile stress distribution is determined by equating Eqs. (15a) and (15b):

$$
R = \frac{4\,\frac{a}{b}\,(1 + 2\,\frac{a}{b})}{(1 + \frac{a}{b})^2} \ . \tag{16}
$$

For the case of a cylindrical cavity, $a = b$ and $R = 3$; for the case of a surface crack (or a liquid grain boundary), $a/b \rightarrow \infty$ and $R = 8$. Intermediate values of R are obtained for other a/b ratios. The position of the highest local tensile stress will change as the ellipse is rotated with respect to the directions of applied stresses.

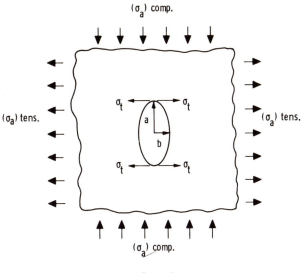

Fig. 5

Thus, it can be concluded that if boundary separation limits and governs the deformation behavior of a polycrystalline material, a much larger compressive stress is required (between 3 to 8 times) to produce the same deformation behavior that would be produced by an applied tensile stress.

DEFORMATION BEHAVIOR OF POLYCRYSTALS

Up to this section, the main concern has been the deformation behavior of individual and small groups of grain pairs located at specific positions throughout the polycrystalline material. Based on the theory evolved for these localized events, the collective behavior of a polycrystalline body will be discussed.

Simple Polycrystals

The simplist case to examine is the polycrystalline material consisting of uniform, cube-shaped grains which are initially separated from one another by a liquid phase of thickness s_0 (see Fig. 1). Three cases will be viewed:

(1) Each boundary contains a vapor bubble of diameter d_0.

(2) Occasional boundaries contain bubbles of diameter d_0. Other boundaries have either smaller diameter bubbles or do not contain bubbles.

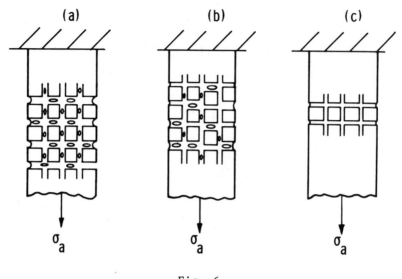

Fig. 6

(3) No vapor bubbles are present.

In each case, the long polycrystalline body is fixed at both ends
and placed in tension by external forces. Elastic deformation will
be neglected.

The first case is illustrated in Fig. 6a. When the applied
tensile stress $\sigma_a > \sigma_c$ ($\sigma_c = 4\gamma/d_0$, see Section 4), all the boundaries
perpendicular to the applied tensile stress will separate at a rate
given by Eq. (5). The strain in the direction of applied stress as
a function of time* is given by:

$$\epsilon \simeq \frac{2\pi \, x^3 \, t \, \sigma}{(x + 1) \, (3\eta - 8\pi \, x^2 \, t\sigma)} . \tag{17}$$

---

*It is assumed that the tensile stress across the boundaries is σ_a,
i.e., the presence of bubbles and their growth can be neglected in
determining the stress at the boundary. This assumption is only
valid when $d_0 \ll D$ and ϵ is small.

For this case, boundary sliding does not occur during deformation,
i.e., the body will separate into rows of grains shown in Fig. 6a.
The situation where $\sigma_a < \sigma_c$ will be examined as part of case 3.

For the second case, when $\sigma_a > \sigma_c$, only those boundaries
perpendicular to the applied tensile stress which contain vapor
bubbles of size d_o will separate. The deformation equation for this
case is the same as given for case 1 (Eq. (17)), but the mechanics
of deformation require that some boundaries slide as shown in
Fig. 6(b).

In the third case (or if $\sigma_a < \sigma_c$ in the previous cases) only
the boundaries at the surface, which have a vapor-liquid interface,
would be free to separate if it were not for the end constraints
which cause the strain throughout a cross-section to be uniform.
Because vapor-liquid interfaces are only present on the surface
of the body, the separation of the body's cross-section can only
occur by the movement of the vapor-liquid interface from its
surface to the interior as shown in Fig. 6c. This is the case of
the two grains shown in Fig. 2b. The volume of liquid between the
separating surfaces is $V = L^2 s_o$, where L is the width of
the body, $L = mD$, and m = the number of grains of dimension D.
Substituting these relations into Eq. (4), it can be shown that
the deformation also depends on m. Since m depends on the cross-
sectional size of the body, it can be shown that the deformation
behavior for this case depends on the size of the body. For very
large bodies, $m \to \infty$, and the body will not exhibit any non-elastic
deformation. The third case could be further extended to the
condition where only a few of the boundaries contain vapor bubbles.

Summarizing these three cases for the simplistic polycrystal
shown in Fig. 6, it can be seen that the deformation behavior
depends on several microstructural features and the flow characteristics of the liquid phase. The microstructural features are the
volume fraction of the liquid phase, which governs the initial
boundary separation thickness, s_o (see Appendix A), and the size of
vapor bubbles which governs the critical (threshold) stress below
which non-elastic deformation will not occur in a practical sense
for large bodies. It should be pointed out that the largest vapor
bubble size is related to grain size, i.e., $d_{largest} = s_o =$
$[V_{\ell}/(3(1-V_{\ell}))]D$. Thus, for equivalent materials except grain size,
the material with the smallest grain size will have the largest
threshold stress for non-elastic deformation.

The flow characteristics of the boundary phase that also govern
the deformation behavior depends on the type of phase present.
For the case of a Newtonian boundary phase, the viscosity is the
single parameter of interest; for non-Newtonian liquid, Eq. (17)
would be modified to include the two parameters, η and n (see

Appendix B). The effect of temperature on the deformation
behavior can be brought into Eq. (17) through these flow parameters.

Real Polycrystals

The microstructures of real polycrystals are much more
complex than that treated above. Within a polycrystalline material,
the grains have different geometries and sizes. The thickness of
the liquid boundary will vary from boundary to boundary. Other
microstructural features may be present, viz., voids, solid
inclusions and cracks. The effect of each of these on the deformation
behavior will be briefly discussed.

Different grain geometries will effect the mechanics of
deformation. For example, the tensile stress normal to each
boundary and the shear stress parallel to each boundary will be
different than that assumed above due to different boundary
orientations. Without rigorous proof, it might be expected that
the effect of grain morphology and size distribution would alter
some of the variables in the deformation equation (Eq. (17)), e.g.,
the relation between s_0 and D will be different from that given in
Appendix A, without changing its general form.

Different boundary separations within the same polycrystal
will have an effect during the initial period of deformation.
Boundaries widely separated by a liquid can act as reservoirs for
adjacent, narrow boundaries that are favorably oriented for
separation. The flow of liquid from thick to thin boundaries
precludes the need for an unstable vapor bubble. Because the thick
boundaries have a potential for separating at a fast rate if they
were not constrained by the thin boundaries, the applied tensile
stress will be unevenly distributed. That is, the thinner
boundaries will carry much of the tensile load, and initially, they
will separate at a much greater rate than that given by Eq. (17).
Once all the boundaries have reached an equilibrium separation
distance to produce a uniform tensile stress distribution, the
flow of liquid from boundary to boundary will stop as discussed in
Section 3. After this initial period, the deformation behavior
should be similar to that given by Eq. (17).

Microstructural features such as large voids, solid inclusions
and pre-existing cracks might be expected to result in the largest
deviation from the deformation behavior given by Eq. (17). These
heterogeneously distributed positions of high tensile stress will
result in zones of higher deformation than found in volume elements
remote from these positions. As discussed in Section 6, the voids
and solid inclusions will form internal cracks. These cracks,
together with pre-existing cracks, will slowly grow to change the
compliance of the body and thus, contribute to the total non-elastic
strain.

Figure 7 illustrates that cracks do grow during deformation and contribute to the total strain in materials that contain a viscous boundary phase. The 4-point flexural creep specimen shown in this figure was made from hot-pressed Si_3N_4, which is believed to contain a viscous boundary phase* at high temperatures. The experiment was conducted at 1400°C with a constant moment of 1.03 mN/m (corresponding to an initial stress of 105 MN/m^2). After a period of 6 hrs. in the range where the specimen exhibited a constant deformation rate, it was cooled. When the oxide surface layer was removed, many large cracks could be observed extending from the tensile to the compression surface. Small cracks which could not be photographed, were also present. Such large cracks would significantly effect the compliance of the specimen. The large separation distance between the crack surfaces at the tensile surface is evidence for a large deformation zone at the crack front.

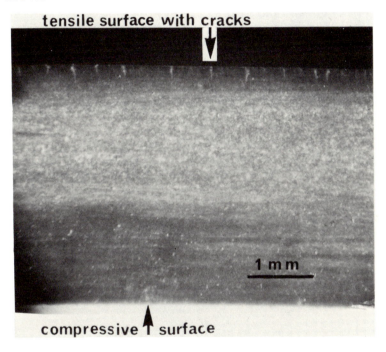

Fig. 7

---

*As discussed elsewhere, [17] impurities such as CaO are believed to strongly effect the viscosity of the boundary phase. The Si_3N_4 material used for this experiment was relatively impure and thus exhibited greater deformation than purer forms of Si_3N_4.
†Norton Co., Grade NC 132.

In summary, the deformation behavior of a polycrystalline material with a viscous boundary phase is governed by the rate that boundaries can separate. Boundary sliding is important to the mechanics of deformation, but it is not a limiting step. The important parameters of the boundary phase are its flow character- istics and its volume content. Important microstructural features of the polycrystal are (1) vapor bubbles within the boundary phase, which are necessary for boundary separation; (2) grain size, which governs the largest bubble size and thus, the threshold tensile stress below which deformation will not occur; and (3) inhomogenei- ties, viz., voids, solid inclusions and cracks, which result in deformation zones and the slow growth of cracks that change the compliance of the body and add to the total non-elastic strain. Polycrystalline materials with a viscous boundary phase will exhibit a much greater rate of deformation in tension than in compression. The different deformation behaviors in tension and compression is the best and most convenient way of differentiating grain separation due to a viscous boundary phase from other mechanisms (viz., diffusion and dislocation phenomenon) that can control deformation. It should also be noted that deformation by boundary separation is not a plastic phenomenon since volume is not conserved.

APPENDIX A: RELATION BETWEEN BOUNDARY THICKNESS, GRAIN SIZE AND VOLUME CONTENT OF LIQUID PHASE

If it is assumed that the polycrystalline body is composed of cube-shaped grains with an edge dimension, D, uniformly separated by a liquid phase of thickness s_o, the volume fraction of the liquid (V_ℓ) is given by

$$V_\ell = \frac{(D + s_o)^3 - D^3}{(D + s_o)^3} . \qquad (1a)$$

By letting $s_o = xD$, this expression can be rewritten as

$$V_\ell = \frac{3x + 3x^2 + x^3}{1 + 3x + 3x^2 + x^3} . \qquad (2a)$$

Since the liquid is considered a minor phase, viz., $x < 0.1$, higher order terms of x can be neglected resulting in

$$x = \frac{V_\ell}{3(1 - V_\ell)} . \qquad (3a)$$

APPENDIX B: DEFORMATION RATE OF GRAIN PAIRS SEPARATED BY
EITHER A NON-NEWTONIAN LIQUID OR A BINGHAM SOLID

The deformation rate ($\dot{\varepsilon}$) of a quasi-plastic phase is defined
by

$$\dot{\varepsilon} = \frac{(\tau - \tau_o)^n}{\eta} ,$$

where τ = an applied shear stress, τ_o = a yield shear stress,
η = viscosity and n = a positive numerical constant. When $n = 1$
and $\tau_o = 0$, the phase is a Newtonian liquid and when $n > 1$, $\tau_o = 0$,
a non-Newtonian liquid. A Bingham solid is defined by $n = 1$ and
$\tau_o > 0$.

The separation rate of parallel plates initially separated by
a quasi-plastic phase of thickness s_0 has been analyzed by Scott. [18]
He obtained solutions for the cases of a non-Newtonian liquid and a
Bingham solid. Using the definitions established in Section 2 of
the text, the rate of deformation of a grain pair containing either
of these phases is given by

(a) Non-Newtonian liquid separating grains:

$$\dot{\varepsilon} = \frac{(3n+1)^n \; \pi^{n+1/2} \; x^{n+2}}{(2n\eta)^n \; (n+2) \; (x+1)} \; (\frac{s}{s_o})^{5n+5/2} \; \sigma^n$$

(b) Bingham solid separating grains:

$$\dot{\varepsilon} = \frac{2\pi \; x^3}{3\eta \; (x+1)} \; (1 + 1/2 \; (\frac{s}{s_L})^{-15/2} \;) (\frac{s}{s_o})^5 \; \sigma$$

$$- \frac{2\pi^{1/2} \; x^2}{3\eta \; (x+1)} \; (\frac{s}{s_o})^{5/2} \; \tau_o$$

where

$$s_L = (\frac{4 \; s_o \; \tau_o^2}{4\pi \; x^2 \; \sigma^2})^{1/5}$$

Scott [18] also points out that boundaries separated with a Bingham
solid cannot be made to approach one another by any distance $< s_L$.

ACKNOWLEDGMENTS

Great thanks is due to Professor W. R. Bitler who patiently reviewed the concepts in this article. One of his criticisms resulted in a better conceptual view of the stress concentration at a crack as related to the finite radius of curvature of the vapor-liquid surface between the grain pairs at the crack front and the change of this curvature during crack growth (see Section 6). Thanks are also due to Dr. A. M. Diness who suggested my involvement in deformation.

REFERENCES

[1] Mechanical Beahvior of Materials at Elevated Temperatures, Ed. by J. E. Dorn, McGraw-Hill (1961).

[2] G. R. Terwilliger and K. C. Radford, "High Temperature Deformation of Ceramics: I, Background", Bul. Amer. Ceram. Soc. 53, 172 (1974).

[3] R. Raj and M. F. Ashby, "On Grain Boundary Sliding and Diffusional Creep", Trans. Met. Soc. AIME 2, 1113 (1971).

[4] For example, F. W. Sears and M. W. Zemansky, College Physics, Addison-Wesley, p. 254 (1952).

[5] N. A. DeBruyne, "The Measurement of the Strength of Adhesive and Cohesive Joints", Adhesion and Cohesion, Ed. by P. Weiss, Elsevier, p. 55 (1962).

[6] D. Tabor, "Friction and Adhesion Between Metals and Other Solids", Adhesion, Ed. by D. D. Eley, Oxford, p. 118 (1961).

[7] J. Hoekstra and C. P. Fritzius, "Rheology of Adhesives", Adhesion and Adhesives, Ed. by N. A. DeBruyne and R. Jouwink, Elsevier, p. 63 (1951).

[8] A. Healey, Trans. Inst. Rubber Ind. 1, 334 (1926).

[9] J. Stefan, Sitzungsberichte Kaiserl, Akad. Wiss. Wien, math. naturn. klasse, 69, 713 (1874).

[10] J. C. Fisher, "The Fracture of Liquids", J. Appl. Phys., 19, 1062 (1948).

[11] A. A. Griffith, "Phenomena of Rupture and Flow in Solids", Phil. Trans. Roy. Soc. London, Ser. A. 221, 163 (1920).

[12] G. W. Morey, "The Property of Glass", p. 197-211, Reinhold (1938).

[13] C. E. Inglis, "Stresses in a Plate due to the Presence of Cracks and Sharp Corners", Inst. Naval Archit. 55, 219 (1913).

[14] H. Neuber, Theory of Notch Stresses, 2nd Edition, Springer (1958). (English Trans. available from Office of Technical Service Dept., Commerce, Washington, D.C).

[15] I. N. Sneddon, "The Distribution of Stress in the Neighborhood of a Crack in an Elastic Solid", Proc. Roy. Soc. Lond. 187, 229 (1946).

[16] H. W. Babel and G. Sines, "A Biaxial Fracture Criterion for Porous Brittle Materials", Trans. ASME, J. Basic Eng., 285 (1968).

[17] F. F. Lange and J. L. Iskoe, "High Temperature Strength Behavior of Hot-Pressed Si_3N_4 and SiC: Effect of Impurities", Proc. Army Mat. Tech. Conf. on Ceramics for High Performance Applications (in press).

[18] J. R. Scott, "Theory and Application of the Parallel-Plate Plastimeter", Trans. Inst. Rubber Ind. 7, 169 (1931).

ANALYSIS OF THE TIME DEPENDENT FLEXURAL TEST

C. A. Andersson, D. P. Wei and Ram Kossowsky

Westinghouse Research Laboratories
Pittsburgh, Pennsylvania   15235

ABSTRACT

The flexural test was analyzed for the general case of a time dependent arbitrary law of deformation. This analysis relates the outer fiber stresses to measured values of moment and time. In lieu of knowledge of the exact relationship between the tensile and compressive stresses, two boundaries were established within which the analyses for most materials will fall. A test method to obtain the additional values required by the analysis is also given.

INTRODUCTION

The flexure test is a common tool used to determine strengths of ceramics. Its advantages lie in the simplicity of the specimen configuration and in the ease of testing. As long as the stress-strain function is linear and is the same for tension and compression, the well established elastic beam theory can be used to determine those stresses and strains. However, if the stress-strain function is non-linear and/or the tensile behavior differs from the compressive, the elastic treatment leads to erroneous results. In these cases, a more complex analysis must be used.

The theory for bending of bars having a time-independent arbitrary law of deformation was given by Nadai.[1] The present work develops a more general theory; i.e., that for the bending of bars having a time-dependent arbitrary law of deformation. This paper also suggests an alternate method of flexural testing required to obtain the data for the analysis. Finally, a comparison between the results for the various analyses is made.

TIME-DEPENDENT DEFORMATION

Time-dependent deformation can be represented by a surface schematically shown in Fig. 1 which is bordered by OANBCDE. Intersecting this surface with various planes results in the curves for the various tests that are conducted on materials. For instance, intersection with a plane of constant stress (GMBL) results in the creep curve ANB, GA being the instantaneous strain. Intersection with a plane of constant strain results in the stress-relaxation curve. A partial stress-relaxation curve (FC) in plane FJKCI is shown. The constant strain rate curve (OFB) is obtained by intersecting the surface with a plane such as OGBH, in which strain is linearly proportional to time.

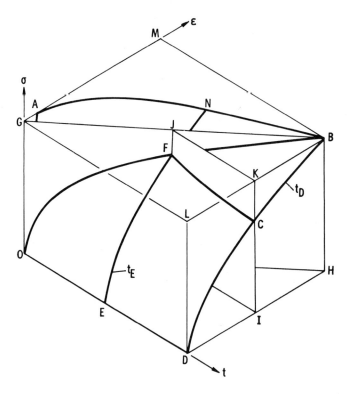

Fig. 1.   Schematic diagram of a typical stress, σ-strain, ε-time, t-surface

For the flexural test, one more curve must be considered. The instantaneous stress-strain curve within the specimen (from the neutral axis to the outer fiber) is described by the intersection of the stress-strain-time surface with a plane of constant time. The curve DCB is one such curve in the DLBH plane. Point B represents the stress and strain condition at the outer fiber, point D the condition at the neutral axis, and point C the condition at a fiber halfway between. The curve EF was similarly generated by a previous constant time plane.

The differences between the instantaneous stress-strain behavior and the behavior of the outside fiber is illustrated in Fig. 2, which is a projection of these curves onto the stress-strain plane. It is the instantaneous stress-strain curve that controls the instantaneous force and moment balances. The fact that they differ from the outer fiber curve disallows the time independent Nadai treatment from being used.

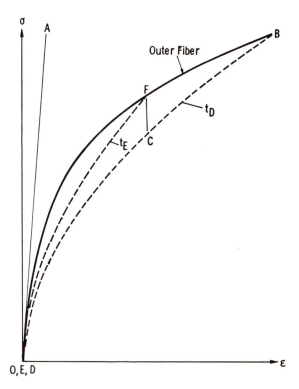

Fig. 2.   Schematic diagram of the stress-strain behavior showing
          the time dependent outer fiber curve, OFB, and two instan-
          taneous curves EF and DCB

ANALYSIS OF THE TIME-DEPENDENT FLEXURE TEST

The analysis of the time-dependent flexure test follows the procedures set down by Nadai [1] for the bending of bars with an arbitrary law of deformation. However, since his analysis is for the time-independent case, it is a special case of the more general one given here. The assumptions are the same: (a) the bar has a constant cross-section (rectangular, for simplicity); (b) it is loaded by forces directed perpendicular to its longitudinal axis; (c) the cross-sectional dimensions are small relative to the length in order to neglect shear stresses; (d) the rate of deflection is constant; and, finally, (e) the cross-sections remain plane during bending. The last assumption implies that the strains, ε, are linearly proportional to the distance, y, from the neutral axis:

$$\varepsilon = \frac{y}{R} \tag{1}$$

where R is the radius of curvature due to bending. The total strain, ε^*, across the height, h, of the specimen is therefore

$$\varepsilon^* = \frac{h}{R} = \varepsilon_t + \varepsilon_c \tag{2}$$

where ε_t is the maximum tensile strain at the outer tensile fiber and ε_c is the maximum compressive strain at the outer compressive fiber.

The equilibrium force equation is:

$$F = \int_{-h/2}^{h/2} \sigma(\varepsilon,t) \, dA_r = \frac{bh}{\varepsilon^*} \int_{\varepsilon_c}^{\varepsilon_t} \sigma(\varepsilon,t) \, d\varepsilon = 0 \tag{3}$$

where $\sigma(\varepsilon,t)$ is the stress, A_r is the area, and b is the specimen width. The applied moment, M, must be resisted by the specimen resulting in the equilibrium:

$$M = \int_{-h/2}^{h/2} \sigma(\varepsilon,t)y \, dA_r = \frac{bh^2}{\varepsilon^{*2}} \int_{\varepsilon_c}^{\varepsilon_t} \sigma(\varepsilon,t)\varepsilon \, d\varepsilon \tag{4}$$

For the four point flexure test in Fig. 3,

$$M = \frac{Pa}{2} \tag{5}$$

for all points within the inner span; P is the applied load, and a is the moment arm. Also, the strain, ε^*, is proportional to the applied deflection, D:

$$\varepsilon^* = cD = \frac{8h}{(\ell - 2a)^2} D \qquad (6)$$

where ℓ is the outer span and D is the center deflection.
Since the test is being conducted at a constant deflection rate, the strain rate is also constant, and

$$\varepsilon^* = kt \ . \qquad (7)$$

Differentiating Eq. (3) with respect to time, t, results in:

$$\frac{1}{bh} \frac{d(F\varepsilon^*)}{dt} = 0 = \gamma + \sigma_t \frac{d\varepsilon_t}{dt} - \sigma_c \frac{d\varepsilon_c}{dt} \qquad (8)$$

where σ_t is the stress at the outer tensile fiber, σ_c is the stress at the outer compressive fiber, and

$$\gamma = \int_{\varepsilon_c}^{\varepsilon_t} \frac{d\sigma(\varepsilon,t)}{dt} d\varepsilon \ . \qquad (9)$$

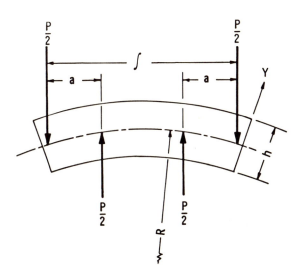

Fig. 3. Bending of a bar in four point flexure

Similarly, differentiating Eq. (4) with respect to time results in:

$$\frac{1}{bh^2} \frac{d(M\varepsilon^{*2})}{dt} = \lambda + \sigma_t \varepsilon_t \frac{d\varepsilon_t}{dt} + \sigma_c \varepsilon_c \frac{d\varepsilon_c}{dt} \tag{10}$$

where:

$$\lambda = \int_{\varepsilon_c}^{\varepsilon_t} \frac{d\sigma(\varepsilon,t)\varepsilon}{dt} \, d\varepsilon \; . \tag{11}$$

Also the differential of Eq. (2) is

$$\frac{d\varepsilon_t}{dt} + \frac{d\varepsilon_c}{dt} = \frac{d\varepsilon^*}{dt} \; . \tag{12}$$

By substitution, Eq. (10) can be written as:

$$\frac{1}{bh^2} \frac{d(M\varepsilon^2)}{dt} = \lambda + \frac{\sigma_t \sigma_c}{(\sigma_c + \sigma_t)} \varepsilon^* \frac{d\varepsilon^*}{dt} + \frac{(\sigma_c \varepsilon_c - \sigma_t \varepsilon_t)}{\sigma_c + \sigma_t} \gamma \; . \tag{13}$$

For the constant deflection rate test, the total strain rate $d\varepsilon^*/dt$ is also constant and Eq. (13) may be written as:

$$\frac{1}{bh^2} (2k\varepsilon^* M + \varepsilon^{*2} \frac{dM}{dt}) = \lambda + \frac{\sigma_t \sigma_c}{(\sigma_t + \sigma_c)} k\varepsilon^* + (\frac{\sigma_c \varepsilon_c - \sigma_t \varepsilon_t}{\sigma_t + \sigma_c}) \gamma \tag{14}$$

If deflection and therefore the total strain, ε^*, is held constant, then the stress relaxation version of Eq. (13) becomes

$$\frac{1}{bh^2} (\varepsilon^{*2} \frac{dM}{dt} \Big|_{\varepsilon^*}) = \lambda + (\frac{\sigma_c \varepsilon_c - \sigma_t \varepsilon_t}{\sigma_t + \sigma_c}) \gamma \; . \tag{15}$$

where $dM/dt \big|_{\varepsilon^*}$ is the rate of moment change due to stress relaxation. Subtracting Eq. (15) from Eq. (14) results in

$$\frac{\sigma_t \sigma_c}{\sigma_t + \sigma_c} = \frac{1}{bh^2} [2M + \frac{\varepsilon^*}{k} (\frac{dM}{dt} - \frac{dM}{dt} \Big|_{\varepsilon^*})] \; . \tag{16}$$

This is the general equation relating the outer fiber stresses to the moment and time. In the Nadai treatment dM/dt at constant strain is zero.

It is necessary to have another function relating σ_t and σ_c in order to obtain exact values. Without knowledge of this function, it is possible to establish boundaries within which most materials will

fall.   It can also be argued that material properties will usually fall
closer to one or the other boundaries, and furthermore it is easy to
establish which.   For materials which undergo true plastic deformation,
the tensile and compressive stress-strain curves will be equivalent,
i.e., σ_t will equal σ_c.   Therefore the first boundary is easily
established.   A large class of materials are highly resistant to
deformation in the compressive mode.   The second boundary occurs by
letting the compressive stress be time-independent linear elastic.
A simple compression test establishes which of the two conditions
hold and in the elastic case it also establishes the compressive
stress function, i.e., the modulus of elasticity.

The following are the methods used to obtain the boundaries.

Equivalent Time Dependent Arbitrary Laws of Deformation in Tension and Compression

In order to obtain the additional moment-time relationships
in Eq. (16), either additional stress-relaxation tests must be
performed or the usual constant deflection rate test must be altered.
One suggested procedure will now be outlined.   The specimen is
initially deflected at the chosen rate, k_d, to some small deflection
(point A on Fig. 4).   The test machine is stopped and the deflection
is held constant for some time increment, Δt (to Point B).
Deflection is then continued for the same time increment, Δt (to
point C), but at double the initial rate, $2 k_d$.   Repeating this
cycle to failure results in the moment-time curve in Fig. 5, the
points A, B and C being equivalent to those of Fig. 2.   The slope,
dM/dt, is then the tangent to the moment-time curve which would have
been generated if the test were not interrupted (curve OACEG).   The
slope $dM/dt \big|_{\varepsilon}*$ is the tangent to the constant deflection curve
where it intersects with the above curve.

When tensile and compressive stress-strain relationships are
the same, then Eq. (16) becomes:

$$\sigma_t = \sigma_c = 2 \phi = \frac{2}{bh^2} \left[2M + t \left(\frac{dM}{dt} - \frac{dM}{dt} \bigg|_{\varepsilon}* \right) \right] . \tag{17}$$

Also

$$\varepsilon_t = \varepsilon_c = \frac{\varepsilon^*}{2} = \frac{k}{2} t \tag{18}$$

where k is proportional to the original deflection rate.

C. A. ANDERSSON, D. P. WEI, AND R. KOSSOWSKY

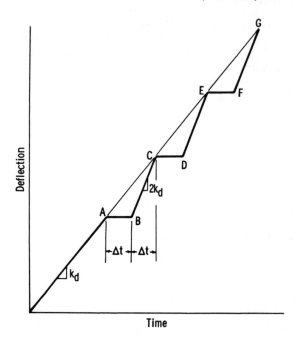

Fig. 4.   Deflection-time relationship for proposed time dependent
          flexure test

The corresponding stress-strain curves are shown in Fig. 6.
The dashed curves are the time-independent stress-strain curves which
represent the instantaneous condition within the specimen. Had the
specimen, initially at point A, been allowed to relax to point X and
then had it been instantaneously loaded, it would have followed
curve OXC.

Time Dependent Arbitrary Deformation Law in Tension and Elastic
Time Independent Deformation in Compression

The interrupted deflection-time test outlined in Fig. 4 results
in a moment-time curve similar to Fig. 5 for the non-linear tensile
and linear compressive case. The corresponding stress-strain curves
are shown in Fig. 7. During the relaxation process the compressive
strain, ε_c, decreases slightly while the tensile strain, ε_t, increases.

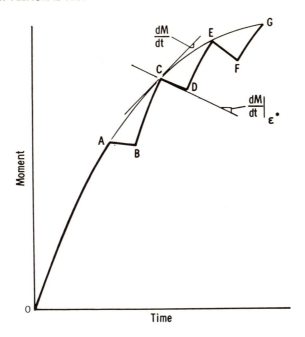

Fig. 5.   Moment-time relationship for the proposed time dependent
          flexure test

This is necessary to maintain both the force balance and the constant
total strain, ε^*. Once again, the dashed lines represent the
instantaneous stress-strain relationships within the specimen.

     The analysis of the present case is dependent on knowledge of
the rate of change of momentum with total strain at constant time.
The moment-strain curve equivalent to the moment-time curve of
Fig. 5 is shown in Fig. 8. The dashed lines represent the
instantaneous behavior.

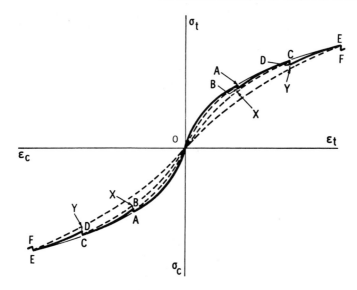

Fig. 6.   The stress-strain relationships for the proposed flexure
test.   Both tension and compression have the same time
dependent deformation behavior.

The differential of force with respect to total strain is:

$$\frac{1}{bh} \frac{d(F\varepsilon^{*})}{d\varepsilon^{*}}\bigg|_{t} - 0 = \int_{\varepsilon_{c}}^{\varepsilon_{t}} \frac{d\sigma(\varepsilon,t)}{d\varepsilon^{*}} d\varepsilon + \sigma_{L} \frac{d\varepsilon_{t}''}{d\varepsilon^{*}} - \sigma_{c} \frac{d\varepsilon_{c}''}{d\varepsilon^{*}} \tag{19}$$

where

$$\int_{\varepsilon_{c}}^{\varepsilon_{t}} \frac{d\sigma(\varepsilon,t)}{d\varepsilon^{*}} d\varepsilon = 0 \tag{20}$$

since σ is not a function of ε^{*}. The differential of the moment at
constant time with total strain is

$$\frac{1}{bh^{2}} \frac{d(M\varepsilon^{*2})}{d\varepsilon^{*}}\bigg|_{t} = \int_{\varepsilon_{c}}^{\varepsilon_{t}} \frac{d(\sigma(\varepsilon,t)\varepsilon)}{d\varepsilon^{*}} d\varepsilon + \sigma_{t} \varepsilon_{t} \frac{d\varepsilon_{t}''}{d\varepsilon^{*}} + \sigma_{c} \varepsilon_{c} \frac{d\varepsilon_{c}''}{d\varepsilon^{*}} \tag{21}$$

where

$$\int_{\varepsilon_{c}}^{\varepsilon_{t}} \frac{d(\sigma(\varepsilon,t)\varepsilon)}{d\varepsilon^{*}} d\varepsilon = 0 \ .$$

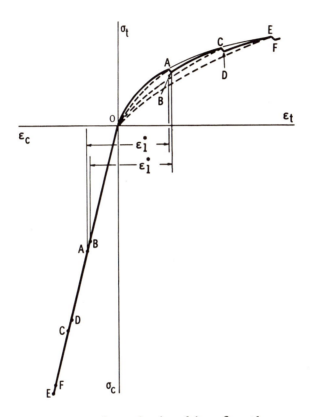

Fig. 7.   The stress-strain relationships for the proposed flexural
          test for time dependent tensile and elastic compressive
          deformation

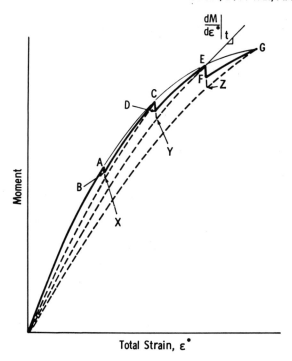

Fig. 8.   The moment-strain relationship for the time dependent
         flexural test

Solving as before

$$\frac{\sigma_t \sigma_c}{\sigma_t + \sigma_c} = \frac{1}{bh^2} \frac{1}{\varepsilon^*} \left. \frac{d(M\varepsilon^{*2})}{d\varepsilon^*} \right|_t = \phi \tag{22}$$

This is the same ϕ which was previously determined in Eq. (16).

The moment-deflection data can now be incrementally analyzed
under conditions of constant time by noting that for small changes:

$$\sigma_{t_i} \Delta\varepsilon_{t_i} = \sigma_{c_i} \Delta\varepsilon_{c_i} \tag{23}$$

where

$$\Delta\varepsilon_{t_i} = \varepsilon_{t_i} - \varepsilon_{t_{i-1}} \tag{24}$$

$$\Delta\varepsilon_{c_i} = \varepsilon_{c_i} - \varepsilon_{c_{i-1}} = \varepsilon_i^* - \varepsilon_{t_i} - \varepsilon_{c_{i-1}} \tag{25}$$

and

$$\sigma_{c_i} = \frac{\varepsilon_{c_i}}{A} \tag{26}$$

where A is the elastic compliance (the reciprical of the elastic modulus). By proper substitution into Eq. (22):

$$\varepsilon_{t_i}^2 + (\varepsilon_{c_{i-1}} - 2\,\varepsilon_i^*)\,\varepsilon_{t_i} + [A\,\phi\,\varepsilon_{t_{i-1}} + (\varepsilon_i^* - A\,\phi)$$

$$(\varepsilon_i^* - \varepsilon_{c_{i-1}})] = 0. \tag{27}$$

The quadratic equation can be solved for ε_{t_i}. By substitution, ε_{c_i}, σ_{c_i} and σ_{t_i} can in turn be determined.

EXAMPLES OF ANALYSIS

To demonstrate the differences in results from the three types of analyses (elastic, time-independent arbitrary deformation law, and time-dependent arbitrary deformation law), moment-time values were generated, and the outer fiber stress-strain relationships determined from these. The stress-strain-time surface used in the synthesis was determined from an equation of the form:

$$\varepsilon = A\sigma + B\,\sigma^n\,t^m \tag{28}$$

and the following parameters were chosen:

$$A = 2 \times 10^{-7}$$
$$B = 2.2 \times 10^{-16}$$
$$n = 3$$
$$m = 1/2$$
$$\text{strain rate} = .01$$

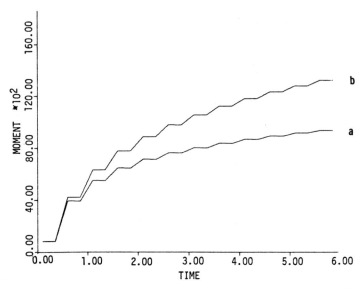

Fig. 9.   Moment-time curves for a) equivalent time-dependent ten-
          sile and compressive deformation, and b) time-dependent
          tensile and elastic compressive deformation

The computed moment-time curves for the two boundary
conditions are illustrated in Fig. 9. For clarity, not all the
calculated points are shown. Results of the tensile analyses for
equivalent tensile and compressive deformation are shown in Fig. 10.
For the given deformation equation and parameters, the elastic beam
analysis results in a maximum stress that is 25 percent too high.
The time independent analysis underestimates the stress by 7 percent.

Results of the tensile analyses for time dependent tensile and
elastic compressive deformation are shown in Fig. 11. In this case
the elastic beam theory not only results in a higher maximum stress,
78 percent, but the maximum strain is diminished by 26 percent.
The time independent analysis reduces the maximum stress by 7 percent.

CONCLUSIONS

The flexural test was analyzed for a time dependent arbitrary
law of deformation. The resulting equation which relates the outer
fiber stresses to the moment, deflection and time is the general
relationship. Both the elastic beam analysis and the time independent
arbitrary deformation law analysis are special cases of the time-
dependent one.

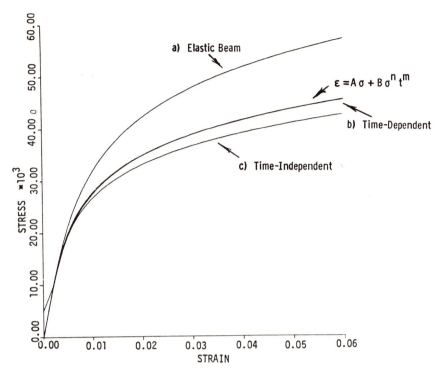

Fig. 10.   Outer fiber tensile stress-strain relationships for
equivalent tensile and compressive deformation as determined from: a) elastic beam theory, b) time-dependent
deformation theory, and c) time-independent deformation
theory

In order to complete the analysis, an additional equation
relating the tensile stress to the compressive stress must be
established. In lieu of this expression, however, two boundaries
can be established. Stresses for most materials will fall within
these boundaries, and generally closer to one or the other.

A test method is proposed to obtain the additional values
required by the analysis. The method consists of a combined constant
deformation rate and stress relaxation test.

The use of elastic beam analysis results in stress values which
are too high, while the use of the time-independent deformation law
analysis results in low stress values.

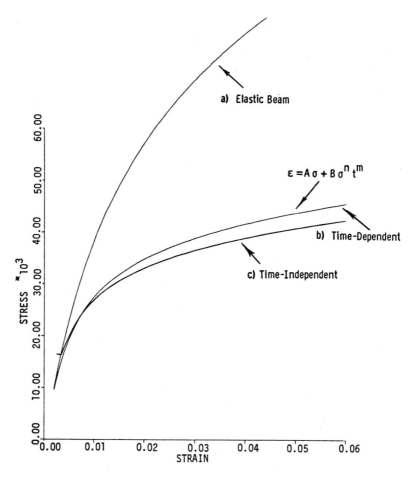

Fig. 11.   Outer fiber tensile stress-strain relationships for non-
linear tensile and linear compressive deformation as determined from: a) elastic beam theory, b) time-dependent
deformation theory, and c) time-independent deformation
theory

BIBLIOGRAPHY

1.   A. Nadai, "Theory of Flow and Fracture of Solids", McGraw-Hill
Book Co., New York, 1950, pp. 353-359.

PRELIMINARY CREEP STUDIES OF HOT-PRESSED SILICON NITRIDE

E. M. LENOE AND G. D. QUINN

Army Materials and Mechanics Research Center

Watertown, Massachusetts

ABSTRACT

Time-dependent response of silicon nitride has been studied by several investigators via flexure and tension experiments. This paper attempts to review the available literature concerning creep behavior of silicon nitride. Preliminary observations of the tension and torsion response of hot-pressed silicon nitride are also reported. The ceramic specimens were heated over their straight-line gage length portions by means of an oxyacetylene torch, allowing rapid heating and cooling, as well as minimizing deformation in the gripping fixtures.

Conventional contoured tension and torsion specimen configurations and mechanical extensometers were used under creep conditions. The majority of the tests consisted of single-step applications of stress. Tension response was explored for stresses ranging up to 3500 psi and the steady state creep rate appeared to be less than 4×10^{-5} inches per inch per hour. Torsion behavior was studied for stress levels up to 7400 psi. Shear deformation appeared to occur at fairly low stress levels and the initial or short-time creep strain rate was approximately 6.1×10^{-4} inches per inch per hour.

INTRODUCTION AND BACKGROUND

Silicon nitride has been largely a British development[1], started in about 1955.* Therefore, in discussing mechanical response

---

*Patent literature dealing with Si_3N_4 dates back to the 1890's.

of the nitrogen ceramics, it is appropriate to contrast current
results to these earlier achievements. Strength, fatigue, creep
and thermal shock behaviors of various types of reaction-sintered
and hot-pressed silicon nitride were described by Glenny and
Taylor[2]. To evaluate creep characteristics, four-point loaded
flexural experiments were performed on beams (0.11X0.11X3.25 inches)
and tension stress rupture data was obtained using tapered cylindrical specimens (0.153" diameter, 1.0" gage length). All specimens
were surface ground to finishes of 5 to 10μ and those containing
obvious defects were not tested. The creep performance of the
various types of silicon nitride had markedly different thermal-
dependent behavior. Reaction-sintered materials generally exhibited
poor creep resistance, although one of the low density grades
showed transient response similar to the early version of hot-
pressed silicon nitride. Type A (containing 5% silicon carbide
(SiC)) as well as Type C materials exhibited higher creep-resistance
at 1200°C as compared to 1000°C experiments. Formation of a glaze
at 1200°C was thought to strengthen the surface of the specimens.
Subsequent studies, in which the Type C ceramic was preheated for
100 hours at 1200°C, appeared to demonstrate enhanced creep resistance
at 1000°C, although the treated specimens did not perform as well
as the as-received ceramic tested at 1200°C.

While the creep strength of the porous grades at 1200°C was
higher than that at 1000°C, the dense hot-pressed silicon nitride
exhibited lower creep strength at 1200°C than at 1000°C. Creep
resistance was noted to vary with the purity of the starting silicon
powder, there being considerable reduction in creep as the silicon
purity was increased. For all the materials, scatter in data was
considerable, leading to difficulty in distinguishing stress level
influences. It was concluded that further developments in manufac-
turing technique were required. The deliberate impurity addition of
magnesium oxide to achieve high hot-pressed densities was thought to
contribute to excessive creep in the hot-pressed material. While
impurities were suggested as being the controlling influence, the
relative importance of various possible impurities were generally
unknown. During the course of the experiments, primary and secondary stages of creep were distinguished, although occasionally the
creep curves were parabolic in shape and it was not then possible
to delineate these as separate stages. Various flexural stress
levels were applied ranging from 4 to 32 ksi for the reaction-sintered and up to 52 ksi for the hot-pressed material. Due to scatter
in the data, only gross trends for increased creep at the higher
stresses could be stated with certainty. For the lower stress levels, occasionally there was an apparent reversal of behavior, no
doubt due to properties variability on a specimen-to-specimen basis.

During the early 1970's the Ford Motor Company undertook
development of reaction-sintered silicon nitride for potential use

in a ceramic gas turbine engine. For the past few years extensive
materials development efforts[3] have been directed toward turbine
components capable of operating at 2300°F under stresses of 10 ksi
for at least 200 hours while undergoing less than 0.5% creep. Five
grades of reaction-sintered silicon nitride have been evaluated[4].
The Ford samples were fabricated by injection molding of silicon
metal using an organic binder. The organic molding compound was
baked out of the rectangular preforms and nitrided using the time-
temperature cycles and nitriding atmospheres shown in Table I.
While little variation was observed in resulting densities, the use
of hydrogen additions in the nitriding cycle modified the microstructure and lowered the aluminum and calcium content. X-ray
diffraction analysis suggested all samples had been completely
nitrided and had achieved similar phase compositions of 65 weight
percent α silicon nitride and 35 weight percent β silicon nitride.

Typical impurity analysis of silicon nitride creep samples is
shown in Table II. Figure 1 compares typical flexural creep response for the Ford materials (Types A, B, and C) with the earlier
data[2].* It is of interest to note that the Type C Ford material
exhibited creep characteristics similar to the Type B ceramic of
Glenny and Taylor[2]. However, the Type C material is less dense
(2.35 g/cc versus 2.5) and sustained 10 ksi at 2300°F as compared
to 4 ksi at 1832°F. (Refer to curves numbered 1 and 2 in Figure 1.)
Apparently the increased purity with respect to Ca as well as a
finer grain size distribution are factors leading to superior creep
resistance of the Type C reaction-sintered silicon nitride.

Comparison of the creep behavior of reaction-sintered (RBSN) and
hot-pressed silicon nitride is made in Figures 2 and 3. The range
in improvement in creep resistance of the RBSN material is illustrated by the lower band of curves. Tension creep of the hot-pressed
silicon nitride is shown by the upper set of data in Figure 2, where
there is about a 1-1/2 to 2 orders of magnitude increase in steady
state creep rate. Two additional matters are illustrated by Figure
2. First, note that the creep strain determined via flexure is considerably less than that observed in tension. Secondly, note the
controlling influence of impurities. For instance, the difference
between samples identified as NC-130A and NC-130B corresponds to the
impurity levels shown in Table III, with NC-130B having generally
lower impurity content.

Reduction in Ca and other impurities significantly reduces the
creep rate. Also, as shown in Figure 3, the stress rupture life is

---

*Reference [2] did not state the load spans used; these were in-
ferred from the statement that 0.0235-inch midspan deflection corresponded to outer fiber tensile strains of 0.125% [2, page 167],
and by application of the beam deflection equations.

Table I. PROCESS HISTORY REACTION-SINTERED SILICON NITRIDE[4]

Material Type	Nitriding Cycle*	Atmosphere	Density (g/cc)	Average Flexure Strength† (ksi)
A	24/24	N_2	2.30	17
B	36/24	N_2	2.36	19.1
C	36/24	$1.8\%H+N_2$	2.35	24.6
D	36/24	$1.8\%H+N_2$	2.34	24.3

*Cycle is — hours at 2300°F and — hours at 2660°F.
†Measured in 4-point bending (1/8×1/8×1-1/2" specimen) with 1-1/8"
outer span and 3/8" inner span, crosshead speed at 0.10 in./min.

Table II. TYPICAL IMPURITY ANALYSIS OF SILICON
NITRIDE CREEP SAMPLES (WT%)[4]

Material	Fe	Al	Ca	Mg	Ni	W	Co	V
Type A	0.50–0.70	0.35–0.48	0.40	0.03	0.02	0.01	0.01	0.01
B	.66–.76	.38–.60	.02–.08	.02	.02	.01	.01	.01
C	.66–.76	.38–.60	.02–.08	.02	.02	.01	.01	.01
D	.60–.94	.10–.23	.01–.05	.02	.04	–	–	.03

Table III. CHEMISTRY OF HOT-PRESSED SILICON NITRIDE BILLETS

Ford-Westinghouse Data[3]
(Measured by Spectrographic Methods (±50%))

Material	Ca	Na	K	Al	Fe	Mg
			Impurities (Weight %)			
NC-130A	0.06–0.08	0.006–0.01	0.004–0.008	0.1–0.2	0.5	0.3–0.4
NC-130B	.03–.05	.004–.006	.004–.006	.2	.5	.3–.4

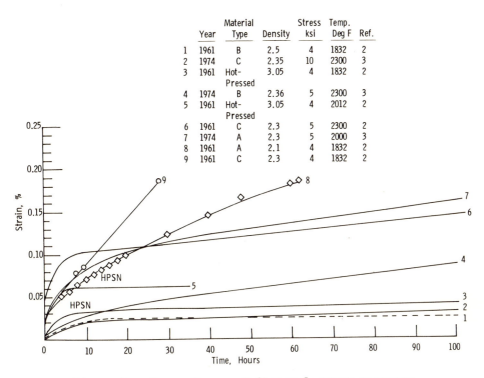

	Year	Material Type	Density	Stress ksi	Temp. Deg F	Ref.
1	1961	B	2.5	4	1832	2
2	1974	C	2.35	10	2300	3
3	1961	Hot-Pressed	3.05	4	1832	2
4	1974	B	2.36	5	2300	3
5	1961	Hot-Pressed	3.05	4	2012	2
6	1961	C	2.3	5	2300	2
7	1974	A	2.3	5	2000	3
8	1961	A	2.1	4	1832	2
9	1961	C	2.3	4	1832	2

Figure 1.   Comparisons of flexural creep response

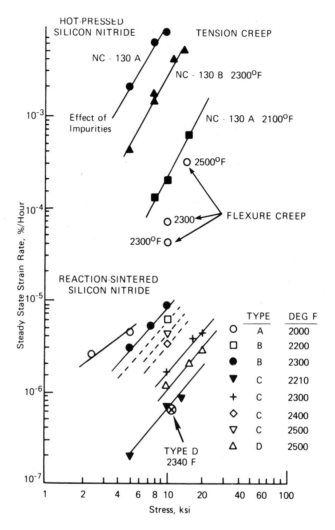

Figure 2.   Comparison of steady state creep response

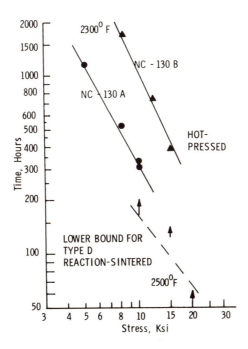

Figure 3.   Stress rupture response of silicon nitride

improved in the higher purity material. A lower bound on the stress
rupture behavior of Type D RBSN is shown for 2500°F. The dashed
curve indicates that flexure creep results have been attained such
that failures were not observed within the time durations at the
indicated stress levels. On the basis of current information, it
can be concluded that the general creep properties of the reaction-
sintered forms are superior to those of the hot-pressed silicon
nitride.

As for basic creep mechanisms, in the case of hot-pressed silicon nitride, grain boundary sliding presumably predominates and
creep rate has been expressed by the equation[5],

$$\partial\varepsilon(t) = A\sigma^n \exp^{\Delta H/RT}$$

where the activation energy has been reported as

$$\Delta H = 150\pm5 \text{ Kcal/mole}$$

for stresses of 10 ksi and temperatures ranging from 2300°F to
2500°F for the 0.04 weight percent Ca material and the exponent
n=2.0, denoting grain boundary sliding.

While dislocations have been identified in silicon nitride,
their densities were very low and there was no detectable change in
dislocation density as influenced by creep deformation. It has been
estimated that the contribution of dislocation motion to creep
strain is a few orders of magnitude less than the observed transient
response.

According to the presumption that the grains cannot accommodate
the deformation, grain boundary sliding leads to the initiation
and steady state growth of triple-point grain boundary wedges which
dominate the second stage of creep. Therefore the viscosity of the
glass phases at the grain boundary controls strain rates.

The grain boundary phase in low calcium content hot-pressed
silicon nitride is estimated to contain about 12 mole % of Ca and
good agreement has been obtained between the activation energies
for creep and the activation energy for viscous flow of Ca-doped
silicate. This agreement, as well as the predominating influence
of Ca on creep rate, suggests that the creep behavior of silicon
nitride is indeed controlled by the properties of the boundary
glass phase.

As far as basic creep mechanisms are concerned, the marked effect of atmosphere on creep rupture is of interest. Total elongation in helium rarely exceeds 1%, while 2.5 to 3.0% failure strain
is typical of the hot-pressed material in air.

The compression creep behavior of silicon nitride has also been studied[6] and the apparent creep activation energy at 10 ksi from measurements at two consecutive temperatures is about 190 Kcal/mole.

Finally, activation energies ΔH of 50 to 55 Kcal/mole and exponent n=1.6 have been calculated from flexure creep experiments on Ford Type D reaction-sintered silicon nitride for 10 ksi stress levels over temperatures of 2200 to 2500°F.

SHORT TERM TENSION AND TORSION RESPONSE

Available creep data on these ceramics is predominated by flexure experiments, although limited tension and compression studies have been completed. Torsion response remains virtually undetermined. For this reason, short time, rapid heating and cooling tests were of interest. The procedures used in our creep investigation are briefly reviewed next.

EXPERIMENTAL CONSIDERATIONS

Figure 4 illustrates the specimen geometries. The gage length portions of the specimens were polished with an oil-based diamond grit slurry for six to seven hours using a simple lapping machine designed to polish parallel to the longitudinal axis. Final surface finish was of the order of 10 to 15 microinches, as determined on an optical comparator. The specimens were markedly tapered to minimize deformations outside the gage section. Figure 5 is a schematic of the tension creep apparatus. The heating technique is shown in Figure 6, an oxyacetylene torch, using O_2 at 5.5 psig and MAPP gas at 5.2 psi gage pressure. The flame is first ignited and then directed over the specimen gage length. Times to achieve a stable surface temperature distribution appear to be less than 10 seconds. Typical temperature distributions, measured with an optical pyrometer, are presented in Figure 7. These tests were conducted with no shielding and it is evident that severe thermal gradients exist, although the temperature distribution is fairly uniform over a substantial portion of the mid-section gage length. Furthermore, reproducibility from test-to-test is approximately ±25°F. Better uniformity results at 2500°F than at lower temperatures. Figure 8 indicates the precision of the temperature measurements. Errors occur due to sharp thermal gradients, edge effects and nonperfect black body conditions.

Thermal deformations during heating prior to tension creep testing are shown in Figure 9, where the observed elongation versus time at temperature is presented. The apparatus and specimen are allowed to reach thermal equilibrium. Judging by the thermal

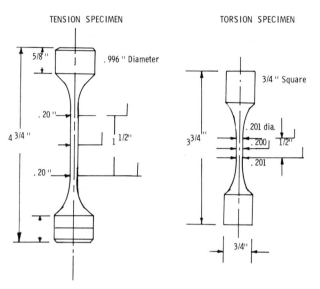

Figure 4.   Typical specimens

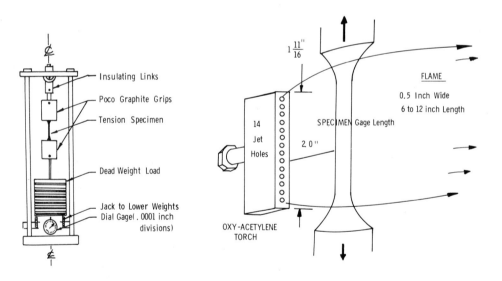

Figure 5.   Schematic of tension creep apparatus

Figure 6.   Specimen heating technique

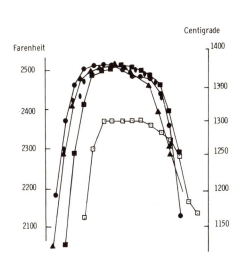

Figure 7. Typical temperature distributions

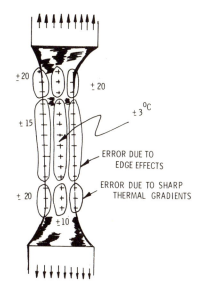

Figure 8. Estimated precision of temperature measurements via optical pyrometers (1380°C)

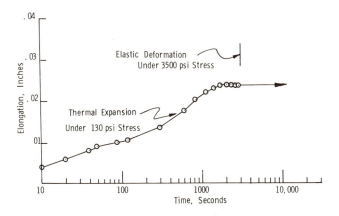

Figure 9. Initial deformation response

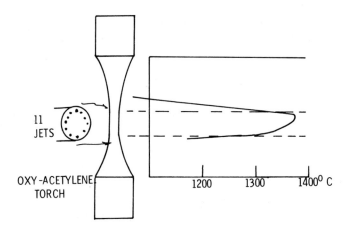

Figure 10.   Temperature distribution in torsion specimen

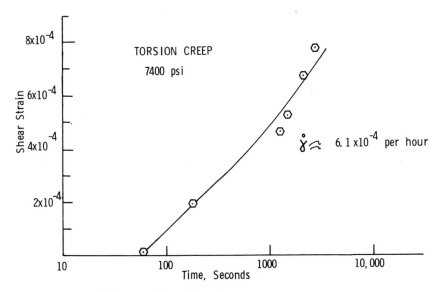

Figure 11.   Torsion creep response

expansion, at 2500°F this requires less than 35 minutes. Immediately thereafter, the dead weight loading was applied by lowering the jack screws. While the specimen is exposed to the oxidizing flame, within a very short time a silicate layer is observed to form over the specimen surface. A few repetitions of the creep experiments on the same specimen suggest diminishing creep for each short test sequence. For very low stress levels, no deformation was observed within an 8-hour period.

For instance, at 3500 psi stress levels initial strain rates 4×10^{-5} inches/inch/hour are suggested. Preliminary torsion creep experiments were also completed. Due to the shorter gage length section in these specimens, a different torch configuration was used. As depicted in Figure 10, a circular configuration with eleven jet holes was used. In this case the temperature distribution was not as uniform and there was an 80°C gradient across the specimen gage length. It is evident that flame deflector shields or additional torch heads are required. Shear stress levels of 1000, 3000, and 7400 psi were studied. The creep rates in shear were larger and apparently were initiated at lower stress levels than for uniaxial tension. A typical torsion creep response is shown in Figure 11.

CONCLUSIONS

The tension creep studies at the low stress levels resulted in lower creep rates than those which might be extrapolated from available information. However, the impurity content of hot-pressed silicon nitride is known to greatly influence creep deformation. There is a possibility that the specimens used in our experiments were not typical of the early versions of this material. In order to develop consistent three-dimensional constitutive equations, it is necessary to characterize the chemical composition and microstructural details of each specimen. This has not yet been accomplished.

These preliminary results have suggested a shear creep dependency which is more pronounced than the tension stress states.

ACKNOWLEDGMENT

The studies were supported by the Advanced Research Projects Agency and the authors are indebted to Mrs. J. C. Riley for preparation of the manuscript.

REFERENCES

1. K. H. Jack, "Nitrogen Ceramics", 17th Mellor Memorial Lecture (1973)

2. E. Glenny and T. A. Taylor, "Mechanical Strength and Thermal Fatigue Characteristics of Silicon Nitride", Powder Metallurgy, $\underline{8}$, pp. 164-195, (1961).

3. A. F. McLean, E. A. Fisher and R. J. Bratton, "Brittle Materials Design, High Temperature Gas Turbine", ARPA-Ford-Westinghouse Interim Report, $\underline{6}$, Contract No. DAAG46-71-C-0162 (Jan. 1, 1974 to June 30, 1974).

4. J. A. Mangels, "Development of a Creep-Resistant Reaction-Sintered Si_3N_4", Ford Motor Company, submitted for publication.

5. R. Kossowsky, Bulletin of the American Ceramic Society, $\underline{53}$, 321, (1974).

6. M. S. Seltzer, A. H. Claver and B. A. Wilcox, "High Temperature Creep of Silicon Nitride and Silicon Materials", to be published.

DEFORMATION TWINNING IN POLYCRYSTALLINE

PEROVSKITES

J. E. Funk, J. Nemeth[*], H. F. Kay[+], and
J. R. Tinklepaugh

New York State College of Ceramics, Alfred Univ.
Alfred, New York

This paper is a review of work done on the mechanical behavior of some perovskite structure ceramics. Most attention was given to $SrZrO_3$ because it has the highest melting point and was the first to exhibit permanent mechanical deformation as a polycrystalline ceramic.

The room temperature stress-strain behavior of polycrystalline $SrZrO_3$ is shown in Figure 1. The primary features of interest in this figure are the non-linear curve, hysteresis, and the remanent strain after unloading the specimen. As $SrZrO_3$ belongs to the same perovskite crystal structure family as ferroelectric $BaTiO_3$ a comparison may explain this mechanical deformation.

It is known that the ferroelectric behavior of $BaTiO_3$ is associated with its domain structure and the movement of these domains under an applied electric stress[1]. Von Hippel[2] stated that mechanical pressure or electric fields can squeeze these domains into or out of a crystal. Forsbergh[8] showed photographs of domain wall motion under electric field stress and described the resulting mechanical distortion of this crystal. Because $SrZrO_3$ is of the same perovskite structure, a similar behavior may be expected for $SrZrO_3$.

*Champion Spark Plug Co., Detroit, Michigan (presently)
+Bristol Univ., Bristol, England (presently)

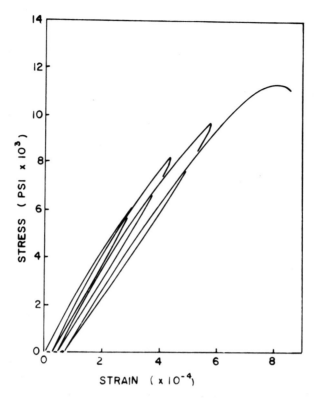

Fig. 1. Stress-strain behavior of hot pressed polycrystalline $SrZrO_3$ at room temperature

Swanson et al[3] measured $SrZrO_3$ as orthorhombic with a_o = 5.814 Å, b_o = 8.196 Å, and c_o = 5.792 Å. The orthorhombic cell contains four primitive monoclinic pseudocubic cells of a' = 4.103 Å, b' = 4.098 Å, and β = 90.217° (90°13'2") as shown in Figure 2. By applying a tensile stress along a_o the primitive cell can twin by changing β from an obtuse to an acute angle (89°46'58"). Upon shearing this angle from the original cube the diagonal in the shear face originally has length a√2 and becomes a√2 + a tan δ/2 so the strain = tan δ/2 or δ/2 for small shear angles, and δ_m = 0.38% in the [101] cube diagonal direction. This is the same as interchanging the orthorhombic a_o and c_o axes. This much strain could only occur in one direction of a perfect, untwinned crystal. In a ceramic the random orientation of the grains greatly reduces the possible total strain.

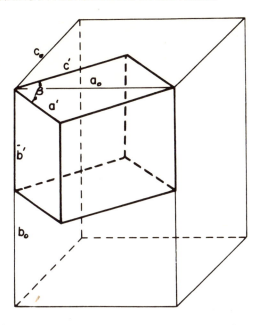

Fig. 2.   Monoclinic primitive cell of $SrZrO_3$ with
the orthorhombic unit cell[3].   Twinning occurs by
compression along c_o.

Assuming that only the $a_o \rightleftarrows c_o$ interchange occurs, a completely random orientation of all a_o vectors, and a unidirectional
tensile stress, only those vectors lying within a cone of semi-
apex angle $\theta = \pi/4$ about the stress direction will produce a
strain proportional to $\delta_m \cdot f(\theta)$.   The a_o vectors in annulus $d\theta$
at $< \pi/4$ is:

$$\frac{2\pi \sin\theta \, d\theta / 2}{3}$$

since there are also an equal number of b_o and c_o vectors.   This
integrates to $0.028\,\delta_m = 1.06 \times 10^{-4}$ maximum strain.   If the
remanant strain at fracture of a polycrystalline ceramic $SrZrO_3$
is less than 1.06×10^{-4}, mechanical twinning is a sufficient explanation for it.

It also seems likely that the behavior of other perovskites
could be predicted according to their symmetry: e.g. cubic
$BaZrO_3$ and $SrTiO_3$ should obviously not twin; tetragonal $BaTiO_3$
may be expected to twin and orthorhombic $CaTiO_3$ is questionable.
According to Kay[4] single crystals of $CaTiO_3$ can be made to twin

under mechanical stress only with great difficulty due to the high
primitive cell shear angle of 0°48'. It is, therefore, unlikely
that a polycrystalline sample would permanently strain. Orthorhombic $CaZrO_3$, with a very large primitive cell shear angle of
1° 43' would not be expected to twin.

$SrZrO_3$ Ceramic

According to Keler and Kuznetsov[5] it is difficult to sinter
$SrZrO_3$ to high density without mineralizers. For this work a
commercial grade of $SrZrO_3$ containing 2.5% SiO_2 and a chemically pure grade were each ball-milled in a steel mill with steel
balls. The iron pickup from the milling operation was dissolved
with HCl and decanted as a soluble chloride. Both were die
pressed at 12,000 psi with no binder additions. The commercial
grade was fired at 1500°C for two hours to 79% of theoretical density. Firing to higher temperatures reduced the density due to
the development of gas bubbles from the SiO_2 present.

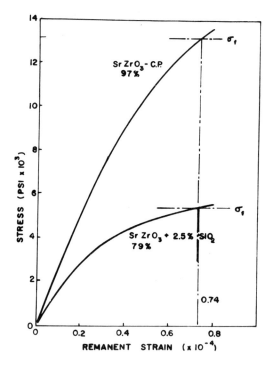

Fig. 3. Stress-remanent strain of $SrZrO_3$ - C.P.
and $SrZrO_3$ + 2.5% SiO_2.

The C. P. grade was fired at 1700°C for four hours to 97% of theoretical density. X-ray fluorescence and spectrographic analysis showed both materials to be SrO deficient ($\sim$ 1.5%) and containing residual iron ($\sim$ 0.1%).

These specimens were tested on a four-point flexure apparatus measuring stress vs. strain cyclically in increasing stress increments to fracture. Figure 1 shows the typical stress-strain behavior of these ceramics. It can be seen that there is a remanent displacement along the strain axis. Experiments were performed to determine the remanent strain at fracture characteristic of this material by a series of load-unload cycles imposed on each sample. These data are shown in Figure 3 where σ_f = average fracture strength.

Comparing the experimental value of saturation strain of $\sim$ 0.74 x 10^{-4} with the estimated 1.06 x 10^{-4} shows reasonable agreement and can account for the mechanical deformation of polycrystalline $SrZrO_3$ at room temperature with no other mechanism necessary.

$SrZrO_3$ Single Crystal

Small crystals with a cubic morphology ($\sim$0.5 mm on a side) were grown in molten KF. One part of chemically pure $SrZrO_3$ was dissolved in 20 parts of KF at 1150-1250°C for 20 hours and then slowly cooled at $\sim$ 7.0°C/hour. The crystals were identified as $SrZrO_3$ by x-ray but were not oriented to determine the relationship of the pseudocubic axes to the orthogonal crystal axes.

Under a polarizing microscope a domain configuration was observed as shown in Figure 4 where the domain walls are approximately parallel to the crystal face diagonal. A device was built to apply stress across corners of the crystals while observing domain motion through a polarizing microscope. No stress or strain measurements were made, but the movement of the domains was observed as the stress was alternated from one diagonal to the other. Figures 5, 6, and 7 are photographs of a $SrZrO_3$ crystal before and after stressing in both directions.

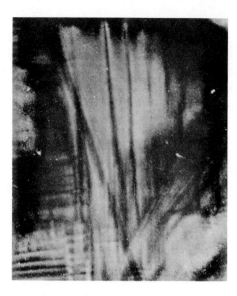

Fig. 4.   Domains in $SrZrO_3$ single crystal-
orthogonal axes at ~ 45° to domain walls (400x)

Some Other Perovskites

Early in this work chemically pure $SrZrO_3$, $BaZrO_3$,
$SrTiO_3$, $BaTiO_3$, $CaTiO_3$ were hot pressed[6] from powders as
received.   These samples were not characterized before or
after pressing.   But as they were hot pressed in a graphite die
at about 1200°C and 3500 psi and were dark grey in color, it
may be assumed they were O_2 deficient.   The stress-strain behavior of these samples with percent of theoretical density are
shown in Figure 8.   Cubic $BaZrO_3$ and $SrTiO_3$, and orthorhombic
$CaTiO_3$ behave as normal Hookeian ceramics while both orthorhombic $SrZrO_3$ and tetragonal $BaTiO_3$ show the characteristic
stress-strain behavior and remanent deformation as expected.

$Sr_xCa_{1-x}SrO_3$

The yield measured in $SrZrO_3$ is quite small and its primitive pseudocell has a shear angle of only about 0°13'.   Tetragonal
$BaTiO_3$ has a shear angle of about 0°35' and orthorhombic $CaTiO_3$
of about 0°48'.   Polycrystalline $BaTiO_3$ behaves mechanically like
$SrZrO_3$ while polycrystalline $CaTiO_3$ behaves as a normal Hookeian
ceramic.

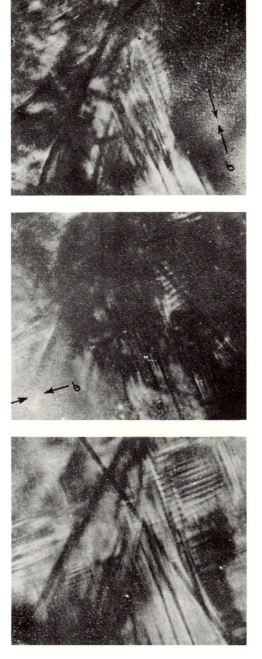

Figures 5, 6 & 7. Effect of compressive stress applied along crystal face diagonal (5). Unstrained (6). After stressing as indicated (7). After stressing at 90° to (6).

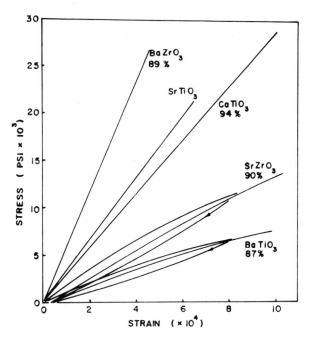

Fig. 8. Stress-strain behavior of several hot pressed perovskites.

        $CaZrO_3$ is also orthorhombic with a_o = 5.587 Å, b_o= 8.008 Å, c_o = 5.758 Å[7]. It is very similar to $SrZrO_3$ with 4 formulas/unit cell and the primitive cell a monoclinic pseudo-cube with a' = c' = 4.012, b' = 4.004 and β = 91°43'. Solid solutions of $SrZrO_3$-$CaZrO_3$ should produce a larger average shear angle than pure $SrZrO_3$. Each $SrZrO_3$ primitive cell is probably dis-torted from its intrinsic shear angle according to the proximity of the nearest $CaZrO_3$ cell but each cell would not have the same average shear angle. Mechanical effects would, therefore, be strongly subject to homogeneous distribution, grain orientation and pore size distribution.

        Batches of $(Sr_{0.95}Ca_{0.05})$ ZrO_3, $(Sr_{0.90}Ca_{0.10})$ ZrO, and $(Sr_{0.85}Ca_{0.15})$ ZrO_3 were prepared as above, sintered at 1700°C for four hours to ~ 92% of theoretical density. Stoichio-metry was not checked on these samples but they were undoubt-edly Sr, Ca deficient due to the acid treatment. They were tested as above and the stress vs. remanent strain at fracture results are shown in Figure 9. It may be noted that: (1) increas-ing additions of $CaZrO_3$ required a threshold elastic strain before permanent deformation occurred; (2) saturation strain was essentially unchanged from $SrZrO_3$; and (3) the fracture strength was significantly increased above $SrZrO_3$.

Other Work

Carlsson[8] did compression tests upon a series of Ca-Sr-Ba ZrO_3 solid solutions and obtained similar results though of a higher magnitude of strain due to the higher stress levels obtainable by compressive loading. The greater internal stress level further allows a change in direction of b_o by $90°$ as discussed by the present authors[10].

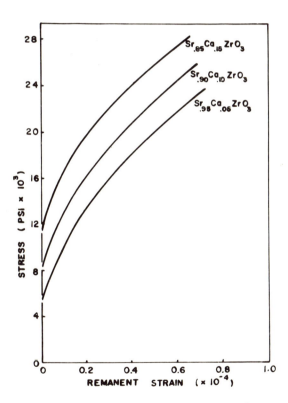

Fig. 9. Stress vs. remanent strain behavior polycrystalline of $Sr_xCa_{1-x}ZrO_3$ at room temperature.

Summary

Although it has been established that mechanical twinning provides a sufficient explanation for the remanent deformation of polycrystalline $SrZrO_3$, $BaTiO_3$ and some non-cubic solid solutions, the effects of porosity and non-stoichiometry in polycrystalline ceramics are of critical importance. Pores provide a non-constraint to at least one side of a grain and non-stoichiometry probably provides both slight lattice distortion and an open structure for stress induced diffusion. Further work on single crystals such as that done by Prasad and Subbarao[9] should be done on $SrZrO_3$ and the solid solutions of Carlsson[8].

REFERENCES

1. a.   W. J. Merz, "The Electrical and Optical Behavior of $BaTiO_3$ Single Domain Crystals," Phys. Rev. <u>76</u> 1221 (1949).

   b.   W. J. Merz, "Domain Properties in BaTiO3," Phys. Rev. <u>88</u> 421 (1952).

   c.   P. W. Forsbergh, Jr., "Domain Structure and Phase Transitions in Barium Titanate," Phys. Rev. <u>76</u> 1187 (1949).

   d.   E. A. Little, "Dynamic Behavior of Domain Walls in BaTiO3" Phys. Rev. <u>98</u> 978 (1955).

   e.   R. G. Rhodes, "Barium Titanate Twinning at Low Temperatures," Acta Cryst. <u>4</u> 105 (1951).

   f.   B. Matthias, A. Von Hippel, "Domain Structure and Dielectric Response of $BaTiO_3$ Single Crystals," Phys. Rev. <u>73</u> 1378 (1948).

2.   A. Von Hippel, <u>Dielectrics and Waves</u>, J. Wiley and Sons, New York (1954.

3.   H. E. Swanson, M. I. Cook, T. Isaacs, E. H. Evans, <u>Standard X-Ray Diffraction Powder Patterns</u>, Nat. Bur. Standards Arc. No. 539 (1960).

4.   H. F. Kay, P. C. Bailey, "Structures and Properties of $CaTiO_3$," Acta. Cryst. 10 (3) 219 (1957).

5.   E. K. Keler, A. K. Kuznetsov, "Synthesis and Physiotechnical Properties of Strontium and Barium Zirconate," J. Appl. Chem., U.S.S.R. <u>34</u> 2044 (1961).

6.   J. Bidwell, "The Stress-Strain Relationships of
        Several Perovskite Compounds," B.S. Thesis, New
        York State College of Ceramics, Alfred, N. Y. (1963).
7.   L. W. Coughanour, R. S. Roth, S. Marzullo, F. E.
        Sennett, J. Research NBS 54 191 (1955).
8.   L. Carlsson, "Non-Elastic Mechanical Behavior in
        $SrZrO_3$ by Reorientation," J. Natl. Sci., 5 335 (1970).
9.   V. C. S. Prasad, E. C. Subbarao, "Deformation Studies
        on $BaTiO_3$ Single Crystals, Appl. Phys. Letl. 22 [8]
        424 (1973).
10.   J. E. Funk, J. Nemeth, H. F. Kay, J. R. Tinklepaugh,
        "The Mechanical Yield of Polycrystalline $SrZrO_3$ at
        Room Temperature," Unpublished; manuscript under re-
        view for Jour. Amer. Ceram. Soc.

TRANSFORMATION PLASTICITY AND HOT PRESSING

A.C.D. Chaklader

Department of Metallurgy
University of British Columbia
Vancouver 8, Canada

ABSTRACT

The transformation plasticity during the phase transition of
quartz to cristobalite, monoclinic $\rightleftarrows$ tetragonal of zirconia,
metakaolin to a spinel phase and brucite to periclase has been
investigated by studying their compaction characteristics.
Viscous flow has been found to be the predominant mechanism of
mass transport (after an initial particle rearrangement stage) in
the case of quartz to cristobalite phase change where the transformation was associated with the formation of an intermediate
amorphous silica phase. The results on the monoclinic $\rightleftarrows$ tetragonal
transformation of zirconia indicated that it is most likely controlled by internal strain induced by the stress associated with
the volume change ($\Delta V/V$) and the flow stress of the weaker phase.
Particle movement and deformation of the weaker phase (possibly
tetragonal) may be the manifestation of this plasticity. The
plasticity in the case of metakaolin to a spinel phase appeared
to start before the exothermic reaction (generally encountered
in a dta plot) and may be diffusion controlled. The plasticity
encountered during brucite to periclase transformation may be
the combined effect of disintegration of precursor particles,
vapor-phase lubrication and some deformability of freshly formed
very fine MgO particles.

INTRODUCTION

The mechanisms of densification during hot-pressing a powder
compact have been attributed to particle rearrangement (without

and with fragmentation), plastic (or viscous) flow and stress
enhanced diffusional creep. Depending on the powder characteristics, temperature and pressure, one or more of the mechanisms
of material transport would predominate at different stages of
hot-pressing. Enhanced densification exhibited by a powder compact during hot-pressing, while the material undergoes a phase
transition, has been attributed to the so-called "transformation
plasticity"(1). In this paper attempts will be made to identify
mechanisms of mass-transport in transformation plasticity, which
give rise to the enhanced compaction. Previous studies indicated
that the enhanced densification during hot-pressing may be
partially attributed to fragmentation of particles(2) and vapor-
phase lubrication(3) (in the case of mass loss with a transformation). Particle movement due to internal stress(4) generated by
a volume change during a phase transition may also give rise to
enhanced compaction. Thus it appears that the overall transformation induced plasticity can be due to several mass-transport
processes. For a particular system one or more of the transport
processes may predominate and this may show enhanced densification
encountered in systems undergoing a phase transformation.

     To deal with the overall transformation plasticity, materials
have been divided into two groups with respect to their phase
change characteristics: phase change without mass-loss and with
mass-loss. Examples of ceramic interest in the former group are
quartz $\rightarrow$ cristobalite, zirconia-monoclinic $\rightleftarrows$ tetragonal,
metakaolin $\rightarrow$ a spinel type phase $\rightarrow$ mullite and γ-alumina $\rightarrow$ α-alumina.
Examples of the latter group are $Mg(OH)_2$ $\rightarrow$ MgO, kaolin $\rightarrow$
metakaolin and aluminum hydroxides to aluminas. The compaction
behavior of powder compacts of four of these materials has been
investigated during their phase transitions and data are analysed
with the existing knowledge of the transformation characteristics
and proposed models. These are quartz $\rightarrow$ cristobalite, zirconia -
monoclinic $\rightleftarrows$ tetragonal, metakaolin $\rightarrow$ a spinel phase and $Mg(OH)_2$
$\rightarrow$ MgO.

EXPERIMENTAL PROCEDURE

     Materials: Reagent grade zirconia was supplied by Koch-Light
Laboratories Limited, England. It is about 99.9% ZrO_2 the average
grain size being $\sim$ 1µm. Optical grade Brazilian quartz was crushed,
ground and fractionated.   A size fraction of $\sim$ 62µm was used for
compaction studies. The ground powder was washed with dilute HF,
rinsed with distilled water and dried before compaction experiments.
Georgia Kaolin was supplied by Ward's Natural Museum and a synthetic brucite, supplied by Alcan Ltd., Arvida, Quebec was used,
the powder characteristics of which have already been reported(2).

Equipment

    The compaction experiments were carried out in a Phillips
induction unit using reactor grade graphite which functioned
both as the susceptor and the hot-pressing component. The die was
heated to the desired temperature in argon to prevent oxidation.
The pressure was applied by graphite rams to the specimen through
a pneumatic ram fed from a gas cylinder. A gas regulator was used
for controlling pressure. A W + 3% Re and W + 25% Re Thermocouple inserted through the top plunger, but 1/16 in. away from
the end, was used to measure the temperature of the specimen. The
linear dimensional change was measured by an inductive displacement transducer connected to a strain gauge bridge and a strip
chart recorder, and also directly by a dial indicator (sensitivity
0.0001 in.) connected in tandem with the transducer.

Procedures

    Except for the compaction studies of zirconia, all other
experiments with kaolin, brucite and quartz were carried out
isothermally for kinetic analysis. Initially, however, a few
experiments were carried out with all materials in which a constant load was maintained while the specimen was heated at an
approximately constant heating rate. This was to determine the
temperature range where the enhanced compaction due to a phase
change was evident. In the case of zirconia, because of the very
fast nature of transformation, the compaction tests were performed in which a constant load was applied before the transition
temperature and the specimen temperature was increased to a point
above this temperature and in some cases cycled through the
temperature range of transformation.

                            RESULTS

(i)   Quartz → Cristobalite

    Figure 1 shows the compaction behaviour ($\Delta L/Lo$) of quartz
powder($\sim$ 62µm) under a constant load of 2000 psi during heating
above the melting point of quartz (i.e. $\sim$ 1750°C). Similar experiments with finer ($\sim$ 45µm) and coarser ($\sim$ 100µm) quartz powder
indicated that the compaction curve shifted to a lower temperature
($\sim$ 25°C) for finer quartz and to a higher temperature ($\sim$ 15°C) for
coarser quartz than that shown in Figure 1. Isothermal compaction
experiments were carried out at 1520, 1580 and 1620°C and under
1000, 2000 and 3000 psi. The die was heated to about 1000°C and
the pressure was applied to reach mechanical equilibrium in the
system. The pressure was then removed and the die temperature was

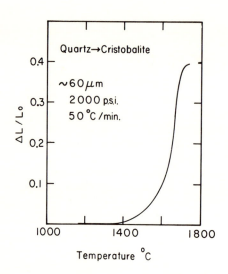

Fig. 1. A compaction curve of quartz powder at a constant heating rate.

increased to the desired level and maintained for 5 minutes, and the same pressure was applied (as used previously to reach equilibrium). The compaction was noted as a function of time. Attempts were made to analyse data in terms of classical hot-pressing equation(5) and liquid-phase-sintering(L.P.S.) models(6). The L-P-S models did not fit the data.

The hot-pressing equation for viscous flow is

$$\frac{dD}{dt} = \frac{3\sigma}{4\eta} (1-D) \qquad 1(a)$$

and in the integrated form is

$$-\ln(1-D) = \frac{3\sigma t}{4\eta} + c \qquad 1(b)$$

The integration constant c becomes $-\ln(1-D_o)$, as at t = o, D = D_o (the density of the compact before hot-pressing).

In order to calculate the relative density of the compact, it was necessary to know the amount of phases present in a compact of quartz powder after different times of hot-pressing (under isothermal conditions). Interrupted experiments were performed in the same hot-pressing equipment and the specimens were analyzed by x-rays for their compositions. From the quantitative estimation of phases, the true density of the compact as a function of time was calculated. For these calculations, the true density values of 2.65 g/cc, 2.32 g/cc and 2.21 g/cc for quartz, cristobalite and the amorphous phase respectively, were used. The true density of a quartz compact at 1520°C was found to be 2.63 g/cc after 3 mins. and 2.52 g/cc after 15 mins. Similarly, these values at 1620°C varied from 2.60 g/cc at 2 mins. to 2.30 g/cc after 15 mins.

Following equation 1(b), log of (1-D) was plotted as a function of time as shown in Figure 2. The experimental data did not follow the theoretical prediction during the first minute or so. The rapid increase in density initially may be attributed to particle movement enhanced by a fluid lubrication and this can

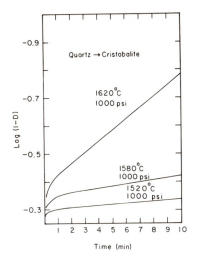

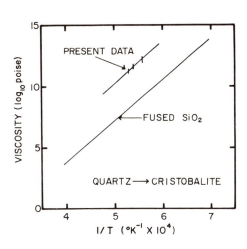

Fig. 2. Log$_{10}$ (1-D) versus time
for quartz powder at different
temperatures.

Fig. 3. Log$_{10}$ viscosity versus
1/T for fused silica glass and
of quartz powder during transformation to cristobalite.

account for the observed behavior. Further justification in using
Equation 1(b) can be found from the calculated slopes of the plot
log (1-D) vs t under different pressures, which bear the same
pressure ratios. These are shown in Table I, which also includes
the calculated viscosity (η) data. The calculated viscosity
values are compared with the viscosity data of pure fused silica(7)
in Figure 3, where log of viscosity vs reciprocal of absolute
temperature is shown. The viscosity values as calculated during
quartz to cristobalite transformation are 2 to 3 orders of magnitude higher than that of pure silica in the same temperature
range. This is not unexpected in view of the fact that the
systems in the present work contained a large fraction of crystalline phases (quartz and cristobalite). It is well known that
the viscosity of a fluid increases with the presence of a non-
viscous component. In the present study, the maximum amount of
the amorphous phase encountered was in the range of 20 to 25% by
volume.

(11) ZrO_2 (Monoclinic $\rightleftarrows$ Tetragonal)

    The powder compacts of zirconia were initially hot-pressed
under a constant load of 2000 psi and a constant heating rate of
25°C/min. The pressure was applied at 800°C and maintained for a

Table I

Slope ratios and viscosity data of quartz compacts

Temperature (°C)	Pressure (psi)	Predicted ratio	Experimental ratio	Viscosity (poise)
1520	1000	1	1.0	1.16×10^{12}
"	2000	2	2.1	$1.02 \times$ "
"	3000	3	3.0	$1.10 \times$ "
1580	1000	1	1.0	7.16×10^{11}
"	2000	2	2.27	$6.56 \times$ "
"	3000	3	4.2	$5.20 \times$ "
1620	1000	1	1.0	$2.07 \times$ "
"	2000	2	1.8	$2.21 \times$ "
"	3000	3	2.3	$2.67 \times$ "

few minutes to reach mechanical equilibrium. The temperature of
the specimen under pressure was increased up to 1250°C and then
cooled to 840°C very slowly and maintained at 840°C for 5 mins.
before reheating again. Thus, the specimen was cycled between
1250°C and 840°C under pressure in order to estimate its flow
properties during both forward and reverse phase change. The
compaction behavior during cycling is represented in Figure 4.
It is quite apparent that almost all densification was encountered during the first heating cycle. Small densification in
the subsequent heating cycles may be simply due to the volume
change associated with the phase transformation. The density
change associated with this transformation is 5.56 g/cc (for
monoclinic) to 6.10 g/cc (for tetragonal), corresponding to about
9% by volume. It was
surprising that no densification was encountered in
the second or later cycles
of heating and cooling in
the powder compact. Microstructural observations of
the fractured compact surfaces showed that no grain
growth occurred during
cycling, as can be seen in
Figure 5. It was thought
increase in the grain-
size during cycling may
affect the plasticity after
the first cycle of heating.

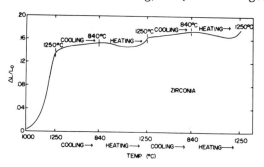

Fig. 4. A compaction curve of
zirconia powder during cycling
through the phase transformation
temperature range.

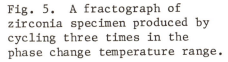

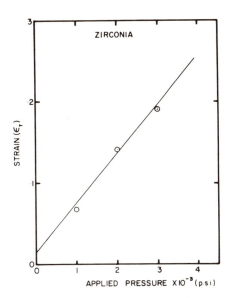

Fig. 5. A fractograph of
zirconia specimen produced by
cycling three times in the
phase change temperature range.

Fig. 6. Total strain (ϵ_T) after
one cycle versus stress.

The effect of varying pressure on the total strain during the
first heating cycle was also evaluated. The results are shown in
Figure 6, where the total strain, ϵ_T, was plotted versus stress, σ.
The results can be expressed by an equation of the form

$$\epsilon_T = A\sigma \tag{2}$$

This equation is very similar to the equation developed by
Greenwood and Johnson(7) to interpret their data of transformation plasticity in metals, which is

$$\epsilon_{(per\ cycle)} = 5(\frac{\Delta V}{V})\sigma/3I \tag{3}$$

This theory suggests that the proportionality between strain
per cycle and stress is mainly controlled by the volume change $(\frac{\Delta V}{V})$
and the flow stress I of the weaker phase at the transformation
temperature. Although powder compaction tests in a die (i.e. in
a constrained system) is not the same as studying the flow behavior
of metals either in compression or in tension tests, the particles
in a powder compact are sufficiently free (with a relative density
of ∿ 0.5 in a die) during transformation to give rise to a pseudoplastic effect, as encountered in the system of zirconia during the

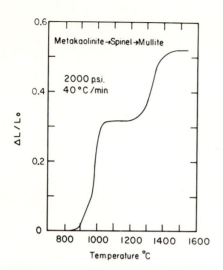

Fig. 7. A compaction curve
during the successive transformation of metakaolin to a spinel
phase and then to mullite.

monoclinic to tetragonal transformation.

(iii) Kaolinite
(Metakaolin to spinel phase)

The interest in the com-
paction behavior in this system
stems from the following:
(i) previous observations(8)
showed that kaolinite compacted
very rapidly during this transformation as compared with the
dehydroxylation reaction;
(ii) very little crystallographic rearrangement is
involved in this transformation
(9); (iii) silica is exsolved
from the structure(9);   and
lastly (iv) an exothermic reaction is associated with the
transformation. The compaction
behavior of metakaolin to a
spinel phase and then to mullite
at a constant heating rate (40°C/
min.) under 2000 psi is shown in
Figure 7. Enhanced compaction
during the transformations is easily apparent from this Figure.
Isothermal compaction experiments were carried out in the
temperature range 850 to 1050°C. The compact temperature was
raised to a desired level and maintained for 2 mins. for equilibration before any pressure was applied. The pressure was usually
maintained for 10 mins. and compaction was recorded. The pressure
was then released and the spring-back was noted. This value was
subtracted from the compaction values recorded in a strip-chart
recorder. The compaction data were normalized with respect to L_o,
the initial length of the specimen.

Figure 8 shows a series of isothermal compaction curves
($\Delta L/L_o$ vs time) obtained at different temperatures. Neither a
semi-log plot (corresponding to the viscous-phase hot-pressing
model(5))nor a log-log plot for the liquid-phase sintering (LPS)
model(6) fitted the data, although a part of the compaction curves
became linear in both of these plots. At short times, when the
compaction was rapid, the data did not fit either of these models.
It was found, however, that all the data can be represented by
an equation of the form

$$\Delta L/_{L_o} = K(1-Ae^{-\alpha t} - Be^{-\beta t}) \tag{4}$$

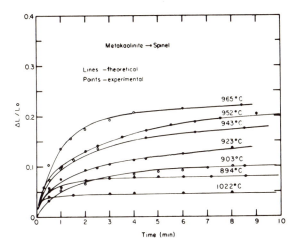

Fig. 8. Isothermal compaction curves during metakaolin to spinel transformation.

Where K, A, α, B and β are constants which can be determined from the experimental plots. $\Delta L/L_O$ is the strain (compaction = ε), t is the time and $K = (\Delta L/L_O)_{t = \alpha}$, i.e. the final compaction. A and B are constants, for which the condition must be satisfied is, at t = o, $\Delta L/L_O = 0$, i.e. A + B = 1.

The values of the constant 'α' were determined by assuming that $\beta \gg \alpha$, and for $t \gg 0$ (say above 3-4 mins.) the contribution from the second exponential term will be approximately zero. Neglecting the second term involving β and taking the derivative of equation (4) and natural log, the final form of equation (4) is

$$\ln(d\varepsilon/dt) = -\alpha t + \ln(KA\alpha) \qquad (5)$$

From the slope of the plot ln slopes (slope of isothermal compaction curves) vs time, the values of α were obtained. From the intercepts and using the values of K from experiments, the values of 'A' were determined. From the boundary condition A + B = 1, the values 'B' were calculated. Then the values of K, A, α, B and experimental values of $\Delta L/L_O$ at different times were substituted in equation (4) and the value of β which produced the best fit of experimental data was chosen. Both the experimental data and the computerized plot of the curves are shown in Figure 8.

It has been shown(10) that a second order differential equation can be developed from equation (4) which describes the hot-pressing characteristics during a phase transformation. The equation is

$$\frac{d^2\varepsilon}{dt^2} + (\alpha+\beta)\frac{d\varepsilon}{dt} + \alpha\beta\varepsilon = K(A\alpha + B\beta)\frac{d\sigma}{dt} + K\alpha\beta\sigma \qquad (6)$$

A viscoelastic model involving springs and dashpots which has a response similar to equation (6) is shown in Figure 9.

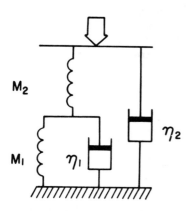

Fig. 9. A visco-elastic model
for describing the compaction
behavior during a phase change.

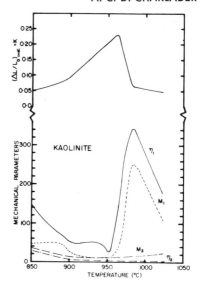

Fig. 10. The values of
mechanical parameters η_1,
η_2, M_1 and M_2 as a function
of temperature (for meta-
kaolin).

The corresponding differential equation for its response is given
by

$$\frac{d^2\epsilon}{dt^2} + \frac{\eta_2(M_1+M_2) + M_2\eta_1}{\eta_1\eta_2} \frac{d\epsilon}{dt} + \frac{M_1M_2}{\eta_1\eta_2} \epsilon = \frac{1}{\eta_2} \frac{d\sigma}{dt} + \frac{M_1+M_2}{\eta_1\eta_2}\sigma \qquad (7)$$

Comparing the two equations it can be seen that the values of the
mechanical parameters η_1, η_2, M_1 and M_2 can be calculated from the
experimental constants K, A, α, B and β. This was done in a computer and the values of the constants as a function of temperature
are shown in Figure 10. The variation of K with temperature is
also shown in the figure. It is quite evident from the figure
that only one of the elastic elements (M_1) and one of the viscous
components (η_1) are temperature sensitive. Thus, these two components can be assigned to the powder characteristics of the
material (kaolin) during the phase change. The maximum compaction,
i.e. the value of K, reached its peak when the value of the viscous component (η_1) is nearly at its minimum. This was not
unexpected, as at this stage the powder compact was most susceptible to deformation.

(iv) $Mg(OH)_2 \rightarrow MgO$

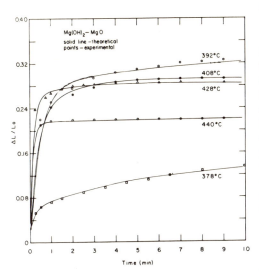

Fig. 11.   Isothermal compaction
curves during brucite to periclase
transformation.

It has been reported
by several workers (1e,8,11)
that during the decomposition
of $Mg(OH)_2$, this material can
be easily densified. The
compaction studies of brucite
both natural and synthetic,
during heating through the
phase transformation temperature range have shown that
enhanced compaction was most
predominant at about 400°C.
For studying isothermal compaction behavior of a synthetic brucite, the temperature range 375 to 450°C was
selected. The experimental
procedure was very similar to
that used for metakaolin →
spinel experiments. A series
of isothermal compaction
curves are shown in Figure 11.
Following the previous proced-
ure, it was found that the data can be expressed as a sum of
exponentials of the form of equation (4). The computerized plots
of the compaction curves following equation (4) are shown as the
solid lines in Figure 8. Using the experimental values of the
constants K, A, α, B and β, the values of the mechanical elements
of η_1, η_2, M_1 and M_2 (corresponding to the viscoelastic model in
Figure 9) were calculated as a function of temperature. This is
shown in Figure 12. Again it is quite evident that only one of
the viscous components (η_1) and one of the elastic elements (M_1)
are temperature sensitive and thus these two elements reflect
the material behavior during the phase change. Contrary to the
case of metakaolin, it can be seen that the value of K reached its
maximum, when the value of η_1 reached a peak. The increase in
viscosity, i.e. decrease in the flow properties of the powder,
may be explained by the fact that very fine particles tended to
coalesce immediately after the reaction was started. This was
observed during heating the powder on a hot-stage in an electronmicroscope. The coalescence of the particles, which can be
referred to as a kind of sintering, resulted in higher density of
the compact, i.e. higher K values. It was also observed while
heating the powder in an electronmicroscope that brucite particles
disintegrated as the reaction proceeded to completion. The
nature of this disintegration is shown in Figure 13. X-ray line
broadening studies indicated that immediately after decomposition

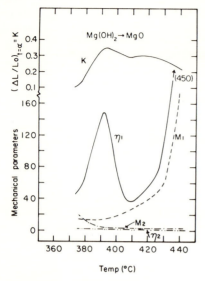

Fig. 12. The value of
mechanical parameters η_1, η_2,
M_1 and M_2 as a function of
temperature (for brucite)

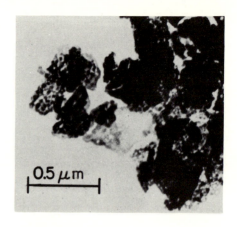

Fig. 13. Disintegration of
brucite particles in a hot-
stage of an electron microscope.

of brucite, the freshly formed MgO particles were of the order of
50 to 100 Å. The fragmentation of particles may give rise to
enhanced fluidity which is reflected in the sharp drop of value of
η_1 at the peak of the decomposition reaction. However, with the
increase in compaction of the powder compact, the mobility of the
particles continuously decreased (i.e. increasing values of η_1)
and this gradually increased the rigidity of the system as
reflected in the increasing values of the elastic element M_1.
Again the elastic element M_2 and viscous component η_2 were found
to be temperature insensitive.

GENERAL OBSERVATIONS

(i) Quartz $\rightarrow$ Cristobalite

The transformation of quartz to cristobalite has been
reported(13) to be an indirect process involving an amorphous
intermediate phase. The overall transformation can be interpreted
in the form of a consecutive reaction. Studies on the melting
behavior of quartz crystal also indicated(14) that the fusion of

quartz was encountered as low as 1400°C (at an extremely slow rate) but at a faster rate (3×10^{-6} cm per minute) above 1550°C. In the presence of such an amorphous (glassy) phase, it was expected that quartz to cristobalite transformation should produce plasticity associated with a phase change.

The compaction behavior of quartz powder in the transformation temperature range is similar to that encountered by Kingery et al (15) during pressing powder compacts in the presence of a liquid phase. Immediately after the application of pressure, rapid densification was observed in all cases as also found in the present study with quartz. This is mostly as a result of particle movement enhanced by lubrication due to the presence of fluid phase. Subsequent to this rapid initial densification, the compaction behavior of quartz powder followed the Murray et al(5) equation for hot-pressing a viscous phase. In this respect, the densification of quartz powder is similar to that observed during hot-pressing of fused SiO_2 powder(16). Both of these materials follow the rate equation of a first-order Kinetic reaction i.e. $\ln(1-D)$ vs t is linear. From the similarity of their compaction behavior it is postulated that the amorphous phase formed during quartz to cristobalite transformation is essentially fused silica. The calculated viscosity of this amorphous phase, which was found to be higher than fused silica by 2 to 3 orders of magnitude in the same temperature range, was most likely due to the presence of a large amount of crystalline (non-fluid) phase in the system.

(ii) Zirconia

It appears that the transformation plasticity associated with this phase change is mainly due to particle arrangement controlled by the volume change ($\Delta V/V$) and the applied stress. However, as it has been possible to produce high density specimens (> 0.96 of relative density) by hot-pressing up to 1250°C, it is mostly unlikely that without any particle deformation such a ·high density compact can be produced. The deformation of particles (i.e. the weaker phase) may also be an essential component of the total deformation process. Thus, it essentially follows the Greenwood-Johnson theory. It has not been possible to test accurately their equation with the present data as the flow stress of either the monoclinic or the tetragonal phase at the transformation temperature range is not known. However, Amato et al(17) calculated the critical shear stress of ZrO_2 using hot-pressing data at 1250°C to be 2300 psi. Incorporation of this value (for I) and ($\Delta V/V$) = .09 in equation (3) showed that $d\varepsilon/d\sigma$ is 6.5×10^{-5} as compared to the experimental value of 6.8×10^{-5}, 7.0×10^{-5} and 6.4×10^{-5} (for $d\varepsilon_T/d\sigma$) at 1000, 2000 and 3000 psi, respectively.

(iii)   Other Systems (Kaolin and Brucite)

    (a)   Metakaolin

        The data of the compaction during metakaolin to spinel transformation and $Mg(OH)_2$ to MgO can be interpreted using a viscoelastic model. The exsolution of SiO_2 from the metakaolin structure during this transformation was expected to give rise to a viscous flow behavior, but it has been shown recently[18] that the exothermic reaction during this phase transition is most likely due to the crystallization of amorphous silica. During compaction studies of metakaolin both under isothermal conditions and at a constant heating rate (see Figure 7) it was observed that densification was encountered below 900°C, whereas dta traces did not show any reaction until a temperature of 980°C was reached. It is possible that metakaolin to spinel transformation occurs just before the exothermic reaction generally encountered in a dta curve. The analysis of the compaction data using a viscoelastic model (Figure 9) as carried out in this study also indicated that the value of the viscous component η_1, started to decrease below 900°C and reached its lowest value at 955°C. The densification continued (increasing K's) up to $\sim$ 960°C. Once the reaction was over, there was a steep rise both in the viscous component η_1 and elastic component M_1, indicating a drastic reduction in the flow properties of the material

    (b)   Brucite

        A slight variation of the same behavior was encountered with brucite (Figure 12). Instead of a continuous decrement of the value of η_1 as obtained with Kaolin, there was an initial rise of η_1, indicating coalescence of particles after the commencement of the decomposition reaction. The newly formed particles tend to react with one another to reduce the surface energy of the system. On the other hand, fragmentation of brucite particles increased the flow properties (i.e. decrease in η_1) of the powder compact. The creep behavior of fully dense MgO specimens of about 1000 Å grain size has recently been investigated[19]. The specimens were fabricated by reactive hot-pressing a synthetic brucite up to 750°C. A fractograph of such an MgO is shown in Figure 14. The specimen contained about 4.5% $Mg(OH)_2$ as indicated by x-ray analysis. Fully dense $Mg(OH)_2$-free MgO specimens having grain-size below 1000 Å could not be produced by this technique (i.e. R.H.P.). Creep studies of these materials indicated that a steady state creep could be produced in this material between 500 and 700°C, indicating that very fine grain-size MgO bodies can be deformed at 0.21 to 0.25 Tm. This suggests that the overall transformation plasticity in this system can be attributed to fluid-phase lubrication[3] and deformability of very fine MgO particles, as encountered in the creep studies.

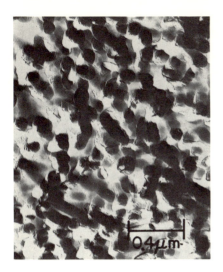

Fig. 14. A fractograph of a
fully dense MgO specimen having
about 1000 Å particle size
used for creep studies.

(c) Flow Properties

The flow characteristics of
metakaolin, brucite and boehmite
(A100H) are compared, in order to
estimate their compactibility
during their phase transition.
All three materials showed reactive hot-pressing behavior
which could be interpreted with
the help of a viscoelastic model
(Figure 9). It has been shown
for metakaolin and brucite in the
present study and for boehmite
previously(10) that the viscous
element η_1 appeared to be closely
related to the flow properties of
the powder compact. That is the
ease of densification was inversely
related to the value of η_1. The
compact density was higher when
the value of η_1 was small and vice
versa. Thus the extent of the
fluidity can be estimated from η_1
in these systems. A plot of the
fluidity $(1/\eta_1)$ was made as a
function of ΔT, where ΔT stands for the difference in temperature
between the temperature of maximum fluidity and any other temperature. This is shown in Figure 15. It is evident from this figure

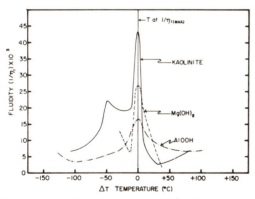

Fig. 15. The fluidity $(1/\eta_1)$ as a
function of ΔT for metakaolin, brucite
and boehmite. (ΔT is the temperature
difference between the temperature
of maximum fluidity and any other
temperature).

that metakaolin is more
fluid than either $Mg(OH)_2$
or A100H at their highest
fluid state. However, the
effect of temperature
should also be considered,
as the reaction tempera-
tures (in Tm) are significantly different in these
materials. For example,
the metakaolin, brucite
and boehmite transfor-
mation temperatures are
0.61 Tm, 0.22 Tm and 0.32
Tm, respectively (where
Tm is the melting point of
the final products of
transformation mullite,
MgO and $\alpha - Al_2O_3$). On
the other hand, the above

approach helps to generate a compactibility index for different
types of materials under reactive hot-pressing conditions.

CONCLUSIONS

Transformation plasticity of several systems involving either
a polymorphic phase change or a decomposition reaction (with
mass loss) has been evaluated by studying their compaction
characteristics. During the transformation of quartz to cristobalite, the formation of an intermediate glassy phase controlled
the flow properties of the system. Thus viscous flow can be
considered to be the predominant mechanism of mass transport
in the system, after an initial particle rearrangement stage.
The transformation plasticity during monoclinic to tetragonal
transformation of zirconia is most likely controlled by the
internal strain induced by the stress associated with the volume
change ($\Delta V/V$) and the flow stress by the weaker phase. Particle
movement and deformation of the weaker phase (probably the
tetragonal phase) produced the plasticity during its phase change.
The compaction behavior during metakaolin to a spinel phase and
brucite to periclase transformation can be interpreted with
a viscoelastic model. The temperature involved in the metakaolin
transformation is 0.62 T_m, thus it is possible that diffusion
control mass transfer mechanism was involved in the transformation
plasticity in this system, although the exact mechanism could
not be determined at present. In the case of brucite transformation to periclase, vapor phase lubrication will enhance
particle rearrangement and this will result in some densification.
This, in combination with some deformation of very fine particle
(~ 100 Å), may introduce enough plasticity in the system to
explain observed behavior encountered during reactive hot-pressing.

Acknowledgement: The experimental data were mostly generated
   by Mr. P. Wenman. Financial support from The National Research
   Council of Canada to conduct this research is gratefully
   acknowledged.

References

1. (a) A.C.D. Chaklader and L.G. McKenzie, Bull. Am. Ceram. Soc.,
       43, 892 (1964).

   (b) A.C.D. Chaklader and L.G. McKenzie, J. Am. Ceram. Soc.,
       49, 477 (1966).

(c)   T.G. Carruthers and T.A. Wheat, Proc. Brit. Ceram. Soc.,
      No. 3, 259 (1965).

(d)   P.E.D. Morgan and E. Scala, p.861 in 'Sintering and
      Related Phenomena', eds. G.C. Kuczynski, N.A. Hooton and
      C.F. Gibbon, Gordon Breach, N.Y. 1967.

(e)   A.C.D. Chaklader, Proc. Brit. Ceram. Soc., 15, 225 (1970).

2. P.W. Sunderland and A.C.D. Chaklader, J.Am. Ceram. Soc., 52,
    410 (1969).

3. A.C.D. Chaklader and G.T. Beynon, J. Am. Ceram. Soc., 53,
    577 (1970).

4. P. Murray, D.T. Livey and J. Williams, p.147 in Ceramic
    Fabrication Processes, ed. W.D. Kingery, Technology Press
    of Massachusetts Institute of Technology and John Wiley and
    Sons, Inc., N.Y. 1958.

5. W.D. Kingery, J. Appl. Phys., 30, 301 (1959).

6. (a)   M.P. Volarovich and A.A. Leonteva, J. Soc. Glass Tech.,
          20, 139 (1936).

    (b)   J.F. Bacon, A.A. Hasapis and J.F. Wholley, Jr., Phys.
          Chem. Glasses, 1, 90 (1960).

    (c)   G. Hetherington, K.H. Jack, and J.C. Kennedy, Phys. Chem.
          Glasses, 5, 130 (1964).

    (d)   S.A. Dunn, Bull. Am. Ceram. Soc., 47, 554 (1968).

7. G.W. Greenwood and R.H. Johnson, Proc. Roy. Soc., A 283,
    403 (1965).

8. A.C.D. Chaklader and R.C. Cook, Bull. Am. Ceram. Soc., 47,
    712 (1968).

9. G.W. Brindley and M. Nakahira, J. Am. Ceram. Soc., 42, 314
    (1959).

10. (a)   R.S. Bradbeer and A.C.D. Chaklader, p.395 in "Materials
           Science Research", Vol.6, ed. G.C. Kuczynski, Plenum.
           Publ. Corp., N.Y. 1973.

     (b)   R.S. Bradbeer and A.C.D. Chaklader, p.553 in "Science
           of Ceramics", Vol.7, Published by "Societe Francaise
           de Ceramique", 1973.

11. (a) P.E.D. Morgan, p.251 in "Ultrafine-Grain Ceramics", Eds. J.J. Burke, N.L. Reed and V. Weiss, Published by Syracuse University Press, Syracuse, N.Y. 1970.

    (b) T.A. Wheat and T.G. Carruthers, p.33 in "Science of Ceramics", Vol.4, Published by Brit. Ceram. Soc., England, 1968.

12. (a) D.T. Livey, B.W. Wankyln, M. Hewitt and P. Murray, A.E.R.E. Harwell, U.K., M/R, 1957 (1956).

    (b) S.J. Gregg and R.K. Packer, J. Chem. Soc., p.51 (1955).

    (c) R.I. Razouk and R.S. Mikhail, J. Phys. Chem., $\underline{63}$, 91 (1964).

13. (a) A.C.D. Chaklader and A.L. Roberts, J. Am. Ceram. Soc., $\underline{44}$, 35 (1961).

    (b) A.C.D. Chaklader, ibid, $\underline{46}$, 66 (1963).

    (c) A.C.D. Chaklader, ibid, 46, 192 (1963).

14. (a) D. Turnbull and M.H. Cohen, J. Chem. Phys., $\underline{29}$, 1049 (1958).

    (b) N.G. Ainslie, J.D. Mackenzie and D. Turnbull, J. Phys. Chem., $\underline{65}$, 1718 (1961).

15. W.D. Kingery, J.M. Woulbroun and F.R. Charvat, J. Am. Ceram. Soc., $\underline{46}$, 391 (1963).

16. T. Vassilos, ibid, $\underline{43}$, 517 (1960).

17. I. Amato, R.L. Colombo and M. Ravizza, J. Nucl. Mater., $\underline{22}$, 97 (1967).

18. (a) P.S. Nicholson and R.M. Fulrath, J. Am. Ceram. Soc., $\underline{53}$, 237 (1970).

    (b) G.R. Blair and A.C.D. Chaklader, J. Thermal. Anal., $\underline{4}$, 311 (1972).

19. J. Crampton, B. Escaig and A.C.D. Chaklader, to be published.

TRANSFORMATIONAL SUPERPLASTICITY IN PURE

Bi_2O_3 AND THE Bi_2O_3-Sm_2O_3 EUTECTOID SYSTEM

C. A. Johnson[*], J. R. Smyth[†], R. C. Bradt and J. H. Hoke

[*]General Electric Co., Research and Dev., Schenectady, NY
[†]Iowa State Univ., Ceramic Engr. Dept., Ames, Iowa
The Pennsylvania State University, University Park, Pa.

INTRODUCTION

Transformational superplasticity is a unique type of deformation phenomenon that is associated with an allotropic phase transition. It was first observed by Sauver in 1924 [1]. This deformation can be induced by thermally cycling a material about its phase transition while under a small externally applied stress. The superplasticity manifests itself by a large discontinuous increment of strain at the transformation temperature. If the deformation process remains active, multiple cycling about the transition yields additive strains that can lead to total strains far in excess of the strain to fracture experienced under normal conditions. This phenomenon has received considerable metallurgical attention during the past decade, and as such has been the subject of several excellent review articles [2-4]. It should not be confused with isothermal or structural superplasticity which depends on a fine grain size or a duplex microstructure rather than a phase transformation.

Various mechanisms have been proposed to explain the phenomenon, including (i) a momentary loss of coherency at the transition phase boundary [5], (ii) ease of deformation of the newly formed phase [6], and (iii) enhanced diffusional processes at the transformation interface [7,8]. None of these mechanisms have been accepted without reservation, rather a process that is independent of the actual mechanism proposed by Greenwood and Johnson [9] is generally favored to describe the phenomenon. The concept is based on the premise that the internal stress generated by the volume change of the phase transition activates the deformation process, which in turn is biased by the small external stress. Their analysis yields the equation:

$$\varepsilon_{trans} = A \frac{\sigma(\Delta v/v)}{S}$$
(1)

where ε_{trans} is the transformational strain, A is a constant, σ is the applied stress, $\Delta v/v$ is the volumetric strain of the transition, and S is the strength parameter. This deformation mechanism may be any of the commonly observed deformation processes and need not be a special one, peculiar to a phase transition. The predicted linear strain-stress relation has been confirmed in numerous metallurgical studies and is now accepted as a criteria for the phenomenon. The general form of the Greenwood and Johnson equation has been experimentally substantiated[10] by the extensive metallurgical study of DeJong and Rathenau.

References to the existence of superplasticity in ceramics have been made for the processes of pressure calcintering or reactive hot pressing[11-14], both of which exhibit discontinuities in the vicinity of their decomposition temperatures. These do not appear to be the true superplastic phenomenon. Also ascribed to transformational superplasticity, but not demonstrated to adhere to equation (1) are observations on the monoclinic to cubic phase transition in zirconia[15] and also the alpha to beta quartz transition[16] Possibly related are two hot pressing studies, one on zirconia[17] and another on alumina[18]. Deformation characteristics in agreement with equation (1) have been reported for Bi_2WO_6[19].

There have not been any studies of, or references to, superplastic deformation of eutectoids in ceramic systems, except for the Bi_2O_3-Sm_2O_3 work reported here[20]. However, numerous studies in metallic systems[21-27] indicate that there is a relationship between the eutectoid reaction in a material and its superplastic behavior. The same linear relationship exists between transformational strain and applied stress that occurs for pure materials. These lines do not usually intersect the origin, indicating either a strain at zero stress, if the strain axis is intercepted; or the necessity of a threshold stress level to induce deformation, if the stress axis is intersected. Also observed is a strain rate sensitivity approaching unity during the thermal cycling, indicative of a viscous like deformation.

Bi_2O_3 AND THE Bi_2O_3-Sm_2O_3 EUTECTOID

Levin and Roth[28] have extensively studied Bi_2O_3 and a large number of Bi_2O_3 binary systems, including Bi_2O_3-Sm_2O_3[29]. Pure Bi_2O_3 transforms from a low temperature monoclinic structure to a high temperature cubic form at 730°C, about 90% of the absolute melting point of 825°C. The detailed nature of the crystallography of the transition is not known, but it increases in volume about

seven percent when transforming to the cubic phase. Levin and Roth
also report that two metastable structures may exist in pure Bi_2O_3,
but they have not been observed in any of the studies here.

The reported Bi_2O_3-Sm_2O_3 equilibrium diagram is shown in Figure
1, with the two hypoeutectoid and two hypereutectoid compositions of
interest clearly labeled. Although Levin and Roth[29] report the
eutectoid to be at four mole percent Sm_2O_3 and 690°C, observations
during these studies revealed the true eutectoid to be nearer to
five percent Sm_2O_3. The structure of the eutectoid microconstituent
is extremely fine as shown in Figure 2. It is lamellar, analogous
to the classical ferrous eutectoid, pearlite. The eutectoid reaction
on cooling is that of a cubic solid solution of Sm_2O_3 in Bi_2O_3 transforming to the monoclinic Bi_2O_3 phase and a rhombohedral phase of
approximately 88 mole percent Bi_2O_3 and twelve mole percent Sm_2O_3.

Dense billets of pure Bi_2O_3 and the Bi_2O_3-Sm_2O_3 compositions
were consolidated from reagent grade powders by hot pressing below
the phase transition, the former at 650°C and 3000 psi for 45
minutes and the latter at 675° and 3000 psi for 30 minutes. The
Bi_2O_3-Sm_2O_3 compositions were prereacted at the (x) temperatures in
Figure 1 prior to hot pressing. Actual 1/2" diameter specimens were
diamond core drilled from the 3/4" thick billets parallel to the
pressing direction. These were subsequently tested in compression
parallel to the pressing direction. As a point of information, the
pure Bi_2O_3 was typically 95% dense with a ten micron average grain
size. Larger grain sizes were grown by extensive annealing at 700°C.
The Bi_2O_3-Sm_2O_3 compositions were comparably dense, with a submicron
lamellar eutectoid and proeutectoid grains varying from ten to twenty
microns in diameter.

Measurement of the transformational superplastic strain of these
materials was carried out in compression using a lever arm system in
a standard creep rig equipped with an LVDT calibrated to monitor the
deformation. All tests were in air. When these materials were
subjected to only small stresses and concurrently heated through
their phase transformations, large discontinuous strains were observed, similar to the superplastic behavior of metals. Details
of the thermal cycling are discussed elsewhere [19,20]. It is sufficient to say here that the resulting transformational strains were
substantial, exceeding all other thermal expansion and creep contributions combined, and in the opposite direction of the volume change
on heating. In pure Bi_2O_3, as will be discussed later, the strain
occurred only on the heating portion of the first cycle because of
excessive grain growth in the cubic phase. However, in a hypereutectoid composition where the second phase inhibited grain growth,
multiple cycling yielded additive strains.

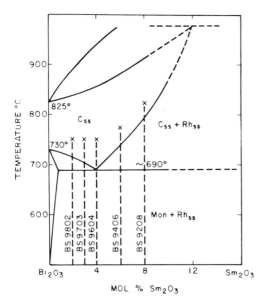

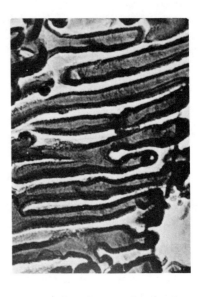

Figure 1. The reported Bi_2O_3-Sm_2O_3 phase diagram[29].

Figure 2. The Bi_2O_3-Sm_2O_3 eutectoid microconstituent. (40,000X)

TRANSFORMATIONAL STRAIN IN PURE Bi_2O_3

The resulting transformational superplastic strain for a 94% density, ten micron grain size body during the heating portion of the first cycle about the monoclinic/cubic transition is shown in Figure 3. No transformational strain was observed on cooling or on subsequent cycles in pure Bi_2O_3. The linear behavior is that predicted by Greenwood and Johnson; however, the magnitude of the strain is about two orders of magnitude larger than that reported for most metal systems.

Observations related to the rate of heating through the transition, plus the effect of varying the initial grain size may be applied to explain some of these aspects of the phenomenon. Figure 4 illustrates the effect of heating rate through the transition, while Figure 5 indicates the pronounced grain size effects. Clearly the deformation mechanism producing this transformational superplasticity is a time dependent, grain size dependent one. In fact an extrapolation of the grain size data of Figure 5 suggests that there should not be any transformational strain for pure Bi_2O_3 at grain sizes greater than about twenty microns. The grain size effect is what inhibits transformational strain on subsequent cycles, as Figure 6 illustrates the grain growth that results from heating to the high temperature phase. Relating to Figure 5, the grain size after a

single transformation is well in excess of the twenty micron limit
for transformational superplasticity. Recalling the conventional
strength parameter in Greenwood and Johnson's equation (1) and considering it in light of the grain size effect as well as the heating
rate effect, suggests that the deformation mechanism is a time dependent, grain size dependent one, probably grain boundary sliding.

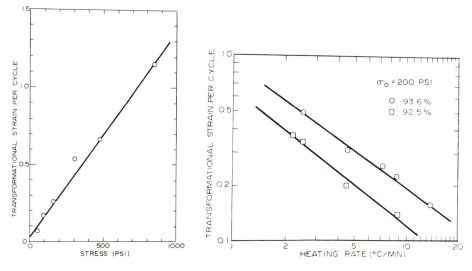

Figure 3. The effect of stress on ε_{trans} of pure Bi_2O_3.

Figure 4. The effect of heating rate on ε_{trans} of pure Bi_2O_3.

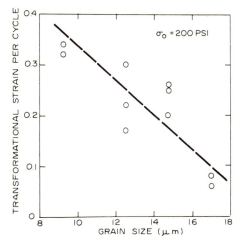

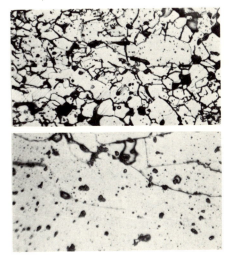

Figure 5. The effect of initial grain size on ε_{trans} of pure Bi_2O_3.

Figure 6. The grain sizes of pure Bi_2O_3 before and after transformation.

THE Bi_2O_3-Sm_2O_3 EUTECTOID

The pronounced grain size effect in pure Bi_2O_3 suggests that if a two phase eutectoid system were employed, the second phase may inhibit grain growth and permit multiple cycling, leading to additive strains. This is one reason for studying a eutectoid system. Figure 7 illustrates the transformational superplastic phenomenon in the reported eutectoid as it compares with the phenomenon in pure Bi_2O_3. Two points are worth noting. First, the transformational strains are slightly less in the two phase system, and second, the intercept suggests a threshold stress for activation of the superplastic process. As previously mentioned the four percent Sm_2O_3 is actually slightly hypoeutectoid, as is illustrated in Figure 8 where the proeutectoid Bi_2O_3 are the light grains. Figures 9 and 10 further illustrate the off-eutectoid compositions as they affect the superplastic transformational strain. The proeutectoid microconstituent apparently not only reduces the amount of strain in the first cycle, but it also appears to raise the threshold stress level. Although it is not clearly resolved because both the volume fraction and the grain size of the proeutectoid microconstituents vary in these compositions, it appears virtually certain that the level of the threshold stress is directly related to the presence of the proeutectoid phase.

The results of a multiple cycling study at three different stress levels for an eight percent Sm_2O_3 hypereutectoid composition is illustrated in Figure 11. Figure 12 is the microstructure after four cycles at 200 psi, which compared with Figures 6 and 8 clearly reveals the structural stability, for no grain growth has occurred, nor did any microcracking. Similar to metal systems, this composition can be multiply cycled to yield additive strains. However, the lines do not extrapolate to zero and the subsequent strains seem to be slightly reduced. However the resulting transformational superplastic strains are still considerably larger than reported for metal systems.

TRANSFORMATIONAL SUPERPLASTICITY IN CERAMICS

Table I summarizes data for some of the metals and ceramics demonstrated to have superplastic transitions. From this an important point is evident, namely that the ceramics have much greater strains per cycle at a given applied external stress level than the metal systems. The reasons for this are apparent in the Greenwood and Johnson equation, where the ceramics have a much larger volume change and very likely a much smaller strength parameter too. As for the strength parameter, in ceramics it is probably indicative of a grain boundary sliding mechanism of creep, which is likely very rapid at homologous temperatures of >90% of the melting point. Some indication of the importance of this melting point fraction

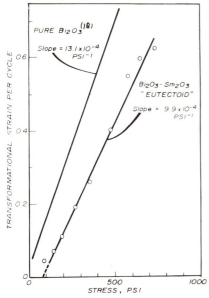

Figure 7. The ε_{trans} for the "eutectoid".

Fig. 8. Microstructure of the "eutectoid", 9604. (750X)

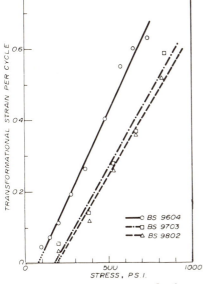

Figure 9. The ε_{trans} of the hypoeutectoids.

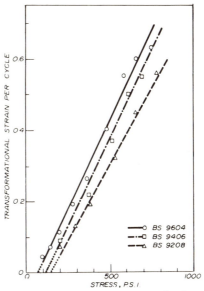

Figure 10. The ε_{trans} of the hypereutectoids.

TABLE I

COMPARISON OF MATERIALS WHICH EXHIBIT TRANSFORMATIONAL SUPERPLASTICITY

Material	Transition	$\Delta V/V$	Temperature (°C)	T/T_{MP}	ε_{Trans}/cycle/100 psi	Reference
Bi_2O_3	mon→cubic	≈0.07	730	0.91	1.4×10^{-1}	(19)
$Bi_2O_3-Sm_2O_3$	eutectoid	≈0.03	690	0.74	0.8×10^{-1}	(20)
Bi_2WO_6	-	≈0.05	940	0.91	6.1×10^{-1}	(19)
Zirconium	α-β	0.007	860	0.53	6.2×10^{-3}	(9)
Titanium	α-β	-	882	0.59	5.8×10^{-3}	(30)
Uranium	β-γ	0.007	775	0.75	5.5×10^{-3}	(9)
Iron	α-γ	0.012	910	0.65	1.7×10^{-3}	(31)
Cobalt	ε-α	0.004	417	0.39	$2.0 \times 10{-5}$	(32)

effect can be gained by comparing the cobalt transition with those of zirconium and uranium. There is no apriori reason to believe the constant, A, preceding these factors in equations (1) should be any different for ceramics than metals.

These results on Bi_2O_3 and the Bi_2O_3-Sm_2O_3 eutectoid clearly indicate that ceramics, under the proper microstructural conditions, can exhibit transformational superplasticity analogous to that reported for metals. Results are in tentative agreement with the Greenwood and Johnson theory. Within the framework of that theory, it appears that additional criteria must be met for ceramic systems to exhibit transformational superplasticity. They are (a) a large transformational volume change to induce large internal stresses and (b) the occurrence of the transformation at $T/T_{mp} \geq 0.75$, so that a substantial deformation process can be activated. These Bi_2O_3 system studies suggest that the deformation process of transformational superplasticity in ceramics is a grain boundary sliding mechanism.

ACKNOWLEDGEMENT

The authors would like to acknowledge the support of the United States Atomic Energy Commission for these studies.

REFERENCES

1. A. Sauver, Iron Age, 113, 581-583 (1924).

2. E. E. Underwood, J. Metals, 14, (12), 914-919 (1962).

3. R. H. Johnson, Met. Rev., 15, Review 146, 115-134 (1970).

4. G. J. Davies, J. W. Edington, C. P. Cutler, and K. A. Padmanabhan, J. Mat. Sc., 5, 1091-1102 (1970).

5. M. G. Lozinsky and I. S. Simeonova, Acta Met., 7, (11) 709-715 (1959).

6. L. F. Porter and P. C. Rosenthal, Acta Met., 7, 504-514 (1959).

7. F. W. Clinard and O. D. Sherby, Acta Met., 12, (8), 911-919 (1964).

8. F. W. Clinard and O. D. Sherby, Trans. AIME, 233, (11), 1975-1983 (1965).

9. G. W. Greenwood and R. H. Johnson, Proc. Roy. Soc., A283, 403-442 (1965).

10. M. DeJong and G. W. Rathenau, Acta Met., $\underline{9}$, (8), 714-720 (1961).

11. A. C. D. Chaklader and L. G. McKenzie, Am. Cer. Soc. Bull., $\underline{43}$, (12), 892-893 (1964).

12. A. C. D. Chaklader and L. G. McKenzie, J. Am. Cer. Soc., $\underline{49}$, (9), 477-483 (1966).

13. A. C. D. Chaklader, Proc. Brit. Cer. Soc., $\underline{15}$, 225-245 (1970).

14. P. E. D. Morgan, Chpt. 12, Pg. 251-271, <u>Ultrafine-Grain Ceramics,</u> Syracuse Press, (1970)

15. J. L. Hart and A. C. D. Chaklader, Mat. Res. Bull., $\underline{2}$, (5) 521-526 (1967).

16. A. C. D. Chaklader, Nature, $\underline{197}$, (2), 791-792 (1963).

17. A. C. D. Chaklader and V. T. Baker, Am. Cer. Soc. Bull., $\underline{44}$, (3), 258-259 (1965).

18. C. J. P. Steiner, R. M. Spriggs, and D. P. H. Hasselman, J. Amer. Cer. Soc., $\underline{55}$, (2), 115-116 (1972).

19. C. A. Johnson, R. C. Bradt, and J. H. Hoke, "Transformational Plasticity in Bi_2O_3" (accepted for publication by the J. Am. Cer. Soc.)

20. J. R. Smyth, "Transformational Superplasticity in Bi_2O_3-Sm_2O_3", Ph.D. Thesis, The Penna. State Univ., November (1974).

21. L. F. Porter and P. C. Rosenthal, Acta Met., $\underline{7}$, (7), 504-514 (1959).

22. W. A. Backofen, I. R. Turner and D. H. Avery, Trans. ASM, $\underline{57}$, 980-990 (1964).

23. F. W. Clinard and O. D. Sherby, Acta Met., $\underline{12}$, (8), 911-919 (1964).

24. D. Oelschlagel and V. Weiss, Trans. ASM, $\underline{59}$, 143-154 (1966).

25. O. A. Ankara and D. R. West, JISI, $\underline{205}$, (1), 36-37 (1967).

26. R. A. Kot and V. Weiss, Met. Trans. $\underline{1}$, (10), 2685-2693 (1970).

27. G. R. Yoder and V. Weiss, Met. Trans., $\underline{3}$, (3), 675-681 (1972).

28. E. M. Levin and R. S. Roth, J. Nat. Bur. Std., 68A, (2), 189-
     195 (1964).

29. E. M. Levin and R. S. Roth, J. Nat. Bur. Std., 68A, (2), 197-
     205 (1965).

30. L. S. Richardson and N. J. Grant, Trans. AIME, 215, (2), 18-23
     (1959).

31. O. D. Sherby and J. L. Lytton, Trans. AIME, 206, (8), 928-930
     (1956).

32. P. Feltham and T. Myers, Phil. Mag., 8, 203-211 (1963).

TRANSFORMATION INDUCED DEFORMATION OF CORDIERITE GLASS-CERAMICS

EDWIN BUTLER

PILKINGTON BROTHERS LIMITED
R&D LABORATORIES, LATHOM, NEAR ORMISKIRK, LANCASHIRE,
ENGLAND

INTRODUCTION

During the crystallisation of a glass-ceramic, several phase transformations may take place before the final crystalline product is formed. The phase transformations are often accompanied by changes in volume which can set up internal stresses in the material. The major phases which appear during the crystallisation of a TiO_2 nucleated cordierite glass-ceramic, are magnesium aluminium titanate, a stuffed β-quartz solid solution and cordierite.

Recent work in this laboratory has shown that under certain crystallisation conditions, a severe degradation in strength of the fully crystallised cordierite glass-ceramic can occur. The reduction in strength has been correlated with the appearance of fine micro-cracks in specimens of the crystallised glass-ceramic.

It is the purpose of the present paper to investigate the relationship between the volume changes which can occur during the crystallisation of a glass-ceramic and the onset of micro-cracking.

Evidence for the micro-cracking mechanisms was sought using such techniques as dilatometry, optical and electron microscopy, and density measurement.

EXPERIMENTAL

The material used in this work was a glass of nominal composition 12.5% MgO, 30.5% Al_2O_3, 45.5% SiO_2 and 11.5% TiO_2 by weight.

Batches of two kilogrammes were melted in a platinum crucible in a gas direct-fired furnace at temperatures between 1550 and 1600°C.   The glass was cast into 10 x 100 x 200mm slabs and were annealed by heating for two hours at 720°C and cooling to ambient temperature over a period of two days.

The crystallisation experiments were carried out in an electric muffle furnace with a temperature control of $\pm$ 5°C.   To avoid difficulties in interpreting the data isothermal heating of the specimens was carried out at all times.

The effect of various cordierite growth rates on micro-cracking at constant grain size was studied by transforming specimens at temperatures of 1220, 1260 and 1300°C after a constant nucleation treatment.   The effect of cordierite grain size was studied by transforming the specimens at a constant temperature after nucleation at 800, 850, 900 and 1000°C.

The volume fraction of cordierite formed during the phase transformation, which at constant nucleation rate is related to the cordierite growth rate, the average cordierite grain size and the degree of micro-cracking were all measured using standard metallographic techniques on an optical microscope.   The relative fractions of β-quartz s.s. and cordierite during the transformation were checked by X-ray diffraction.

Electron microscope specimens were thinned using an Edwards ion beam etching unit and examined in an A.E.I. EM7 electron microscope at an accelerating voltage of IMV.

Density measurements were made using a stabilised liquid density column.

RESULTS AND DISCUSSION

The initial part of the study was concerned with identifying the specific phase transformation responsible for the micro-cracking.   Accordingly glass specimens were heated to various stages of the crystallisation process and examined for signs of micro-cracking.   There was no evidence of micro-cracking in specimens which has been heated to the stage where β-quartz s.s. and magnesium aluminium titanate (MAT) had formed.   However at higher temperatures micro-cracking was observed in specimens in where the cordierite grains has grown to the point of impingement on each other.

It is clear that the transformation from β-quartz s.s. to cordierite is responsible for the micro-cracking phenomenon.

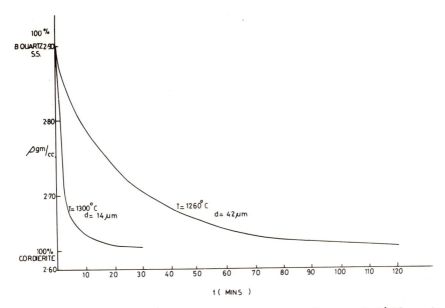

Fig. 1 Plot of density vs time for two specimens of different nucleation rate transformed at 1260 and 1300°C.

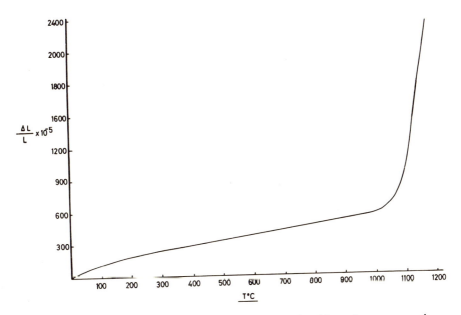

Fig. 2 Linear thermal expansion of a cordierite glass–ceramic on tranforming from β–quartz s.s. to cordierite.

Figure 1 shows how the density of specimens consisting of β-quartz
s.s. changes with time at the transformation temperature.   The
change in density corresponds to a volume expansion of about 10%
on the formation of cordierite.   To verify this expansion, a
specimen, consisting largely of β-quartz s.s. was heated in a
dilatometer from ambient temperature to 1250°C (Figure 2).   The
first section of the curve is due only to the thermal expansion of
the β-quartz s.s., but the steeply rising section, beginning at
1000°C, is a result of the expansion of the specimen on transformation
to cordierite.   The transformation from β-quartz s.s. to cordierite
over this temperature range was confirmed by X-ray diffraction.
Since this large volume strain does not always result in micro-
cracking it may be useful to consider what factors might be
involved in the criteria for micro-cracking.

As the transformation takes place at a relatively high
temperature ($\sim$ 1200°C) it is reasonable to assume that some of
the internal stresses would be relieved by thermally activated
deformation and recovery processes.   Transmission electron
microscopy of the micro-cracked specimens (Figure 3) shows that
extensive plastic deformation takes place during the transformation
from β-quartz s.s. to cordierite.

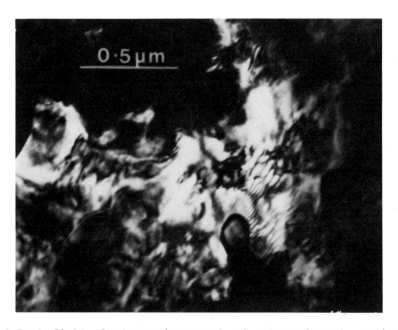

Fig. 3 Dark field electron micrograph of a transformed cordierite
glass-ceramic showing dislocation networks, tangles and sub-
boundaries. IMV.

Dislocation tangles, networks and sub-boundaries are clearly
visible, the latter features indicating that recovery processes
such as climb and intersection are operative during transformation,
The situation, therefore, may be analogous to the hot working
process where a dynamic equilibrium exists between strain and
recovery rates.   The onset of micro-cracking, therefore may
depend on which of these two processes is dominant.   In the present
case the strain rate is imposed by the rate of growth of the
cordierite grains.

     The growth rate studies at various transformation temperatures
indicated that there was no correlation whatsoever between growth
rate of the cordierite grains and the incidence of micro-cracking.
Specimens could be tranformed at both high and low growth rates to
produce cracks and vice versa.   However, when the effect of grain
size after transformation was investigated, a clear relationship
between grain size and micro-cracks became obvious.   By varying the
nucleation rate, grain sizes of between 5 and 55um were obtained.
Figure 4 shows the relationship between final cordierite grain
size and the surface area of crack per unit volume transformed at
various temperatures.

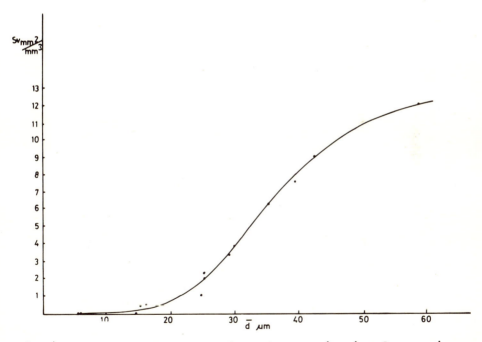

Fig. 4 Plot of surface area of crack vs grain size for specimens
transformed at various nucleation rates and temperatures.

It is clear that the dominating influence on micro-cracking is the cordierite grain size, as determined by the nucleation rate, and that the limiting grain age below which micro-cracking does not occur is about 10μm.   For a given nucleation rate the effect of growth rate was insignificant, a high nucleation rate (i.e. fine grain size) always resulted in a crack-free specimen, whereas a low nucleation rate (coarse grain size) always resulted in micro-cracking irrespective of the chosen growth rate.

It is apparent that the high temperature stress relieving mechanisms which have been observed by electron microscopy, play only a minor role in determining the onset of micro-cracking.

A model system representing the growth of cordierite grains in a cordierite matrix was devised to try to account for the grain size dependency of micro- cracking.   The model consists of a sphere, expanding in a spherical shell of the same material.   The radius of the inner sphere a is varied over the range of grain sizes observed, whilst b, the radius of the outer shell, remains constant at $\sim$ 10a,   Since plastic relaxation effects are insignificant, only the elastic solution is considered.   This model represents an elastic stress criterion approach to the problem of crack propogation.

The increase in radius of the inner sphere Δ due to the change in volume is given by:

$$\Delta = a \ (\sqrt[3]{(1+k)} - 1)$$ (1)

where k is the fractional increase in volume.

The tangential stress on the outer shell σ_t at radius r is,

$$\sigma_t = P_i \ (\frac{a^3 \ (b^3 + 2r^3)}{2r^3 \ (b^3 - a^3)})$$ (2)

where $$P_i = \frac{E\Delta}{a} \ (\frac{2(b^3 - a^3)}{3b^3(1 - \mu)})$$ (3)

where μ = Poisson's ratio.

Combining (2) and (3)
$$\sigma_t = \frac{1}{3} \frac{E\Delta}{(1-\mu)} \ (\frac{a^2 \ (b^3 - 2r^3)}{r^3 \ b^3})$$ (4)

The maximum stress occurs at the interface, r = a   and is independent of a but the rate of change of stress with radius depends critically on the particle radius a.   This means that the stress field extends further from the interface the larger the particle size.

Figure 5 illustrates the relationship between grain size and depth of crack penetration from the interface.   Known values of the observed fracture stress (σ_f) Youngs modulus and imposed elastic strain were used to evaluate the factor ($\frac{\sigma_f}{E\Delta}$) which was then used as a stress criterion for the propogation of a flaw. Several values of ($\frac{\sigma_t}{E\Delta}$) at increasing distance from the interface ($r - a$) were computed from (4), for a range of grain sizes (a). The depth of the crack is then given by the distance from the interface ($r - a$) at which ($\frac{\sigma_t}{E\Delta}$) = ($\frac{\sigma_f}{E\Delta}$)$_c$.   This model can be compared to the energy criterion approach   which Lange and others (1,2) have used to explain the particle size dependency of cracking in two phase materials of differing thermal expansion coefficient. In this case the stored elastic energy available for the extension of a pre-existing flow is related to the particle size, and it is shown that for a given combination of materials, crack extension will not occur until the product σ^2 R (where R is the particle size) is $\geqslant$ a constant.   The constant, however, is difficult to evaluate and the shape of the pre-existing flaw or subsequent crack is most important.   In the present case the volume expansion is expected to give rise to radial cracks and this factor together with application of the stress criterion allows a more simplistic and semi-quantitative approach to be made.

Using observed values of the fracture strength, Youngs modulus and surface energy of cordierite glass-ceramics an inherent flaw size could be postulated allowing the minimum grain size below which micro-cracks are not observed, to be estimated.   However, a value for the surface energy is not available at present.

According to equation (4) an inherent flaw (r_c)would just be stable at a stress corresponding to a critical grain size a_c.   At smaller grain sizes the stress falls off at such a rate that at this critical radius r_c the stress is too low to propogate the crack. Above this critical grain size the stress gradient is much shallower and the crack will propogate up to the radius (r) where the stress falls to the equilibrium value.

This model, therefore, predicts fairly accurately the micro-crack/grain size relationship which has been observed in this system;   If in figure 5, the depth of crack penetration is converted to surface area/ unit vol., the general shape of the curve in figure 4 is reproduced by the dashed curve.

CONCLUSIONS

The micro-cracking phenomenon encountered during the crystallis-ation of a cordierite glass ceramic, has been shown to be associated with the transformation from B-quartz s.s. to cordierite.   The transformation is accompanied by a volume expansion of about 10%

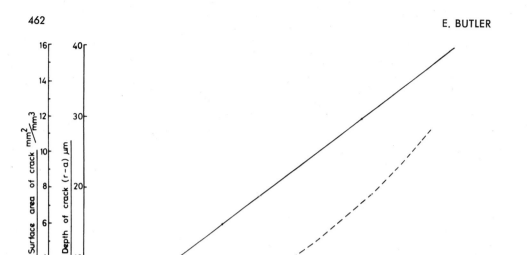

Fig. 5 Plot of crack depth vs grain size (full line) and crack surface area vs grain size (dashed line) as derived from equ. 4;

which is evidenced by a corresponding density change and a linear expansion of about 2%.

Transmission electron microscopy revealed the operation of high temperature stress relieving mechanisms such as plastic deformation and recovery during the change from β-quartz s.s. to cordierite, however, transformation experiments suggest that they have little effect on the criterion for micro-cracking.

The micro-cracking is shown to be a function of the final cordierite grain size, which is related to the cordierite nucleation rate. A model system based on a stress criterion has been devised and can satisfactorily explain the grain size dependency of micro-cracking and also predicts a minimum grain size below which micro-cracking is not expected to occur.

REFERENCES

1. F. F. Lange, "Criteria for crack extension and arrest in residual, localised stress fields associated with second phase particles." Westinghouse technical report No. 9 (N00014-68-C-0323)

2. R.W. Davidge and T.J. Green, "The strength of two-phase ceramic/glass materials", J. Mat. Sci. 3, 629 (1968).

ACKNOWLEDGEMENTS

The author wishes to thank the Directors of Pilkington Brothers Limited, and Dr. D.S. Oliver, Director of Group Research and Development for permission to publish this paper.

He also wishes to thank Dr. G. Nair of Pilkington Brothers and Dr. F.F. Lange of Westinghouse Research for very helpful discussions.

THE VARIED ROLE OF PLASTICITY IN THE FRACTURE OF INDUCTILE CERAMICS

K. R. Kinsman, R. K. Govila and P. Beardmore

Scientific Research Staff, Ford Motor Company

Dearborn, Michigan 48121

ABSTRACT

Even in the most brittle solids there is usually a measure of local microplasticity which influences the temperature dependence of the fracture stress. Deviations in behavior from the usual response of solids which exhibit increasing ductility with temperature have been examined for a range of nominally inductile ceramics including a LAS glass ceramic, Si_3N_4, SiC, and Al_2O_3. The temperature dependence of the fracture behavior has been evaluated using specimens in which cracks of controlled size were introduced by a hardness indentation technique. The effects on the fracture behavior of varying crack lengths by using different indentation loads, of blunting the implaced cracks, and of varying temperature at constant crack size were determined. The results are consistent with the anticipated influence of plasticity on fracture behavior in metals, polymers and more ductile non-metallics.

INTRODUCTION

In structural applications ceramics have little tolerance for flaws. Even though they may be characterized as inductile, most are capable of a modicum of plastic response to applied force. The nature and source of this plasticity determines the fracture resistance of the material. The proper amount of local plasticity, confined to the stress field of the advancing tip of a crack, can be beneficial in preventing the crack from enlarging to failure. Thus, in theory ceramics are innately sound for structural use so long as the incipient flaws are actively prevented from enlarging. In practice, however, occasionally the manifestation of plastic response of the material is to promote crack extension rather than blunting. The dichotomy, therefore is that the same dislocation activity that leads to crack blunting alternatively can act to nucleate or extend a flaw. More significantly, for recent interest sake, the temperature dependence of plasticity typically is positive, thereby raising the possibility of high temperature plasticity of a detrimental kind. From a structural viewpoint, ceramics are usually complex in the sense that they are typically comprised of more than one phase and contain significant "impurity" levels either or both of which may tend to segregate to crystalline grain boundaries. The purpose of this paper is to describe some aspects of the temperature dependence of fracture in certain structural ceramics which recently have been the subject of some attention. We do this by observing the behavior of materials containing strength limiting flaws of a controlled size (1,2) and by noting the plasticity factors at play by fractographic inspection.

MATERIALS

Hot pressed Si_3N_4 (Norton HS-130) and a nominally fully dense lithium-aluminum-silicate (Owens-Illinois C-140) were investigated in some detail, a preliminary study was made of a hot pressed SiC (Norton) and a commercially available Al_2O_3 (G.E. Lucalox). While the nominal structure of hot pressed Si_3N_4 is yet the subject of some discussion (3,4), for the purpose of this investigation we may consider it to be comprised of a mixture of α and β Si_3N_4, and that the strength limiting impurities segregate to the crystalline boundaries in the form of a less refractory phase. Conventional interpretation (5-10) considers that the grain boundary phase is a viscous glass at high temperatures comprised chemically of an unfortunate combination of processing aids, principally MgO and CaO, as well as oxygen residual to the transformation of α to β Si_3N_4 during fabrication (11). Lithium aluminum silicate (LAS) is a crystallized glass ceramic comprised principally of the β-spodumene and β-eucryptite structures amidst which is dispersed intentional and significant impurity fractions initially in the form of processing aids (12). In such glass ceramics, it is safe to assume that crystallization is not entire and that residual glass is located at crystalline interfaces as a viscous layer. Current fabrication practice for hot pressed SiC seems to yield a product less overtly responsive to impurities (13) though the differences are likely more in extent than in kind. Commercially available Al_2O_3 (Lucalox) contains significant grain boundary impurity segregation (14) though it appears that it is not viscous in the sense of the segregant in Si_3N_4 and LAS.

EXPERIMENTAL PROCEDURE

Specimens of dimensions 1/8" x 1/4" x 1" were machined from blocks or bars of the various materials. In those instances where directionality due to hot pressing is a factor to be considered, the bars were fashioned so that the tensile face was perpendicular to the hot pressing direction [i.e. "strong direction" (15)]. The tensile surface was ground in a lengthwise direction to a 220 grit diamond finish and the edges chamfered (16) on those specimens slated for baseline fracture tests. Additional polish to a 3μ diamond finish was done preparatory to controlled cracking of the balance of the specimens. The LAS and Al_2O_3 were then baked out 2 hours at 250°F to remove water entrained from the polishing media.

A single operational crack* was introduced into the center of the polished surface of the specimen by an indentation technique using a diamond pyramid (Vickers) indenter. The technique has been described elsewhere (1,2) and is shown schematically in Fig. 1a; a typical surface crack(s) and subsequent appearance of the fracture surfaces are shown in Fig. 1b and Fig. 1c,d respectively. Preliminary studies directed to establish the relationship between indentation load and crack depth indicated a reproducibly definable correspondence. In most instances (Al_2O_3 the exception) crack depth, CO (Fig. 1), could be measured directly from the fracture face.

The temperature dependence of the fracture strength was tested over the range room temperature to 1550°C in four point flexure (midspan 5 mm.; outer span 15 mm.). Materials were evaluated nominally in the as-received condition. Tests were performed in air in a Centorr furnace affixed to an Instron testing machine operated, unless otherwise indicated, at a crosshead speed of 0.005 in./min. (corresponding to a surface strain rate of 0.0014/min.). The specification of strain rate in terms of crosshead rate is continued for convenience.

RESULTS AND DISCUSSION

The range of results is most conveniently described by considering each material in turn, emphasizing commonality in behavior. To a degree the procedure is meant to be tutorial with regard to phenomenology, yet heuristic because the phenomena we discuss are not well quantified for the specific systems considered.

---

*Actually two cracks are emplaced which cross at $\sim 90°$ (Fig. 1). Operationally one is located at maximum crack opening stress, the other has been demonstrated to be completely benign in this type of experiment.

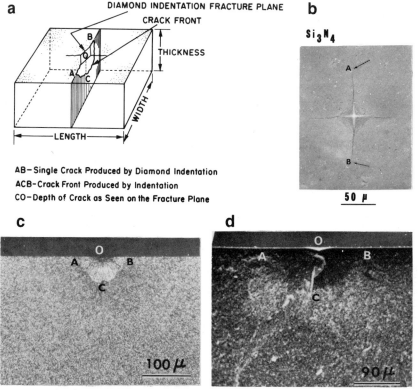

Fig. 1   (a.) Schematic representation of the significant features
of the hardness-indenter-emplaced crack. (b.) The appearance on
the tensile surface of crack(s) emplaced in hot pressed silicon
nitride with 3 kg. load. Appearance of the emplaced crack on the
fracture surface in (c.) lithium-aluminum-silicate and (d.) hot
pressed silicon nitride.

Silicon Nitride

    The simplest comparison of the fracture behavior of material
in the "normal" state versus intentionally flawed hot pressed Si_3N_4
is seen in Fig. 2. The uppermost (nominally unflawed) curve is
representative not only of our own work but of the studies of
numerous investigators (5,9,11) using a range of specimen sizes. The
experimental scatter range has been omitted, the average through
the data sufficing for our purpose. The temperature dependence of
fracture stress is empirically simple, it remains relatively invariant to about 1100°C and then drops off smoothly to around half
strength at 1400°C. Throughout the entire temperature range, fracture occurs at the grain boundaries through the "glassy" impurity

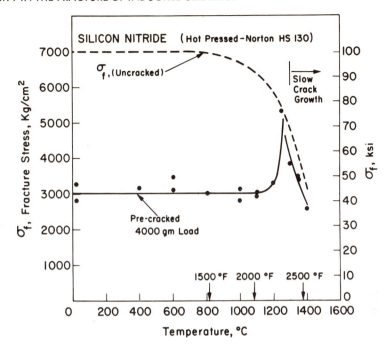

Fig. 2   Temperature dependence of fracture stress for the base-line material and for a precracked (90μ cracks) material.

layer which becomes less viscous as the temperature is raised. The first sign of weakness in the grain boundary impurity layer is manifest as a depression in fracture stress at about 1100°C as the viscous layer becomes unable to support the stress field associated with sharp cracks. The microplastic response of the grain boundary layer is mirrored also by a concomitant and proportional increase in ultrasonic attenuation (17) with temperature above 1100°C as well as the onset of detectible stressing rate dependence of frac-ture stress (18). While the former is a viscous response, the latter can be taken to auger a subcritical crack growth mechanism (19), though not conclusively.

When the material is intentionally flawed, as with a 90μ radius surface flaw (4000 gm indenter load), the fracture stress is both lower to an extent consistent with the fracture mechanics des-cription of the problem, and remains essentially temperature invar-iant, as would be expected for an invariance in the fracture process in this material. Between 1200°C and 1250°C there occurs a sharp peak in fracture stress. At that peak and at higher temperatures,

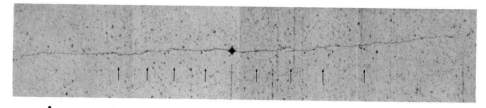

1350°C   **0.5 mm**

Fig. 3   Surface crack formed at 1350°C by incrementally loading
a small emplaced crack.

subcritical crack growth is included in the fracture process. The
peak in fracture stress indicates that the emplaced crack finds
it more difficult to propagate as the temperature is raised through
a narrow range. Such behavior is typical of brittle materials
experiencing the onset of some plasticity in the fracture process.
While the true intensity of the peak may be actually slightly less
than measured due to the testing configuration (20), it is a real
feature of the fracture behavior and is not testing mode dependent (21). Since the crack path is confined to the grain boundary
the plasticity process at play is likely not a classical crack tip
blunting in the sense that crystal dislocation slip accomodation
occurs, because the Taylor-Von Mises criteria is unlikely to be
met in these circumstances. Rather a displacement and general
disruption of the continuity of the crack tip out of the emplaced
crack due to viscous flow at the grain boundary is a more likely
mechanism of "blunting". That viscous flow is a detriment is
additionally supported by the observation (Fig. 2) that the fracture process is profoundly influenced by the higher stresses
supported by the smaller natural flaws in the baseline material.
This is manifest in Fig. 2 as a departure from temperature
invariant fracture stress at a lower temperature for the material
in the baseline condition than in the precracked condition. In
precracked material this is further substantiated in the systematic
shift to lower temperature (and higher stresses) of the fracture
stress peak for smaller artificial flaw sizes (22).

     The above discussion sets out conditions favoring subcritical
crack growth at loads below the normal fracture stress. However,
proof must be by observation and Fig. 3 shows a crack which has
been grown slowly in a controlled way by incrementally increasing
the stress on a crack originally the size shown in Fig. 1b. Crack
advance was monitored by observation between stress increases.

     Correlation with the features of the fracture surface was
accomplished by loading the specimen to failure at room temperature,
an acknowledged regime of catastrophic crack advance. There was a
one to one correlation between the extent of slow crack growth on

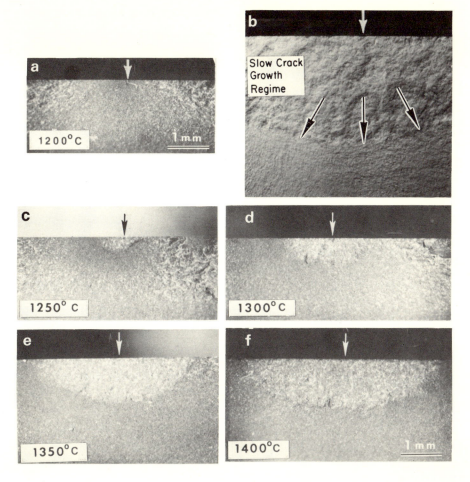

Fig. 4   Fracture appearance by optical ceramography typical of failure initiated at emplaced flaws (arrow) at temperatures (a.) below and (c.-f.) throughout the regime of subcritical crack growth. (b.) is a higher magnification scanning electron micrograph showing the surface roughness correlated with the slow growth fracture path at 1250°C.

the specimen surface and the feature so identified on the fracture surface on the scanning electron micrograph in Fig. 4. Consistent with the operational mechanism for crack "blunting" in this material, the slow crack growth path is intergranular but with significant excursions out of the mean fracture plane so that the optical reflectivity of the specimen is altered. This is strikingly apparent in the sequence (a) to (f) in Fig. 4. At 1200°C, and at a fracture initiating surface crack of radius ~ 90μ (indicated by arrow) the fracture path is intergranular and

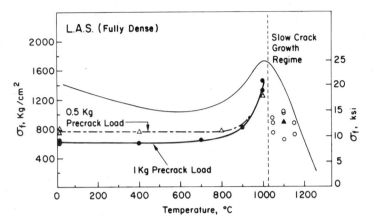

Fig. 5   Temperature and crack size dependence of fracture stress
in lithium-aluminum-silicate. The precrack loads indicated yield
30μ and 50μ radius surface cracks.

otherwise comparatively featureless, indicative of catastrophic
failure. At 1250°C and above (Fig. 2) the extent of slow crack
growth (as indicated by the change in reflectivity) prior to
catastrophic failure increases with temperature. At a constant
initial crack size the stress to initiate viscous flow controlled
crack growth decreases with increasing temperature. Concomitant
with the lower stress, the crack size required for the onset of
catastrophic failure as dictated by the Griffith criteria is
proportionally larger. Parenthetically, conventional fracture
mechanics tests (9) imply that subcritical crack growth is not a
factor below approximately 1250°C. Our results indicate that
this observation may be influenced by the operative "crack size".

     Hot pressed Si_3N_4 is an example of a brittle material which
experiences a viscosity related change in fracture mode (subcritical
crack growth) wherein the fracture path remains essentially
invariant.

Lithium-Aluminum-Silicate

     In LAS the temperature dependence of the fracture stress of
the baseline material is in a more classical mold as indicated by
the uppermost curve enveloping the data in Fig. 5. A slight
negative slope is followed by a peak in fracture stress at about
1000°C. Fractographic inspection indicates that the fracture mode

up to this temperature is transgranular cleavage; above 1000°C
intergranular slow crack growth occurs. There is evidence for
local microplastic response in this material even at room temperature (23). Since fracture is transgranular up to the peak stress,
the mechanism of increasing resistance to fracture more nearly
approximates the classical dislocation slip accomodation.

Fracture stress dependence on crack size is as anticipated
for a 30μ crack compared to a 50μ crack (Fig. 5) (24). The temperature dependence is similar, reflecting an invariance in fracture
mechanism and the onset of significant plastic "blunting".

The existence and fractographic features of subcritical crack
growth were confirmed directly in the manner applied to Si_3N_4, and
Fig. 6 shows examples of the behavior associated with naturally
occurring surface flaws geometrically similar to those artificially
emplaced. At a temperature below the peak, fracture occurs
entirely by cleavage; at high temperatures, the crack first extends
slowly along the viscous grain boundaries until it reaches a
critical size beyond which catastrophic cleavage can no longer be
suppressed. The grain boundary path is very rough and effects a
useful contrast in light optics (Fig. 7). A systematic temperature
dependence of this response is demonstrated by the montage in
Fig. 7 depicting the fracture surface appearance in association
with emplaced flaws of a uniform size. As in Fig. 6a, cleavage is
the sole fracture mode at 20°C and 900°C. But at 1000°C and 1150°C
a proportionately larger rim of grain boundary fracture occurs
around the initial crack as it grows "slowly" to the critical size
for catastrophic (cleavage) failure. The change in fracture path
associated with change in fracture mode is displayed in Fig. 8
wherein sections of the crack front within regions AC in Fig. 7(a,
c) are observed at high magnification.

Deformation of viscous media is strain rate sensitive. Even
though the viscous media here is presumably confined in a thin
layer at crystalline boundaries, the response in the regime where
the viscous layer behavior controls fracture should also be strain
rate sensitive. Thus the variables of temperature and strain rate
can be arranged to extend or contract the region of slow crack
growth along viscous boundaries. In Fig. 9 the strain rate
dependence of fracture stress at 1100°C is plotted for crosshead
speeds ranging from 0.002 in./min. to 0.5 in./min. At this
temperature the standard crosshead speed for our experiments,
0.005 in./min., yields significant development of viscous fracture
control in the form of subcritical crack growth (Fig. 7) prior to
catastrophic cleavage failure. The initial increase in fracture
strength with increasing strain rate reflects an approach to the
limiting rate at which the material can accomodate strain viscoplastically. At low rates, in a manner of speaking, the crack

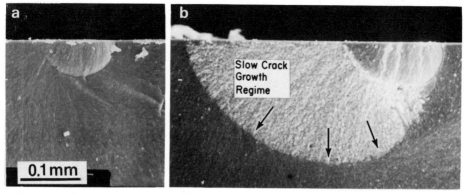

Fig. 6   Scanning electron micrographs of fracture initiated at natural half-penny shaped surface flaws at (a.) 950°C, crosshead rate 0.5 in./min. and (b.) 1100°C, crosshead rate 0.002 in./min.

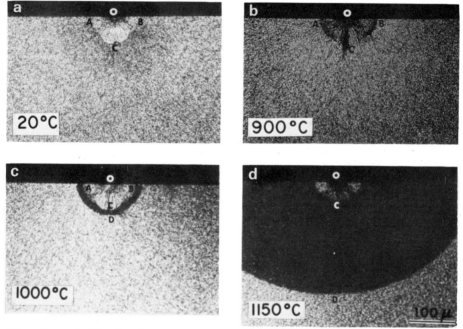

Fig. 7   Optical micrographs of the appearance of fracture initiating at emplaced flaws (1 kg. indenter load) of a uniform size at a series of temperatures. c. and d. exhibit a rim of slow crack growth enlarging the initial flaw.

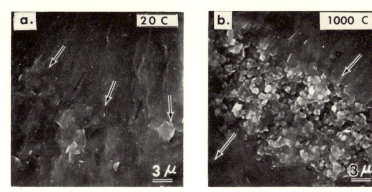

Fig. 8   High magnification scanning electron micrographs of
(a.) the crack front demarking the emplaced crack and continued
cleavage at 20°C, and (b.) a rim of intergranular fracture defining
the extent of slow crack growth at 1000°C.   1 kg. indenter load.

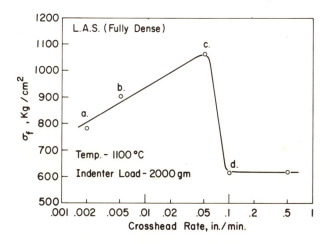

Fig. 9   Strain rate dependence of fracture stress at 1100°C for
LAS precracked with a 2000 gm. indenter load.

creeps to critical size. At the peak, the more limited viscous
response to deformation accomodation gives rise to optimal
"toughness". Beyond the peak the imposed strain rate exceeds the
ability of the material to conform viscoplastically at the grain
boundaries. While there is no doubt some bulk crystal plasticity
occurs, as discussed in relation to Fig. 5, this also cannot track

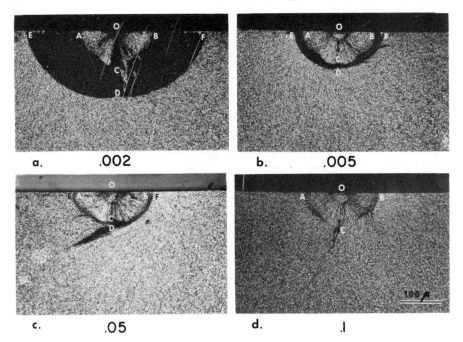

Fig. 10   Appearance of the fracture surface in association with
site of fracture initiation at emplaced crack as a function of
crosshead speed.   The indicated crosshead speeds and the frame
identification correlates with Fig. 9.

the rate of loading.   The material thereafter behaves in a brittle
manner, gains nothing in the way of plasticity augmented fracture
resistance, and fails at a lower and strain rate invariant stress.
Fractographic confirmation of the mechanisms at play throughout
this range for points (a.-d) in Fig. 9 are shown in Fig. 10 (a.-d.).
As expected, increasing strain rate, and increasing fracture
resistance, correlates with a decreasing extent of subcritical
crack growth (plastic zone) to the point (Fig. 10d) where it
vanishes altogether.

Even the slightest amount of subcritical crack growth serves
not only to alter the nature of the crack front but also changes
the size of the crack.   Cracks grown at high temperatures and
loaded to failure at room temperature yield the normal inverse
crack size dependence of fracture stress (24).   However, in the
regime anticipatory to crack extension along viscous boundaries
(as in the range 950-1000°C in Fig. 5) where plastic accomodation
apparently occurs by bulk crystal deformation, it is possible to
relax the stress concentration at the crack tip by bulk crystal
plastic deformation.   By imposing the proper load on a specimen

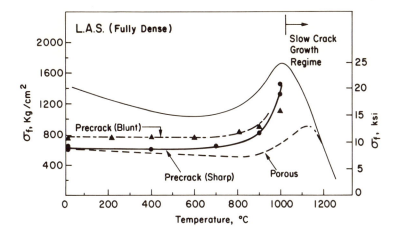

Fig. 11   Temperature dependence of fracture stress for lithium-
aluminum-silicate emphasizing the role of the sharpness of emplaced
cracks of uniform size (50μ radius).

containing an emplaced crack, for example at 1000°C, it is possible
to "blunt" effectively the crack without changing its size. That
this can be accomplished structurally has been confirmed ceramo-
graphically (24). The strength differential between sharp and
blunt cracks of the same size, as shown in Fig. 11, is reproducible
and of the expected sign.

Fully dense, nominally crystalline, lithium-aluminum-silicate
provides an example of a brittle material which experiences a viscosity related change in fracture mode coincident with a change in
fracture path.

Silicon Carbide

Temperature dependence of the fracture stress of hot pressed
silicon carbide is virtually featureless in the case where the site
of fracture initiation is determined by precracking. Modest temperature dependent behavior is found in the baseline state of the
sort generally consistent with the onset of some local plastic
response. However, the form of the dependence shown in Fig. 12a is
not firmly agreed upon (13,16). Fracture at all temperatures is a
mixture of transgranular and intergranular paths, attesting to the
relative temperature insensitivity in the fracture process of any
grain boundary segregants. A modest loading rate sensitivity of
fracture stress has been reported above 1200°C (11), consistent
with the tendency for a decrease in fracture stress in Fig. 12a, both
of   which could be grain boundary impurity related. No evidence
for subcritical crack growth was observed in our experiments

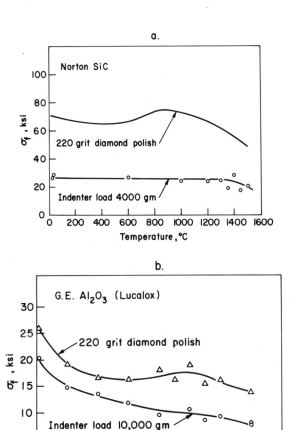

Fig. 12   Temperature dependence of fracture stress for (a.) baseline hot pressed Norton SiC compared to that precracked with a 4 Kg.
indenter load, and (b.) baseline Al₂O₃ (Lucalox) compared to that
precracked with a 10 kg. indenter load.

at 0.005 in./min. crosshead rate. However Prochazka and Charles
(13) report qualitative evidence for this phenomena at 1500°C in
the form of fractographic evidence akin to that discussed for Si_3N_4
(Fig. 4). Our experiments at this stage certainly do not preclude
the occurence of this phenomena at higher temperatures, lower strain
rates and higher stresses (at the emplaced crack tip) than currently
obtained in the precracked specimens.

Alumina

Preliminary investigation of Lucalox by the above procedures
has yielded a temperature dependence of a fracture stress more in
keeping with that observed at low temperatures in lithium-aluminum-
silicate though it is not unlike SiC (Fig. 12a). The data in
Fig. 12b is meant only to confirm feasibility of such a study, the
"baseline" 220 grit diamond finish being relatively far too harsh
a finish to pass as such. The temperature dependence of site
initiation controlled fracture, unlike others discussed, is
significant, possibly in excess of that accounted for by the
temperature dependence in the elastic modulus. As in SiC, no
evidence was found for subcritical crack growth though the same
disclaimer applies.

CLOSING REMARKS

Principally in Si_3N_4 and LAS, significant plastic response has
been observed in association with the fracture process. It is
clear that plastic accomodation involves a thin viscous layer remaining at crystalline boundaries and that subcritical crack growth
accompanies a lowering of viscosity with increase in temperature.
While in the case of Si_3N_4 this change in fracture mode involves
no change in fracture path, in the glass ceramic (LAS) a change in
fracture path and fracture mode are coincident. An observation of
broad implication, subtle in Si_3N_4 and obvious in LAS, is that the
nature of the crack front during subcritical crack growth is different than that during catastrophic failure.

It is instructive to schematicize the balance of our observations, in Fig. 13, as they provide a framework for discussion of
the evident trends. In parts (a.) and (b.) the upper curve is
taken to be that of a typical response of a brittle baseline material containing a distribution of natural flaws. Differences in
shape as occur, for example in Si_3N_4, are inconsequential to the
following remarks. The experimental approach used allows defini
tive commentary on the role of fracture initiating flaws of various
sizes and the trade-offs available by varying strain rate and
temperature:

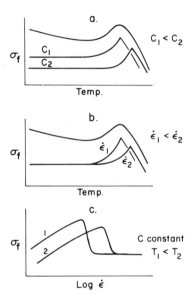

Fig. 13   Schematic representation of (a.) crack size dependence of σ_f at constant strain rate, (b.) strain rate dependence of σ_f at constant crack size, and (c.) strain rate dependence of σ_f as influenced by temperature at constant crack size.

Crack size dependence of σ_f at constant strain rate
(Fig. 13 a) - the smaller crack sustains a higher applied
load but in so doing causes plastic flow at a lower
temperature.

Strain rate dependence of σ_f at a constant crack size
(Fig. 13 b) - a lower rate of loading permits plastic
accomodation in the vicinity of the crack tip at a lower
temperature. A much higher rate of loading can wipe out
the peak in σ_f altogether, carrying the temperature
invariance of σ_f to its intersection with the σ_f
decrease due to gross plastic flow.

Both of these effects form a logical extension of the results reported in this paper and have been confirmed experimentally (22,24)
both with Si_3N_4 and LAS. The response of a material with a distribution of naturally occurring flaws to imposed temperature/strain
rate variables is akin to the effects described. An instructive
example is the strain rate response of a lithium-zinc-silicate
glass ceramic (only partially crystallized) studied by Lyall and
Ashbee (25). Their results are best described as a combination
of the effects portrayed in Fig. 13a,b against the realization
that the size of their fracture initiating flaws was not controlled
and likely (though not necessarily) constant.

Strain rate dependence of σ_f (Fig. 13c) - the mechanistic
nature of this dependence already has been discussed.
Additional functional dependence on crack size at constant
temperature and on temperature at constant crack
size are envisioned consistent with the character of the
controlling visco-plastic (diffusive) processes. While
the schematic Fic. 13c retains the convenience of Log $\dot{\varepsilon}$
used in Fig. 9, no significance is attached to that specification at this juncture except that it is consistent
with the envisioned diffusive processes.

The ability to control the size and site of fracture initiation
in an otherwise standard fracture test has permitted the evolution
of a different and useful vantage of the role of plasticity in
the fracture of nominally inductile materials.

ACKNOWLEDGMENTS

This work was supported in part by the Advanced Research Pro-
ject Agency under contract DAAG 46-71-C-0162. We thank
Mr. G. Easley for assistance in mechanical testing and
Mr. D. Janowski for the scanning electron microscopy.

REFERENCES

1. K. R. Kinsman, M. Yessik, R. K. Govila and P. Beardmore, submitted to Bull. Am. Cer. Soc.
2. R. K. Govila, Acta Met., 20, 447 (1972).
3. S. Wild, P. Grieveson and K. H. Jack, Special Ceramics V, P. Popper (Ed.), The British Ceramic Research Assoc., Stoke-on-Trent, 329 (1972).
4. H. F. Priest, F. C. Burns, G. L. Priest and E. C. Skaar, J. Am. Ceram. Soc., 56, 395 (1973).
5. D. W. Richerson, Bull Am. Ceram. Soc., 52, 560 (1973).
6. R. Kossowsky, J. Am. Ceram. Soc., 56, 531 (1973).
7. J. A. Mangells, Ceramics for High Performance Applications, Army Materials Technology Conf. Series, Cape Cod (1974).
8. A. G. Evans and J. V. Sharp, J. Mater. Sci., 6, 1292 (1971).
9. A. G. Evans and S. M. Wiederhorn, J. Mater, Sci., 9, 270 (1974).
10. S. D. Hartline, R. C. Bradt, D. W. Richerson, and M. L. Torti, J. Am. Ceram. Soc., 57, 190 (1974).
11. F. F. Lange, MCIC Rev. Ceram. Tech., No. 29, 1 (1974).
12. J. A. Mangells, personal communication (1973).
13. S. Prochazka and R. J. Charles, General Electric Rpt. 73 CRD 169 (1973).
14. W. C. Johnson, D. F. Stein and R. W. Rice, Fourth Bolton Landing Conf., (1974) in press.
15. A. F. McLean, E. A. Fisher and R. J. Bratton, Brittle Materials Design, High Temperature Gas Turbine, A.M.M.R.C., Contract No. DAAG 46-71-C-0162, Interim Rpt. CTR 72-19, pg. 97 (1972).
16. Ibid, Interim Rpt. CTR 73-9, pg. 147 (1973).
17. W. A. Fate, Scientific Research Staff, Ford Motor Company, unpublished research.
18. F. F. Lange, J. Am Ceram. Soc., 57, 84 (1974).
19. R. J. Charles, J. Appl. Phys., 29, 1657 (1958).
20. C. A. Anderson, D. P. Wei and R. Kossowsky, Plastic Deformation of Ceramic Materials, Pennsylvania State University, Plenum Press, in press.
21. R. W. Davidge and A. G. Evans, Mater. Sci. Eng. J., 6, 281 (1970).
22. K. R. Kinsman, R. K. Govila and P. Beardmore, Scientific Research Staff, Ford Motor Company, unpublished research.
23. K. R. Kinsman and P. Beardmore, Scientific Research Staff, Ford Motor Company, unpublished research.
24. R. K. Govila, P. Beardmore and K. R. Kinsman, Scientific Research Staff, Ford Motor Company, unpublished research.
25. R. Lyall and K. H. G. Ashbee, J. Mater. Sci., 9, 576 (1974).

DEFORMATION TEXTURE AND MAGNETIC PROPERTIES

OF THE MAGNETOPLUMBITE FERRITES

M. H. Hodge*, W. R. Bitler** and R. C. Bradt**

*TRW, Inc., Philadelphia, Pa.

**The Pennsylvania State University, University Park, Pa.

INTRODUCTION

The barium and strontium hexaferrites, $BaFe_{12}O_{19}$ and $SrFe_{12}O_{19}$, are ferrimagnetic magnetoplumbites with a hexagonal crystal structure favoring the [0001] easy magnetic direction. Commercial utilization of these materials has lead to the development of sintered poly-crystalline bodies in which a substantial degree of preferred orientation or texture of the individual crystallites has been accomplished by powder pressing in a magnetic field [1]. The uniaxial compaction involved, however, has a randomizing tendency, one that might be avoided if the crystallite alignment process were concurrent with or following the sintering stage. Several researchers have recently achieved a high degree of preferred orientation and correspondingly enhanced magnetic properties through hot forming [2-4]. This paper addresses the effects of hot working upon orientation in these ferrites.

EXPERIMENTAL PROCEDURE

Stoichiometric and commercial hexaferrites with both isotropic and wet-oriented processing histories were studied. The stoichiometric materials were consolidated at the exact 6:1 ratio, while the commercial bodies were non-stoichiometric and contained dopant grain growth inhibitors. In preparing the stoichiometric ferrites a dry blended mixture of Fe_2O_3 and the relevant carbonate was calcined at 1320°C followed by ball-milling to submicron particle size. Isotropic specimens were uniaxially pressed using a binder while the wet-oriented bodies were pressed from a slurry subjected to a magnetic field of 4000 Oe. After sintering at 1215°C for two hours, the stoichiometric ferrites had the properties listed in Table I.

Table I

Properties of Sintered Ferrites

Ferrite type	%ρ	(B_r) gauss	(H_c) oers	$(BH)_{max}$ X10^{-6}
Isotropic $BaFe_{12}O_{19}$	82	1850	1200	0.70
Wet Aligned $BaFe_{12}O_{19}$	87	2500	950	1.10
Commercial $BaFe_{12}O_{19}$	94	2450	1950	1.25
Isotropic $SrFe_{12}O_{19}$	69	1600	1300	0.51
Wet Aligned $SrFe_{12}O_{19}$	77	3350	2100	2.50
Commercial $SrFe_{12}O_{19}$	93	3700	3100	3.10

The commercial ferrites were included in this study as a contrast to the model stoichiometric materials. They were processed in a similar manner, but the details are of a proprietary nature.

Cylindrical samples with axial ratios ranging between 1:1 and 2:1 were compressively deformed at elevated temperatures in a creep mode and also by rapid press forging using a commercial testing machine* at crosshead speeds of between $2x10^{-3}$ and $5x10^{-3}$ in/min. In both the creep and forging modes, stresses from 500 to 6000 psi were employed. The temperature range between 1000°C and 1200°C was dictated by the sample deformation rates at the lower temperature and by extensive grain growth at the high temperature.

Preferred orientation was determined by X-ray diffraction of the mid-sections cut perpendicular to the deformation axis. The degree of orientation, (R), was then described from the integrated intensities by:

$$\underline{R} = \frac{\{\frac{\Sigma 00.\ell}{\Sigma hR.\ell}\}\ \text{deformed}}{\{\frac{\Sigma 00.\ell}{\Sigma hR.\ell}\}\ \text{random}} \tag{1}$$

"$\underline{R}$" is the ratio of basal peaks to all of the diffraction peaks. It was originally derived by Taft[5].

---

*Instron Corp., Canton, Mass.

   Bulk densities were measured by the Archimedes method and grain
sizes were measured from photomicrographs by Heyn and Grossman's
linear intercept technique[6]. Magnetic measurements were made using
a modified ballistic galvanometer equipped with integrators to
measure the remanence, (B_r), and high output Hall generators to
measure the coercivity, (H_c).

   Activation energy determinations were carried out both by plotting log (strain rate) vs. log (reciprocal temperature) and also by
the differential temperature technique. The latter involved recording the gradient of the time-strain curve at one temperature, adjusting the temperature, recording the new curve, and then employing the
standard Ahrrenius relationship:

$$Q_c = -RT_1 T_2 \{ \frac{\ln(\dot{\epsilon}_1) - \ln(\dot{\epsilon}_2)}{T_2 - T_1} \} \qquad (2)$$

In actual practice, the load was removed during the temperature
change, which required a finite length of time.

   Activation volumes were measured by the analogous differential
stress technique whereby the ratio of log (strain rate) immediately
prior to and following a stress change were applied through the
relationship:

$$v^* = -RT \{ \frac{\ln(\dot{\epsilon}_2) - \ln(\dot{\epsilon}_1)}{\sigma_1 - \sigma_2} \} \qquad (3)$$

The stress exponent evaluations were derived from standard log-log
plots of strain rate versus stress.

RESULTS AND DISCUSSION

   Compressive deformations in both the forging and creep modes
regularly exceeded strains of 150%. These yielded highly "barrelled"
specimens. Using the mid-length cross sectional area as the basis
for approximate stress calculations, the relationship,[7]

$$\epsilon = B \ln(A_o/A), \qquad (4)$$

was found to describe the dependence of the area (A) upon the strain
(ϵ). Here, (A_o) is the original area and (B) is a constant. Stress
corrections arising from Equation 4 were applied to all of the
forging load strain curves, a typical one of which is shown in Figure
(1). When the area-corrected data were plotted on a log-log basis,
Figure (2), the resulting stress-strain relationship[8] was:

$$\sigma = c\epsilon^n \qquad (5)$$

applied where (σ) is the true stress, (c) is a constant, and (n) is the strain hardening coefficient. In all the cases investigated, the strain hardening coefficient was unity, indicating that neither hardening nor recovery were dominant factors during the forging process.

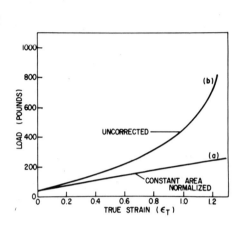

Fig. 1. Area Compensation of Forging Recording

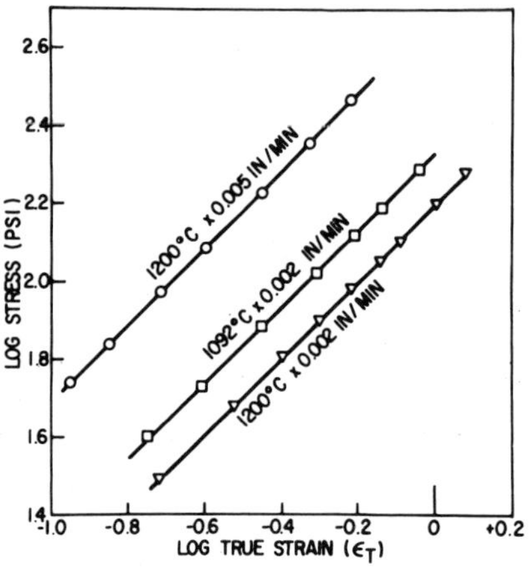

Fig. 2. Forging Strain Hardening Coefficient Graphs

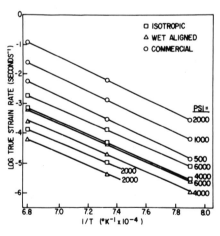

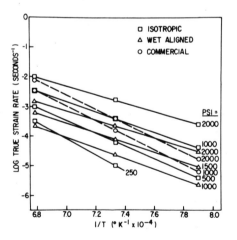

Fig. 3. $BaFe_{12}O_{19}$ Creep Activation Energy Plot

Fig. 4. $SrFe_{12}O_{19}$ Creep Activation Energy Plot

Area corrections to a creep process are somewhat more difficult to apply; thus, instead of the direct approach adopted during forging, the apparent strain-time curves of creep were scrutinized for evidence of "barrel" induced strain-rate changes. A progressive diminution of strain with time occurred. It was suggestive of primary creep and followed the strain-time dependence:

$$\varepsilon = \alpha \ln(t) \tag{6}$$

Here (α) is a constant in the equation corresponding to Cottrell's [9] "alpha" creep. Equation (6) may be written in derivative form as:

$$\dot{\varepsilon} = c\sigma^N t^{-1}. \tag{7}$$

The equivalent steady state equation would be

$$\dot{\varepsilon} = c\sigma_t^N \tag{7a}$$

where (σ_t) is the true stress. By assuming the stress in Equation (7) to be the applied stress, uncorrected for area change, and that in Equation (7a) to be the true stress, the equivalence:

$$[\frac{\sigma}{\sigma_t}]^N = t, \tag{7b}$$

should hold if Equations (7) and (7a) are descriptive of the same process. The stress corollary of Equation (4) is that:

$$\varepsilon \alpha \ \{\ln(\frac{A_o}{A})\} \ \alpha \ \{\ln(\frac{\sigma}{\sigma_t})^N\} \tag{7c}$$

which, along with the relationship in Equation (6) bears out the deduced equivalence at Equation (7b). Thus, from the standpoint of activation analyses, this particular creep deformation process may be said to be independent of time and therefore steady state, thus corroborating the forging findings.

The results of the activation energy, (Q_c), determinations are shown in Figures (3) and (4) for the barium and strontium hexaferrites respectively. The two stoichiometric hexaferrites differed sharply in that the barium hexaferrites proved independent of stress, while the strontium hexaferrites exhibited a pronounced stress dependence. A further disparity was found in the activation volume results which are shown in Figure (5). Here, the strontium hexaferrite was decidedly temperature sensitive while the barium hexaferrite failed to exhibit any such dependence. A rationale for the activation energy and activation volume behavior may be found by using Cottrell and Aytekin's [10] thermally activated steady state creep equation:

$$\dot{\varepsilon} = D \exp - \{(\Delta H - \sigma v^*)/RT\} = D \exp - \{Q_c/RT\}, \qquad (8)$$

where (D) is a constand and (ΔH) is the activation enthalpy for creep. Equation (8) shows that, at zero stress, (ΔH) is equal to (Q_c), the activation energy. Since (ΔH) is usually accepted to be the activation enthalpy for diffusion of the rate controlling species, an attempt was made to apply a suitable activation volume, (v^*), to Equation (8) such that the activation enthalpy was a constant for the apparently stress sensitive strontium ferrite system. This "best-fit" approach yielded a 5% strain activation volume of $6 \times 10^{-21} cm^3$ which compares with the experimentally determined range of 5 to $20 \times 10^{-21} cm^3$ over the temperatures studied. The (ΔH) value corresponding to these conditions is 127 Kcal/mol, Table II. Table III indicates the calculated activation enthalpies for $SrFe_{12}O_{19}$ utilizing the stress differential technique via Equation (3) for activation volume determination at different strain levels. It yields an average of 130 Kcal/mol in satisfactory agreement with the 127 Kcal/mol value. Manipulations of this nature were likewise applied to barium hexaferrite, but because of the extremely small activation volumes involved, only a variation of six Kcal/mol was realized over the entire stress range. This variation is within the experimental error of the activation energy determinations and thus explains the apparent stress and temperature insensitivity of these and the activation volume measurements. The average activation enthalpy of barium hexaferrite was 129 Kcal/mol, strikingly similar to that of the strontium hexaferrite.

Table II

Isotropic Activation Enthalpies by Extrapolation

Material	σ (psi)	σv^* (Kcal/mol)	Q_c (Kcal/mol)	ΔH (Kcal/mol)
$SrFe_{12}O_{19}$	250	8	125	133
$SrFe_{12}O_{19}$	500	16	105	121
$SrFe_{12}O_{19}$	1000	33	88	121
$SrFe_{12}O_{19}$	2000	66	68	134
$BaFe_{12}O_{19}$	2000	3	123	126
$BaFe_{12}O_{19}$	4000	6	123	129
$BaFe_{12}O_{19}$	6000	9	123	132

Table III

Isotropic Strontium Ferrite Activation Enthalpies
Calculated from Activation Energies and Activation Volumes

ε	$\Delta\sigma$ (psi)	Q_c Kcal/mol	v^* $\times 10^{21} cm^3$	σv^* Kcal/mol	ΔH Kcal/mol
0.050	766	95	12	31	126
0.051	746	97	12	31	128
0.124	193	127	9	6	133
0.151	578	121	7	14	135
0.370	1151	107	5	20	127
0.581	796	124	4	11	135

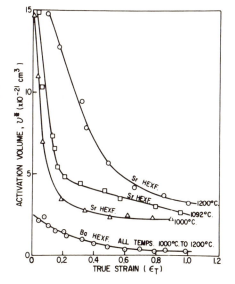

Fig. 5. Activation Volumes
as a Function of Strain

Fig. 6. Density Variation as a
Function of Strain

While no diffusional activation energy data exists for the hexaferrites, it is interesting to speculate which might be the possible
rate controlling ionic species. All diffusional data pertaining to
the energies of Ba^{2+} and Sr^{2+} in appropriately equivalent host compounds [11-15] indicate at least a 15% discrepancy. Sr^{2+} is consistently the lower of the two. In addition [16-22] Fe^{3+} is invariably
much lower in activation energy than 0^{2-} in oxide species diffusion.
It seems reasonable to conclude that the oxygen ion is the rate
controlling species during creep and forging deformation of the
hexaferrites.

The stress exponent findings, summarized in Figures (7) and (8) are directly affected by the activation volume dependence of the activation energy for creep. This obviously complicates the strontium hexaferrite case. Since this effect is minimal in the case of barium hexaferrite, discussion will be confined to this system. Interestingly, the stress exponent of the stoichiometric barium hexaferrite increased with strain at 1200°C, Figure (9), although the commercial material did not show this trend. No single creep model proposed to date is capable of explaining the gradual increase of stress exponent values from N=2 at 5% strain to N=4.6 at 90% strain. However, by assuming that this is the result of the overlapping of two distinct creep regimes, each indicative of an extreme case, it is possible to analyze the creep mechanisms. The two which appear most applicable are the Langdon [23] grain boundary sliding model with a stress exponent of two and the Weertman [24] dislocation loop climb mechanism with a 4.5 stress exponent. While satisfying the stress exponent requirements, Langdon's model does little to explain the activation energy and activation volume observations. Indeed, the ionic diffusion engendered by it would presumably correspond to activation volumes of ionic proportions, that is, three orders at magnitude lower than experimentally measured.

The explanation of the observed creep phenomena may lie in a combination creep-sintering model consisting of Raj and Ashby's [25] viscous flow creep of hypothetical sinusoidal grain boundaries and the van Bueren and Hornstra [26] sintering model which is based upon the same grain geometry. Although Raj and Ashby model requires a stress exponent of one, the disparity between this and experimental values near two may be rationalized by including the appropriate overlapping contribution from the high exponent mechanism. Flow is contingent upon the movement of vacancies toward high stress areas of the grain boundary which, though intrinsically of ionic proportions, are considered by van Bueren and Hornstra as having coalesced into pores. The sizes of these pores can be rationalized more readily with the large, experimentally observed activation volumes and, when taken in light of the bulk density behavior, their variation with strain is readily explicable: The average pore size within a given ferrite may be reasonably taken to vary inversely with its bulk density so that the overall activation volume trend with strain, Figure (5), would be the natural consequence of the densification, Figure (6). Furthermore, the observed temperature dependence of the activation volumes in isotropic strontium hexaferrite could be attributed to the equivalent vacancy migration rate at the different temperatures, Figure (5). Finally, the smaller activation volumes of the isotropic barium hexaferrite, and their apparent temperature insensitivity, can be attributed to the higher density in comparison with the strontium hexaferrite.

Before considering a third creep parameter, the stress exponent, mention should be made of the bulk density increase with strain in

these ferrites. Figure (6) shows that, while the stoichiometric
ferrites rapidly increase in density with strain, the commercial
ferrites undergo little change. The most remarkable facet of this
measurement is that the material densities are, without exception,
independent of the temperature, stress, or even type of deformation.
The densities were only a function of strain.

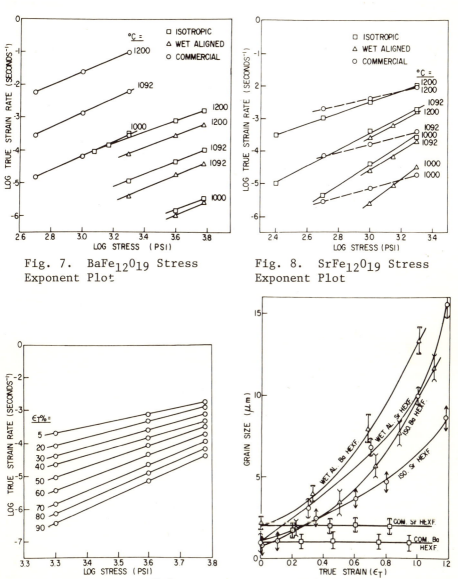

Fig. 7. $BaFe_{12}O_{19}$ Stress
Exponent Plot

Fig. 8. $SrFe_{12}O_{19}$ Stress
Exponent Plot

Fig. 9.   Variation of "N" with
Strain in $BaFe_{12}O_{19}$

Fig. 10. Variation of Grain Size
with Strain

An increasing reliance upon a Weertman-type dislocation climb mechanism at higher strain levels is not unlikely. Initial microstructural changes consistent with grain boundary sliding and densification lead to a situation where grain shape changes associated with dislocation action are probably necessary. Extensive electron microscope transmission studies failed to positively establish the existence of dislocations in highly strained specimens although some deformation twins were observed [7]. However, the absence of dislocations is not too surprising in view of the exaggerated grain growth which accompanied the strain process. Figure (10) depicts the variation of grain size with strain for all of the hexaferrites investigated. Significantly, only the commercial ferrites accommodated the deformation without incurring any substantial grain growth. This change in the microstructure of the stoichiometric ferrites is thought to be the primary reason for the absence of dislocations, since grain growth of this magnitude would correspond to an annealing process. It could also contribute to the rapid densification mentioned earlier.

It is no coincidence then, that texture development, as shown in Figure (11) occurred in the stoichiometric ferrites, but not to any extent in the commercial materials. Apparently, the combined impetus of temperature and uniaxial pressure leads to the enhanced growth of those grains whose basal planes lie perpendicular to the applied stress. During straining where grain growth is inhibited by dopants, as in the commercial materials, the negligible orientation changes attest to the coupling of the grain growth to the texture development processes. Some grain rotation appears to be a factor. The significance of the intensity of preferred texture development can hardly be overestimated. The degree of basal plane alignment far exceeds the present commercial processes and approaches an intensity nearly as well oriented as a single crystal. For example, X-ray patterns taken along the deformation direction, on a midsection, exhibit only the basal plane reflections, some of which are not reported in normal polycrystalline patterns.

The effects of these physical changes upon the magnetic properties are shown in Figures (12), (13), and (14) which illustrate the strain variation of remanence, (B_r), coercivity, (H_c), and the magnetic energy product, (BH_{max}). The remanence vs. strain curves are precisely what would be anticipated from systems undergoing combined densification and grain alignment for the stoichiometric materials, and minor densification only for the commercial materials. With the exception of the isotropic strontium hexaferrite, the coercivity vs. strain trends are similarly predictable on the basis of a hard ferrite system undergoing progressive grain growth. That is, one whose coercive quality is contingent upon retaining single domain grains and whose single domain grain size limit is about one micron [27]. The one exception to this trend is the isotropic

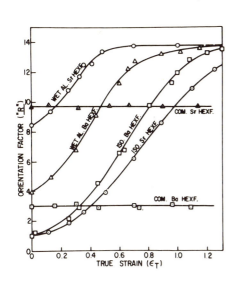

Fig. 11. Variation of
Orientation with Strain

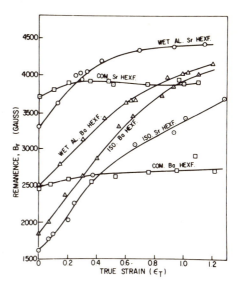

Fig. 12. Variation of
Remanence with Strain

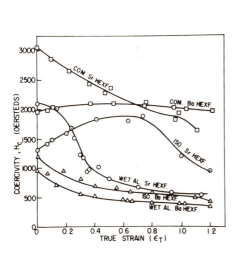

Fig. 13. Variation of
Coercivity with Strain

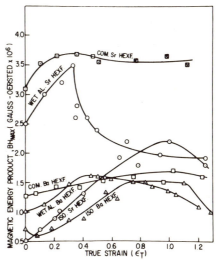

Fig. 14. Variation of the
Energy Product with Strain

strontium hexaferrite. It possesses both the lowest initial bulk
density and least initial orientation, Figures (7) and (11) respectively. This is significant in that the most probable reason for
coercive degradation is the spontaneous nucleation of reverse
domains in erstwhile single domain grains by their neighboring
grains' stray fields. This mechanism, first proposed by Holz [28]
naturally discriminates against polycrystalline ensembles which
exhibit both close mutual proximity and highly preferred orientation.
As indicated previously, the system furthest from this category is
the isotropic strontium hexaferrite. Here, the coercivity actually
increases with strain due to sample densification prior to the onset
of substantial texturing when the relatively close grain-to-grain
contact initiates reverse domain nucleation, thus degrading the
coercivity. Other possible mechanisms of coercive improvements
such as the introduction of defects and ferrite decomposition were
ruled out by annealing and X-ray studies, respectively.

The overall effects of deformation upon the magnetic energy
product of the stoichiometric ferrites, Figure (14), reflects the
rapid rise in remanence initially, followed by the steady fall in
coercivity at higher strain levels resulting in the peaking (BH_{max})
behavior. A distinct improvement in the energy product is achieved,
but the success is somewhat tempered by the derogatory, complementary
grain growth. The slight improvement in the commercial materials may
be accounted for by their increase in bulk density. Undoubtedly,
even more substantially improved energy products could be achieved
if the coercivity decrease could be minimized.

CONCLUSIONS

The barium and strontium magnetoplumbite hexaferrites can be
deformation processed by either a creep technique or a press forging
mode. These can result in increased density and the development of
a very strong basal texture. The deformation appears to be accompanied and accomodated by grain growth, which contributes to the
mutual grain orientation and densification. Activation analysis
suggests that the rate controlling ionic species is O^{2-}, which
probably acts through some grain boundary sliding mechanism in the
early stages of deformation and later via a dislocation climb
mechanism.

Magnetic property changes during deformation processing reflect
the structural changes. With texture development and densification,
the remanence approaches the single crystal value. However, the
attendant orientation and grain growth degrades the coercivity. The
energy product exhibits a maximum with strain.

ACKNOWLEDGMENT

The authors would like to acknowledge the support of the National Science Foundation, and the cooperation of J. Proske and T. Shirk of Stackpole Carbon Co., St. Marys, Pa.

REFERENCES

1. A. L. Stuijts, Trans. Brit. Ceram. Soc., 55, 57-74, 1956.

2. R. M. Haag, AUSD-0248-71, CR, 1971.

3. N. Ichinose and Z. Tanno, J. Elect. Cer. Japan 3, (9), 57-61, 1972.

4. M. H. Hodge, W. R. Bitler, and R. C. Bradt, J. Amer. Cer. Soc. 56, (10) 497-501, 1973.

5. D. R. Taft, RADC-TR-67-614, 1967.

6. E. Heyn and M. A. Grossman, Physical Metallurgy, John Wiley and Sons, Inc., New York, 1925.

7. M. H. Hodge, Ph.D. Thesis, Penn State Univ., 1973.

8. P. Ludwick, Elemente der Technologischen Mechanick, Berlin, Springer, 1909.

9. A. H. Cottrell, J. Mech. and Phys. Solids, (1), 53-63, 1952.

10. A. H. Cottrell and V. Aytekin, J. Inst. Metals, (77), 389-391, 1950.

11. M. E. Baker, Ph.D. Thesis, Cornell University, Thesis Univ. Microf. No. 68-16762.

12. S. P. Murorka and R. A. Swalin, J. Phys. Chem. Solids, 32, (6), 2015-2020, 1971.

13. S. P. Murorka and R. A. Swalin, J. Phys. Chem. Solids, 32, (6), 1277-1285, 1971.

14. M. F. De Souza, Phys. Rev., 188, (3), 1367-1370, 1969.

15. D. C. Freeman and D. N. Stomires, J. Chem. Phys., 35, 799-801, 1961.

16. H. M. O'Bryan and F. V. DiMarcello, J. Am. Ceram. Soc., 53, (7), 413–416, 1970.

17. R. Krishnan, Phys. Stat. Solidi, 32, (2), 695–701, 1969.

18. R. J. Bratton, J. Am. Ceram. Soc., 52, (8), 417–419, 1969.

19. J. R. Keski and I. B. Cutler, J. Am. Ceram. Soc., 51, (3), 440–444, 1968.

20. J. H. Christian and H. L. Taylor, J. Appl. Phys., 38, (10), 3843–3845, 1967.

21. T. Sasamoto and T. Sata, Kogyo Kaguku Zasski, 74, (5), 832–839, 1971.

22. J. Kummer and M. E. Millberg, C. and EN., 90–99, 1969.

23. T. G. Langdon, Phil. Mag., 22, 689–700, 1970.

24. J. P. Weertman, J. Appl. Phys., 26, 1213–1217, 1953.

25. R. Raj and M. F. Ashby, Mat. Trans., 3, 1937–1944, 1972.

26. H. G. van Bueren and J. Hornstra, 5th Int. Symp. React. Solids, Munich, 1964, ed., G. M. Schwab, Elsevier, Amsterdam, 1965.

27. J. Smit and H. P. J. Wijn, Ferrites, John Wiley and Sons, Inc., New York, 1959.

28. A. Holz, J. Appl. Phys., 41, (3), 1095–1096, 1970.

HOT WORKING ALKALI HALIDES FOR LASER WINDOW APPLICATIONS

B. G. Koepke, R. H. Anderson and R. J. Stokes

Honeywell Corporate Research Center
Bloomington, Minnesota 55420

INTRODUCTION

Alkali halide crystals exhibit two properties which make them outstanding candidates for use as optical components in infrared laser systems; low optical absorption and opposing effects of temperature on the indices of refraction and thermal expansion[1,2]. Both properties tend to minimize beam distortion during the transmission of high power through a window or lens. Alkali halide single crystals are, however, too weak to withstand the service stresses expected of these components. Permanent plastic deformation of the window material due to thermal or mechanical stresses would be intolerable since both the shape of the deformed window and the stress induced birefringence would distort the beam. There is a need, then, to strengthen alkali halide single crystals without deteriorating their optical properties.

Strengthening can be achieved in many ways, principally by grain size refinement and by alloying. The conversion of single crystals to polycrystalline material can be carried out quite effectively by hot working. Some time ago Stokes and Li[3] demonstrated that polycrystalline NaCl could be produced from single crystals by extrusion at 350°C, with tensile yield strengths at room temperature over three times those of the starting crystals. Furthermore, the polycrystalline billets were fully dense. Since extrusion does not lend itself readily to the production of large flat shapes, recent emphasis has been on hot press forging[4-8].

It is important to recognize that the fabrication of halides must be carried out at high temperatures and relatively low strain rates to take advantage of extra slip systems. At low temperatures alkali halides slip on {110}<110> systems[9]; these systems provide only two independent slip systems instead of the five needed for a general change of shape[10]. At high temperatures the {100}<110> slip systems come into operation and provide the additional three independent slip systems. The transition temperatures are typically 200°C for NaCl[11] and 250°C for KCl [12] although the exact value is dependent on additional parameters such as strain rate and stress state.

Alloying, of course, can also result in significant strengthening of alkali halide crystals. Strengthening resulting from additions of both monovalent[5,11,12] and divalent[13] ions has been demonstrated. The idea of combining both sources of strengthening to generate polycrystalline alloy material has been considered and the results of hot working evaluated[6].

In this paper we address ourselves both to the techniques used to hot work alkali halide crystals into window blanks, and to the structures and properties of the hot worked materials. From the point of view of high power laser window applications one of the materials with a high figure of merit is KCl[1]. Thus the materials examined here are KCl and alloys of KCl-KBr containing 5 mole % KBr. The fabrication techniques include conventional and constrained press forging, isostatic press forging and hot rolling. Since optical properties are paramount to the ultimate usefulness of these materials, results on the optical properties of the hot worked material are included together with mechanical properties and microstructural data.

HOT WORKING TECHNIQUES

Press Forging

Conventional Press Forging. A schematic diagram showing conventional press forging along with examples of KCl crystals press forged at two temperatures is shown in Figure 1. In this configuration the single crystal billet is simply compressed between two heated platens. The working temperatures (150°C - 350°C) for alkali halides are fortunately low enough that a muffle furnace is not needed. In this temperature range silicone oils (#702) have been used to advantage as platen lubricants. As shown in Figure 1(b) KCl crystals can be given large deformations at 300°C without cracking. To obtain further grain refinement and thereby improved mechanical properties the working temperature is decreased (or the strain rate is increased) as much as possible until eventually edge cracks form and propagate radially inward as forging proceeds. Below temperatures of 200°C or so this becomes a serious problem

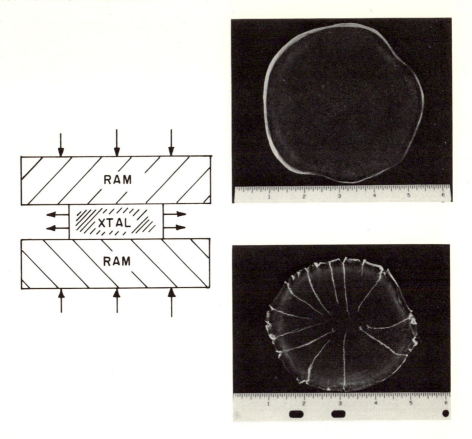

Figure 1.   (a) Schematic of conventional press forging.   (b) KCl
crystal pressed along <100> to 80% reduction at 280°C at constant
ram speed corresponding to an initial strain rate of 10^{-2} min^{-1}.
(c) KCl crystal pressed to 72% reduction at same rate at 190°C.

as shown in Figure 1(c).

   Constrained Press Forging. To minimize this radial cracking
problem in crystals forged at lower temperatures and/or higher
strain rates, the technique of constrained forging was adopted.
In constrained press forging (see Figure 2) the cylindrical single
crystal billet is surrounded with an aluminum or copper ring. As
the billet expands against the ring a compressive hoop stress is
developed around the periphery of the billet thereby inhibiting
radial cracking[7]. As shown in Figure 2(b), constrained forging
allows crack-free material to be forged as low as 150°C.

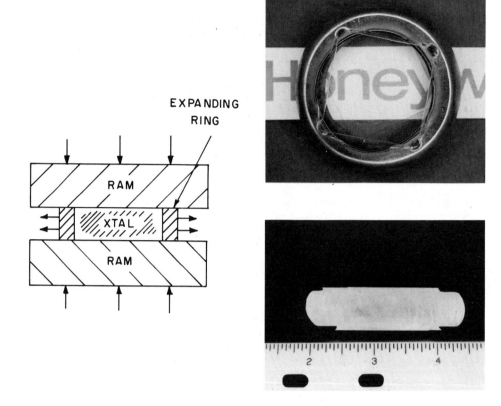

Figure 2.   (a) Schematic of constrained press forging.   (b) KCl
crystal press forged to 60% reduction at 150°C at a constant ram
speed corresponding to an initial strain rate of 8×10^{-3} min^{-1}.
(c) Cross section of billet showing effects of bulged constraining
ring.

     The Double Piston Technique.   One problem with the simple
constrained press forging technique in Figure 2 is that the high
frictional forces on the constraining ring at the platen surface
cause the constraining ring to bulge and wrap itself around the top
and bottom of the deforming billet.   This is evident in the billet
shown in Figure 2(b) and is clearly shown in the cross section of
Figure 2(c).   To minimize this problem and to increase the amount
of useable material the "double-piston technique" illustrated in
Figure 3 has been evolved.   In this case the platens do not move
and the frictional forces opposing expansion of the constraining
ring are kept to a minimum.   Alkali halide material is forced
into the chamber against the constraining ring by action of the
two pistons.   Little or no bulging occurs as can be seen in the

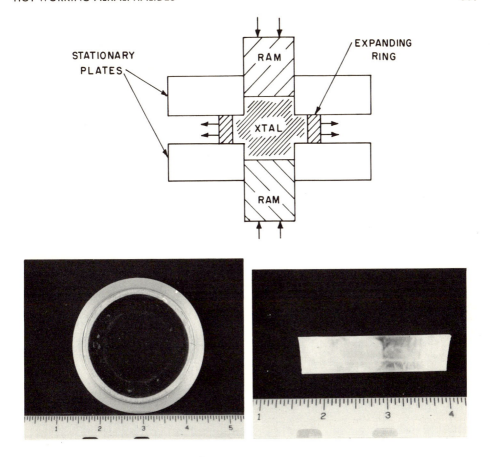

Figure 3. (a) Schematic of double piston forging technique. (b)
KCl crystal forged to 60% reduction at 200°C by this technique.
(c) Cross section of billet.

cross section of Figure 3 (c) and the amount of useable window
material in the final forging is therefore large. This technique
has been used successfully to produce pure polycrystalline KCl at
temperatures down to 150°C and at strain rates of 6 x 10^{-1} per min.

   Isostatic Press Forging. The double piston technique also has
its disadvantages. The forging stresses can exceed 10,000 psi and
the complicated flow patterns can lead to some discontinuities in
structure. The most recent press forging technique we have developed
is "isostatic" press forging. The technique is less complicated
and was evolved particularly to overcome persistent edge cracking
during fabrication of KCl crystals at the lower working temperatures
and during fabrication of the alloy single crystals.

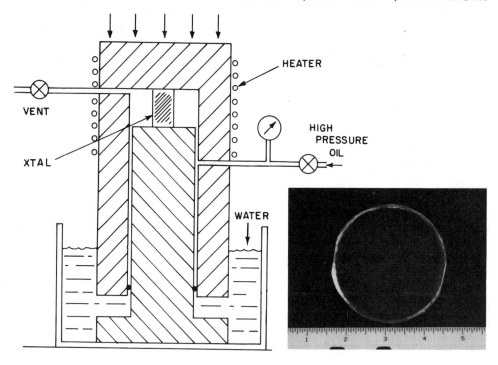

Figure 4.   (a) Schematic of isostatic forging technique. Water
cooling is necessary to maintain high pressure seal. (b) KCl
crystal forged to 73% reduction at 150°C and constant ram speed
corresponding to an initial strain rate of 10^{-2} min^{-1}.

        In the "isostatic" forging technique, shown in Figure 4(a),
press forging between two heated platens takes place in a closed
chamber filled with hot pressurized silicone oil. This technique
has been used to press forge pure KCl crystals to over 70% reduction at 150°C without cracking as shown in Figure 4(b). Crack
free billets have been produced at temperatures as low as 100°C
from KCl crystals and at 250°C from KCl-5 mole % KBr crystals
(strain rate 10^{-2} per min) with this technique.

        The success of the technique is most certainly attributed to
the inhibition of tensile cracking by the superimposed hydrostatic
pressure as shown many years ago by Bridgman[14]. Another beneficial effect of pressurizing the lubricant has been explained by
Backofen[15]. In press forging, it is difficult to keep the outer
edge of the platen-billet interface lubricated because the lubricant in this region is easily ejected by the forging stresses.
The resulting increased frictional forces enhance barreling and
edge cracking. The effect can be minimized by pressurizing the
lubricant. Backofen also suggests that lubrication can be improved

HEATED ROLLS

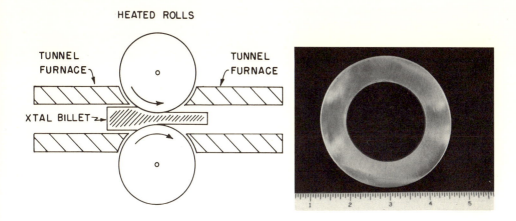

Figure 5. (a) Schematic of rolling mill used to hot work alkali
halide crystals. (b) KCl crystal cross rolled in <110> directions
on {100} face to 60% true compressive strain at 200°C. The mean
strain rate during the last pass was approximately 1.5 min^{-1}.

by periodically unloading the platens during forging[15]. Our best
results to date have benefited from this alternative.

     With the isostatic press forging technique it is possible con-
sistently to produce crack free, fully dense polycrystalline forgings
of both pure and alloyed KCl. However it should be noted that the
turn around time is slow, each forging requiring approximately one
day with our set up.

Rolling

     Rolling is an attractive hot working technique for a number of
reasons. First, the process is rapid and many samples can be made
in a single day. Second, since the deformation in rolling occurs
under conditions approximating plane strain, the rolling direction
is an additional variable parameter. Finally the size of the deforming zone under the rolls depends on the roll and billet
geometries and the reduction per pass. As a result microstructural
gradients can be arbitrarily generated through the thickness of the
sample if desired.

     A schematic of the rolling mill used in this study is shown
in Figure 5. The heated rolls can be operated at temperatures up

to 370°C. Before and after a rolling pass the specimen temperature
is maintained by tunnel furnaces. To produce deformation rates
suitable for hot working alkali halides the mill has been modified
to run very slowly with the roll speed adjustable from 1.0 RPM to
0.012 RPM (corresponding to strain rates in the 5×10^{-2} to 5 per
min. range for a 10% reduction on a $\frac{1}{4}$ to $\frac{1}{2}$ inch thick billet).

Single crystal billets are surrounded with a closely fitting
aluminum constraining ring. The ring has two advantages, first it
inhibits radial cracking and second, and most important, it provides sufficient friction for the rolls to grab the specimen. An
example of a rolled crack free KCl billet fitted with an aluminum
constraining ring is shown in Figure 5(b).

MATERIALS AND PROCEDURES

The starting materials were $1\frac{1}{2}$" dia. x $1\frac{1}{2}$" long pure KCl
crystals[*] and 1" dia. KCl-5 mole % KBr alloy crystals[+]. All unconstrained and constrained press forging used pure KCl crystals;
isostatic forging used both pure KCl and KCl-5 mole % KBr crystals;
hot rolling used $\frac{1}{2}$" high pure KCl crystals.

Crystals were always press forged along the <100> direction,
the ram speeds were maintained constant at values corresponding
to initial strain rates in the 5×10^{-3} to 1 per minute range.
Working temperatures in press forging ranged from 100°C to 400°C
depending on the method and material as noted earlier.

Crystals were rolled on {100}, {110} and {111} faces. All
billets were cross rolled in alternate orthogonal directions to
insure a flat product, the directions were chosen to activate the
highest number of slip systems. Rolling was carried out at temperatures set between 175° and 300°C and at roll speeds set to give
final strain rates ranging from 10^{-1} to 3.5 per minute. In every
case the reduction per pass corresponded to 10% compressive true
strain.

Textures of the materials after hot working were determined
with a Siemens automatic pole figure goniometer using Mo $K\alpha$ radiation. Mechanical properties were determined from 3-point bend
tests on bars cut from the billets with a wire saw and subsequently
mechanically polished on wet silk and finally immersion polished
in water just before testing to remove all damage. At least three
bars were cut from each billet. Most microstructural observations
were made on broken bend test bars etched with a 50-50 solution of
methanol in distilled water.

---

[*] Optovac Inc.
[+] Grown at Honeywell Corporate Research Center

Figure 6. (a) Microstructure of a KC1 crystal press forged to 80%
reduction at 175°C without constraints. The mean grain size is
4µm. (b) Microstructure of a KC1-5 mole % KBr crystal pressed to
67% reduction at 250°C by the isostatic technique. The mean grain
size is 4µm.

The optical absorption at 10.6µm was measured both before and
after hot working to evaluate the effects of the deformation on the
optical properties. These measurements were made with the automatic
laser calorimeter developed by Bernal G.[16].

RESULTS

Microstructures

With few exceptions, the microstructures of the hot worked
KC1 and KC1-KBr alloy crystals exhibited a fine grain structure
with mean grain sizes ranging from 3µm to over 10µm depending on
the working conditions. Figure 6(a) shows a typical microstructure
of a KC1 crystal which had been forged to 80% reduction at 175°C.
In many cases the microstructures exhibited deformation banding
such as shown in Figure 6(a) with the grain size varying considerably between adjacent bands. Forging pure KC1 at higher temperatures produced similar structures but with correspondingly
larger grain sizes. KC1-5 mole % KBr crystals forged at 250°C by
the isostatic technique exhibited grain sizes close to those of
pure KC1 billets forged at 150 to 175°C as shown in Figure 6(b).
In general it may be stated that the overall microstructural features of press forged samples were dependent neither on the
pressing technique nor on the alloy composition up to 5% KBr.

The microstructures of KC1 crystals cross rolled at 250°C on

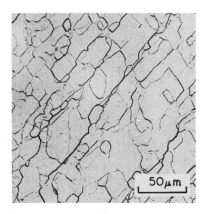 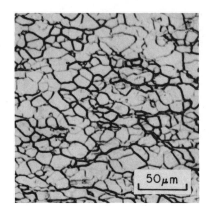

Figure 7. Microstructures of KCl crystals cross rolled two (a) and six (b) passes on {100} faces in <110> directions at 250°C. Note square grain morphology in sample given two passes.

{100} faces in <110> orthogonal directions are reproduced in Figure 7 for 20% and 60% compressive true strain respectively. At the higher strain the microstructure was similar to that developed by hot forging. At the lower strain, however, the grain morphology was rectangular with the boundaries lying roughly perpendicular to {110} slip plane traces in the starting crystal. This indicated the boundaries were dislocation walls formed by simple polygonization. The question remained as to the nature of the more complex boundaries associated with higher strain. Further information concerning this point was obtained by examining the textures introduced by hot working described in the next Section.

Textures

Two distinct textures were observed as a consequence of the fabrication methods. First, the texture developed during unconstrained hot forging was consistent with the lattice rotations predicted theoretically for deformed alkali halide single crystals by Chin and Mammel[17-19]. With five independent slip systems operating and freedom for lattice rotation the deformation texture tended toward stable end points on the [100]-[110] and [100]-[111] boundaries. This is illustrated in Figure 8 in which the recognizable components of the compression textures of five unconstrained KCl crystals pressed to 60% reduction along <100> at 200°C (specimens A & B), 175°C (specimens C & D) and 150°C (specimen E) are plotted on a standard stereographic traingle. The good correspondence between the forging textures and the predicted deformation textures[17-19] represented a strong indication that true recrystallization had not taken place during the forging operation. Instead

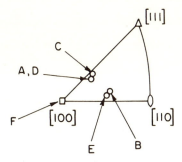

Figure 8. Compression textures of KCl crystals pressed to 60% reduction with and without constraints.

A,B - unconstrained at 200°C.
C,D - unconstrained at 175°C.
E   - unconstrained at 150°C.
F   - constrained at 200°C, 175°C, 150°C and 125°C.

the single crystal material developed the observed boundaries in
Figure 6 through dislocation reactions and they were primarily
small angle grain boundaries.

    The texture developed during constrained hot forging was
markedly different. Because the constraining ring insured axis-
mmetric flow there was very little lattice rotation away from the
symmetric <100> starting orientation. As indicated on Figure 8,
the final textures were sharp and close to those of the initial
crystals. Figure 9(a) shows a (220) pole figure of a KCl crystal
constrained during forging along <100> at 150°C and illustrates
how sharp these textures were. The textures of hot rolled crystals
were even sharper. Figure 9(b) shows a (220) pole figure of a KCl
crystal cross rolled on a {100} face in <110> directions to 60%
true compressive strain. Again the final microstructure (see
Figure 7(b)) is really that of a polygonized single crystal rather
than a randomly oriented polycrystal.

                    Mechanical Properties

    The mechanical behavior of hot worked KCl produced using the
procedures described here can be summarized with the aid of Tables
1 and 2. First and foremost hot working pure KCl single crystals
produced an order of magnitude increase in yield strength from
values of 500 psi for single crystals (under 3-point bending) to
values as high as 6000 psi for polygonized material. Similar
increases in strength have been noted by other workers[4,5,6].

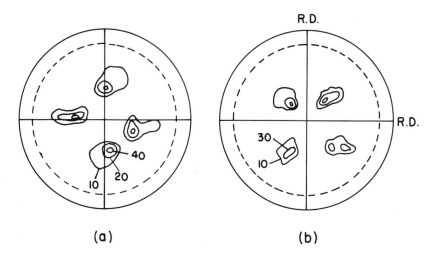

Figure 9. (a) (220) pole figures of KCl crystals (a) pressed to
60% reduction along <100> at 175°C by the constrained forging
technique and (b) cross rolled in {100}<110> orientation to 60%
true compressive strain at 250°C.

     The exact values of the hot worked material were dependent on
a number of factors. First was the effect of grain size. As has
been reported elsewhere[20] the subgrain structure of the hot
worked pure KCl was not necessarily stable and after standing for
a few hours at room temperature abnormal grain growth occurred.
The resulting structure was then extremely course grained in
which case the yield strengths were only 2 to 3 times the value
for single crystals as indicated in Table 1. To record the
mechanical strength of fine grained hot worked pure KCl material
therefore, the measurements had to be made immediately after fabrication was complete. When this precaution was taken, the strength
could be correlated with the subgrain size and generally increased
as the subgrain size decreased. Other workers have found the
strength data to follow an inverse square root relationship under
these conditions[4].

     The mechanical strengths in Table 1 were essentially indepen-
dent of forging method but they were sensitive to purity. For
example, one billet of "pure" KCl exhibited strengths exceeding
6000 psi. Subsequent chemical analysis of this material revealed
over 5000 ppm of sodium present. The effect of deliberate monovalent alloy additions such as 5 mole % KBr appeared to be additive
rather than multiplicative. Thus, both the single crystal and
polycrystalline KCl-5% KBr alloys were approximately 1000 psi
stronger than the pure counterparts. The most important contribution of the alloy addition, however, was that it inhibited abnormal

TABLE 1

Material	Pressing Technique	Temperature (°C)	Initial Strain Rate (min^{-1})	Percent Reduction	Yield Stress (psi)	Fracture Stress (psi)	Mean Grain Size (μm)
KCl Single crystal					500		
KCl	Unconstrained	200	.008	61	4380	6480	6
KCl	Unconstrained	200	.008	82	1670	2500	Several mms. (Due to abnormal grain growth)
KCl	Constrained with Cu sleeve	150	.008	60	4830	7470	7
KCl	Constrained with Cu sleeve	150	.064	61	6670*	9350	4
KCl	Double piston	150	.065	55	4050	6080	6
KCl	Double piston	150	.67	57	5160	8010	4
KCl-5% KBr crystal					1380		
KCl-5% KBR	Isostatic (4500 psi pressure)	300	.012	72	6100	8000	6
KCl-5% KBr	Isostatic (4500 psi with periodic relaxations)	250	.012	67	6370	6640	4

*superior mechanical properties associated with high Na impurity content.

Table 1. Forging parameters and mechanical properties of KCl and KCl-5% KBr crystals press forged by different techniques along <100>.

grain growth. This stabilized the fine grained as-fabricated micro-structure and at the same time retained the high strength of the material.

The general trend to higher strengths with decreasing subgrain size was also evident in rolled KCl crystals (see Table 2). The rolling temperature was found to have a greater effect on the sub-grain size and thus on the mechanical properties than either total strain or strain rate.

The results in Table 2 show initial crystal orientation and thus the resultant deformation texture to be another factor affect-ing the mechanical properties. Chin [17,18] has shown that billets with textures near <110> and <111> should be stronger than those with <100> textures. This was evident in the rolled material and has also been shown by others to hold for press forged KCl [6].

Hot working can also drastically increase the fracture tough-ness of alkali halide crystals. Becher et. al. [21] have demon-strated an order of magnitude increase in the fracture energies of pure and alloy KCl crystals hot worked by both press forging and by rolling.

A final factor affecting the mechanical properties of hot worked alkali halides was the amount of internal stress in the billet. The strongest materials noted in Table 1 also exhibited the lowest ductilities indicating high degrees of residual work hardening. The potential for increasing the strengths even further by working at even lower temperatures and higher strain rates are not practical as a result of the cracking generally encountered.

In summary, hot working methods were very effective in strength-ening alkali halide single crystals. An order of magnitude increase in strength and fracture toughness was readily achieved by the variety of fabrication methods. To retain the strength alloys were used to stabilize the grain structure. Any further increases in strength above 6000 psi call for fabrication conditions which raise the level of internal stress in the material to the point where very limited ductility is observed and brittleness becomes a problem.

Optical Properties

The effects of deformation processing on the optical properties of KCl were determined since the success of the hot working-strengthening operation was wholly dependent on whether the low absorption coefficients of the single crystals could be maintained in the processed billet. With this in mind the absorption co-efficients at $10.6\mu m$ of most KCl and KCl-KBr alloy crystal billets

TABLE 2

Rolling Plane and Directions	Temp. (°C)	No. of Passes	Mean Strain Rate During Last Pass* (min⁻¹)	Yield Stress (psi)	Fracture Stress (psi)	Mean Grain Size (μm)
{100}<110>	250	2	0.10	2360	6630	6
		6	0.15	3580	6920	7
	250	2	1.1	2580	7170	6
		6	1.4	2390	5440	7
	200	2	1.1	3270	6300	4
		6	1.4	3280	7620	3
{110}<110>, <100>	200	2	1.1	4990	7340	3
		6	1.4	4850	8020	3
{111}<110>, <112>	200	2	1.1	3880	6620	3
		6	1.4	4130	7960	4

*Calculated from expression given by G. E. Dieter, Jr. in Mechanical Metallurgy. (McGraw-Hill, New York, 1961) p. 209.

Table 2. Rolling data and properties of KCl crystals cross rolled on {100}, {110} and {111} faces in reductions of 10% true compressive strain per pass.

were determined before and after hot working. The results of a
large number of these measurements on pure KCl samples are
summarized in Table 3. Basically the measurements indicated that
there was a slight increase in absorption due to hot working and
that press forging produced a greater increase than hot rolling.
This could be expected since the working temperatures used in
press forging were typically lower than those used in rolling.
The Table also indicates that the increases in optical absorption
associated with hot rolling are sensitive to the orientation of
the rolled crystals. Rolling in the {100}<110> orientation resulted in essentially no increase in absorption.

The least increase in absorption measured to date in press
forged material was achieved by the isostatic technique.
KCl-5 mole % KBr crystals with absorption coefficients of less
than $5 \times 10^{-4} cm^{-1}$ were isostatically pressed under 4500 psi oil
pressure to 70% reduction at temperatures ranging from $400^\circ C$ down
to $250^\circ C$ and exhibited increases in absorption of less than
$2 \times 10^{-4} cm^{-1}$, i.e. the absorption was virtually unchanged.

TABLE 3

Table 3. Statistics of change in 10.6μm absorption, β, of KCl
as a result of hot working.

Hot Working Method	Number of Samples	Average Increase in Absorption; $\frac{\beta(poly)}{\beta(single)}$
Constrained press forging	14	6.0
Rolled - {100} surfaces	16	1.2
Rolled - non {100} surfaces	8	3.7
All samples	38	3.5

DISCUSSION

The goal of this work has been to prepare laser window
material with the highest possible strength, free from cracks and
internal stress. It has been demonstrated that under suitable
conditions it is possible to take advantage of the limited plasticity of bulk alkali halide crystals and to deform them directly
into prescribed shapes while significantly improving their mechanical properties and retaining the low optical absorption at 10.6µm
characteristic of the starting material. A variety of fabrication
techniques have been developed which provide flexibility depending
on the need. When strength and grain size are not so important
pure crystals hot worked by simple unconstrained press forging at
high temperatures can be used successfully. For high strength
alloys isostatic pressing is more suitable. Although the flat
window cross section we have been fabricating is simple, there
appears to be no reason why more complex shapes cannot be produced
using, for instance, shaped platens and dies.

The structures and properties of hot worked material result
from the competing processes of work hardening and softening that
occur simultaneously during fabrication. The square grain
morphologies of Figure 7, the sharp crystallographic textures of
Figure 9 and our observations that the fracture surfaces of
broken test bars are generally planar and parallel to the original
{100} planes all indicate that the microstructures in the halides
fabricated in this study are subgrain structures. The subgrain
boundaries form during hot working from dislocation reactions such
as those discussed by Amelinckx and Dekeyser[22]. Similar structures have been observed in deformed and annealed NaCl single
crystals by Davidge and Pratt[23] and in NaCl crystals[24,25] and
LiF crystals[26] deformed at elevated temperatures in creep tests.
The inverse square root dependence of the strength on subgrain
size[4] indicates the strengthening is due to the resistance to
dislocation motion by the subgrain boundaries which, in this work,
have dimensions between 3 and 10µm.

Monovalent alloy additions increase the strength of the KCl
crystals but apparently have no effect on the subgrain formation or
dimension. Thus the strength of the hot worked alloy material is the
same as hot worked pure KCl with an equivalent grain size plus about
1000 to 1500 psi corresponding to the increase in lattice friction
stress. In other words the alloying effect is additive. A much
more important contribution of substitutional monovalent additions,
however, is the long term stabilization of the fine grained microstructure.

514   B. G. KOEPKE, R. H. ANDERSON, AND R. J. STOKES

It has been shown that hot working can improve the strength of halides at little expense to their optical properties. The slight increases in optical absorption noted in Table 3 can have a number of origins. Tiny voids and intergranular separations can be induced during fabrication, particularly at lower temperatures and higher strain rates. Void formation can occur, for instance at multiple slip band intersections or intersections of slip bands with subgrain boundaries[27]. In many cases the voids occur in sheets or "veils" situated along major deformation bands at approximately 45° to the billet surface. The simpler localized plane strain deformation involved in rolling crystals of suitable orientation might be less likely to result in void formation than the more complex homogeneous shape change involved in forging. This accounts for the smaller increase in optical absorption noted in rolled {100}<110> crystals. In material produced by isostatic press forging the reduced tensile stress fields result in a lower incidence of void formation and absorption coefficients quite close to those of the starting crystals. For many reasons, then, isostatic press forging has emerged as one of the most promising means to hot work halides for laser window applications.

Conclusions

This work has shown

i) A number of techniques including both press forging and rolling can be used to hot work alkali halide crystals into crack free laser window blanks. One of the most promising techniques is isostatic press forging. KCl and KCl-KBr alloy crystals can be fabricated into window blanks with yield strengths exceeding 5000 psi at little expense to the extremely low absorption ($<5 \times 10^{-4} cm^{-1}$) at 10.6μm of the starting crystals.

ii) The hot worked structures are fine grained and are composed of low angle subgrain boundaries with the subgrain sizes ranging from 3 to 10μm.

Acknowledgments

The authors are indebted to D. W. Woodward for able experimental assistance during the course of this research and to E. Bernal G. for stimulating and critical discussion. Dr. R. B. Maciolek, Honeywell Corporate Research Center kindly supplied the KCl-KBr alloy crystals. The continued interest of Dr. C. H. Li, Director, Honeywell Corporate Research Center and Dr. C. M. Stickley, Advanced Research Projects Agency, is gratefully acknowledged. This work was supported by the Advanced Research Projects Agency.

References

1. M. Sparks, J. Appl. Phys., 42, 5029 (1971).

2. F. C. Horrigan, C. Klein, R. Rudko and D. Wilson, Microwaves,
    8, 68 (1969).

3. R. J. Stokes and C. H. Li, in Materials Science Research,
    edited by H. H. Stadelmaier and W. W. Austin (Plenum, New York,
    1963) Vol. 1, p. 133.

4. P. F. Becher and R. W. Rice, J. Appl. Phys. 44, 2915 (1973).

5. A. F. Armington, H. Posen and D. H. Lipson, J. Electronic
    Mater., 2, 127 (1973).

6. H. K. Bowen, R. N. Singh, H. Posen, A. Armington and S. A. Kulin,
    Mater. Res. Bull., 8, 1389 (1973).

7. R. H. Anderson, B. G. Koepke, E. Bernal G. and R. J. Stokes, J.
    Amer. Ceram. Soc., 56, 287 (1973).

8. see also Proc. Second and Third Conf. on High Power Infrared
    Laser Window Materials, AFCRL-TR-73-0372 (1973) and AFCRL-TR-
    74-0085 (1974) respectively. (Distribution limited.)

9. J. J. Gilman, Acta Met., 7, 608 (1959).

10. G. W. Groves and A. Kelly, Phil. Mag., 8, 877 (1963).

11. R. J. Stokes and C. H. Li, Acta. Met. 10, 535 (1962).

12. N. S. Stoloff, D. K. Lezius and T. L. Johnston, J. Appl. Phys.,
    34, 3315 (1963).

13. G. Y. Chin, L. G. Van Uitert, M. L. Green, G. J. Zydzik and
    T. Y. Kometani, J. Amer. Ceram. Soc., 56, 369 (1973).

14. P. W. Bridgman, Studies in Large Plastic Flow and Fracture
    (McGraw-Hill, New York, 1952) p. 118.

15. W. A. Backofen, Deformation Processing (Addison-Wesley, Reading,
    MA., 1972) p. 170.

16. E. Bernal G., and R. Stryk to be published.

17. G. Y. Chin, Met. Trans., 4, 329 (1973).

18. G. Y. Chin and W. L. Mammel, Met. Trans., 4, 355 (1973).

19. G. Y. Chin and W. L. Mammel, Met. Trans., 5, 325 (1974).

20. B. G. Koepke, R. H. Anderson, E. Bernal G., and R. J. Stokes,
     J. Appl. Phys., 45, 969 (1974).

21. P. F. Becher, R. W. Rice, P. H. Klein and S. W. Freiman,
     these proceedings.

22. S. Amelinckx and W. Dekeyser, Solid State Phys., 8, 325 (1959).

23. R. W. Davidge and P. L. Pratt, Phys. Stat. Sol., 6, 759 (1964).

24. J. P. Poirier, Phil. Mag., 26, 713 (1972).

25. W. Blum and B. Illschner, Phy. Stat. Sol., 20, 629 (1967).

26. G. Streb and B. Reppich, Phys. Stat. Sol. (a), 16, 493 (1973).

27. R. J. Stokes, in Fracture, An Advanced Treatise, edited by
     H. Liebowitz (Academic Press, New York, 1972) Vol. VII, p. 157.

PRESS FORGING BEHAVIOR OF KCl CRYSTALS AND RESULTANT STRENGTHENING

P. F. Becher, R. W. Rice, P. H. Klein and
S. W. Freiman
U. S. Naval Research Laboratory
Washington, D.C. 20375

INTRODUCTION

Use of alkali halides for optical components of laser systems has been limited by the fact that single crystals deform and fracture at low stresses. As dislocation processes are responsible for both the yield and fracture of these materials, one should be able to strengthen by alloying (1) and forming polycrystalline materials (2,3,4) or possibly a combination of both.

In order to produce pore-free polycrystalline materials of high optical quality, powder processing techniques have been avoided. Instead, hot working techniques have been employed where, by polygonization and recrystallization of a single crystal, deformed primarily by $\{110\}<1\bar{1}0>$ slip, a dense polycrystalline material is obtained (5,6). These techniques not only produce stronger ($[\sigma_y>21$ MN/m^2 (>3000 psi)] polycrystalline materials but also yield optical properties such as absorption coefficients that are effectively controlled by the starting single crystal properties (7,8) which are more readily optimized (e.g., 9). This paper will discuss the press forging of pure and strontium doped KCl single crystals with regard to microstructural development and consequent yield strength and fracture toughness.

EXPERIMENTAL PROCEDURE

Many of the aspects of press forging are only briefly presented here since they have been extensively discussed (5-8,10,11). First, single crystal samples are cleaved or wire sawn to the desired geometry (generally, a 2:1 aspect ratio) and orientation. To inhibit cracking during forging, crystals are chemically polished (flowing water bath followed by HCl or 2 acetic acid:1 HCl), rinsed in isopropanol and dried, and also often forged with a constraining ring (a fully annealed, thin-wall copper sleeve) (6).

A Universal test machine*, fitted with graphite rams, was used for forging with accurate strain rate control. Two heating systems were employed, both of which could be programmed for heating and cooling rates and maximum temperature, (1) an externally wound resistance-heated, insulated alumina tube, and (2) immersion heater elements mounted in the ram end faces with the rams enclosed in an insulated alumina tube. The latter was more desirable as the individual ram temperatures could be controlled, giving a more uniform temperature distribution in the forging chamber. Attached to the ram end faces were polished, hard, stainless steel anvils that could be water- or gas-cooled. Lubricants were found to be necessary to both prevent the sample sticking to the anvil and reduce end face friction. Silicone fluids could be used to temperatures up to ~250°C, while graphoil shims were employed at higher temperatures.

Subsequent examination and testing of forgings have also been previously described (7); however, it is noted that HCl or methanol etchants were used for microstructural examination. The mechanical tests consisted of a three point bend test, which for alkali halides gave yield stresses identical to four point tests, and the constant moment fracture energy test (12).

RESULTS AND DISCUSSION

The effects of deformation strain, temperature and alloying on the press forging behavior and microstructure of KCl will be discussed first, followed by the roles of recovery-polygonization and recrystallization processes in the microstructural development.

* Wilson Instrument Div., ACCO, Bridgeport, Conn.

Forging Behavior and Microstructure

The deformation of unconstrained <100> axis KC1 by {110}<1$\bar{1}$0> slip exhibited a decrease in the work-hardening coefficient ($d\sigma/d\epsilon$) with increase in temperature, Fig. 1. The curves were characterized by two regions; namely, an initial rapid hardening stage (up to -0.2 to -0.3 true strain) which decreases and becomes linear (Fig. 1). Strontium alloys exhibited a much more distinct first stage with a higher hardening coefficient which again goes through a transition to a linear hardening stage (with a comparable coefficient to pure KC1) with increasing strain. Also, strontium alloys required higher forging temperatures (>175°C) to maintain stress levels comparable to pure KC1 as a result of increased resistance to dislocation motion from the divalent cation additions. Forging end face friction contributed in the form of an increasing hardening coefficient with increasing strain (generally ≥ -1.0 strain), but in the absence of such effects, the hardening coefficient remained linear, as in Fig. 1.

Initial substructure had little effect on the stress-strain behavior of unconstrained forgings. Forging along one <100> axis to -0.2 to -0.4 strain and annealing to form a polygonized structure, followed by a second forging along an orthogonal <010> axis an additional -1.2 to -2.0 strain, yielded comparable stress-strain curves. The only difference in the re-forging behavior was more rapid initial hardening which, like strontium alloying, gave rise to a more distinct transition to the linear hardening region with increasing strain. It is important to note from the above results that (1) the linear hardening coefficients above -0.3 strain are small [-7.6 to -20 MN/m^2 (-1.1 to -2.9 10^3 psi/in/in)], and (2) there are no obvious instantaneous or gradual stress drops.

The microstructures developed after various forging strains in <100> axis KC1 crystals are represented in Fig. 1. Note that increasing the forging temperature had a minor effect resulting in a distinct "grain" structure at somewhat lower strains (7a). Up to strains of ≤ -0.3, the substructure consisted of irregular dislocation arrays, which gave way to a more defined and homogeneous polygonal-type substructure with increasing strain. The substructure or grain size appeared to be refined from ≥15μm at -0.3 to -0.4 strain to ≤10μm at

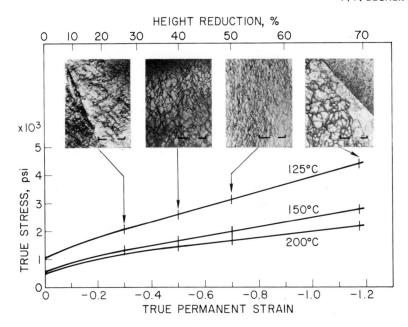

Figure 1. Deformation During Forging of Unconstrained
<100> KCl and Resultant Microstructure. (Bar on micro-
graphs = 100μm.)

~0.8 strain, probably by further network formation within grains, as in Fig. 2a. Above ~ -0.8 strain, a bimodal grain size was observed having large (≧100μm)
grains in a fine-grained matrix. In fact, in pure KCl,
"secondary recrystallization" of this type has often
been observed after forging at room temperature (13).
Often these large grains were idiomorphic, exhibiting
straight boundaries (Fig. 1) which x-ray data showed
generally consisted of {100} surfaces. Koepke et al
(13) have observed that the idiomorphic shapes were
enhanced at low forging temperatures and high strains,
altering to curved boundaries with increasing temperature. Strontium additions modified the microstructure
in two ways, (1) higher forging temperatures were
required to develop a polygonal structure, and (2) large
grain formation was reduced (7a). Preforging,

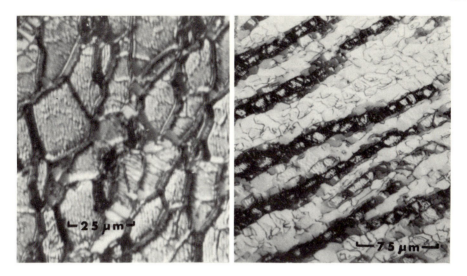

Figure 2.   Microstructure of Unconstrained <100> Axis
KC1 Forgings.   These micrographs exhibit features indicative of polygonization in (left) the formation of
subgrain boundaries within grains (200°C, -1.7 strain)
and (right) formation of "grains" within parallel bands
(260°C, -0.9 strain).

annealing, and reforging resulted in microstructures
comparable to single axis forging in KC1, although a
more homogeneous microstructure was obtained (7b).

        The fine "grain" structure often was observed to
form within deformation bands, Fig. 2b.   In fact, the
cross section of <100> forgings consists of four quad-
rants, each composed of parallel {110} slip (or deformation) bands.   Intersection of these bands was obvious
at cross section center axis, but such slip band intersections were much less distinct within each quadrant.
Thus, flow in each quadrant primarily involved one set
of parallel {110}<1$\bar{1}$0> slip systems with intersecting
slip systems contributing to a lesser extent.   In near
<110> axis forgings, primarily only one set of parallel
deformation bands are observed.

        Laue back reflection x-ray patterns of KC1 forged
at 150° to 200°C consisted of some streaked spots,
indicative of lattice rotation, and multiple spots and
short grainy arcs, from cell formation by polygonization, at ~ -0.6 strain.   With increasing strain (> -0.6),
the arcs predominated and became somewhat longer but

retained some grainy character. X-ray studies of the
crystallographic texture of these forged KCl samples
indicated a perponderance of [100] character (5) and
some [110] component in the surface normal to the <100>
forging axis. Bernal G. et al (10) have shown that the
texture of unconstrained <100> axis KCl forgings at
temperatures >150°C contain components lying along the
[100]-[110] and the [100]-[111] boundaries for strains
of -0.9. At 150°C, only the [100]-[110] boundary component was observed, while constrained <100> forgings
always exhibited a strong [100] component. This strong
texture was consistent with the fact that most forged
or rolled KCl bodies exhibited pseudo-cleavage.

Polygonization-Recrystallization

       The deformation behavior of unconstrained KCl
indicated that recovery-polygonization processes strongly contribute to the forging microstructure. The lack
of rapid workhardening with increasing strain (above
-0.3) was consistent with Mecking and Kirch's analysis
of dynamic recovery-recrystallization (14) where recovery and polygonization make major contributions. They
considered the softening effects of recovery-polygonization and recrystallization during the deformation of
metals and showed that the contribution of softening to
the stress-strain behavior increases when going from
recovery to polygonization to recrystallization as the
dominant processes. In fact, these processes compete
to lower the strain energy during deformation and recrystallization can be suppressed when the stored strain
energy is diminished by recovery processes. The onset
of recrystallization to form strain-free grains, as in
single crystal copper, resulted in an actual rapid
decrease in the true stress after rapid strain hardening
under constant strain rate conditions. Polygonization
can cause greater softening than recovery (by cross slip
and annihilation) and decreases the work hardening rate
more effectively. Thus, in KCl one suspected that
softening by polygonization processes contributed significantly in maintaining a low linear hardening coefficient at strains between -0.3 to -1.0. Initial support for the importance of polygonization was indicated
in limited anisothermal (25° to 400°C) scanning calorimetry studies of several low temperature (100°C to
150°C) KCl forgings where only a uniform evolution of
heat was observed. Detectable exothermic peaks, typical of recrystallization, were not observed.

Polygonization has been generally visualized as predominating when slip occurs only on one plane and results in formation of sub-boundaries by dislocation climb process. Note that the formation of parallel {110} slip bands in n<110> axis forgings and in each quadrant of <100> axis forgings was indicative of this. However, as described by Miekk-OJA and Lindroos (15), polygonization can be a more complex process, not necessarily so limited in the nature of slip or climb. They show that glide dislocations can be rapidly "knitted" into existing dislocation networks or forest dislocation in the presence of an applied stress. Bernal G. et al (16) also have pointed out that reaction of <1$\bar{1}$0> dislocation on 60° {110} slip planes can react to form dislocation networks on {0$\bar{1}$1}.

The microstructure developed with increasing forging strain indicated that the irregular dislocation arrays introduced at low strains could act to form subgrain networks within deformation bands with the addition of further dislocations by glide and climb. This resulted in a regular polygonal substructure which appeared to be further refined by formation of networks within these subgrains with further strain in the absence of recrystallization, as noted by calorimetry studies. The addition of strontium increased the temperature required to produce a "polycrystalline" microstructure, most likely by retarding dislocation climb and thus polygonization.   Bernal G. et al (8) have also observed the breakup of rectilinear "grains" whose boundaries align with the deformation bands with increasing strain during hot rolling of KC1. This and the sharp crystallographic textures (10) and the Laue patterns observed in forged KC1 were consistent with subgrain or grain formation by polygonization.

The increase in the arc lengths on Laue patterns and changes in texture with strains > -0.9 suggested that polygonization processes were leading to recrystallization. Comer et al (17), with the aid of replication electron microscopy, observed that initially ($\leqq$50% height reductions), the forged microstructure was closely related to the slip systems in KC1. But then at >90% height reductions, they found evidence of recrystallization in the form of distinct grain structures. Traskin et al (4) indicated that at very large reductions ($\approx$100%), complete recrystallization in KC1 did not occur at temperatures below 300°C. In the present study, a decrease in the hardness of KC1 forgings

with > -0.9 strain suggested recrystallization, which
was supported by the inhomogeneous formation of large,
often idiomorphic, grains. The microstructural stabilization by strontium additions may result from the
suppression of recrystallization or reduction of the
growth rate of these strain-free grains (18). The
behavior in KCl was not unlike that of aluminum crystals which continuously polygonize with increasing
strain until deformation results in sufficient stored
energy to initiate recrystallization (14). Thus, the
microstructure of KCl forgings resulted from the continued polygonization of dislocations involved in {110}
<1$\bar{1}$0> slip. This led to the formation of well defined
boundaries with increasing strain until local strain
energy became sufficient to initiate recrystallization.

Strengthening in Press Forged KCl

     The major reason for hot working of alkali and
alkaline earth halides was to improve the yield strength
of such materials. Stokes (3) and Carnahan et al (2)
earlier showed that by producing fine-grained bodies by
hot extrusion, one could improve the strengths of alkali
halide materials. Current investigations have confirmed
this for hot worked KCl and various alloys. The
strengthening behavior of press forged, pure, and strontium-doped KCl, as well as the improved fracture toughness of these materials, will be discussed.

     Previously, Becher and Rice (5) showed that press
forged "pure" KCl exhibited increased yield strength
with decreasing grain size following the Petch behavior
(e.g., ~31 MN/m^2 (4500 psi) at 5μm grain size to $\leq$ 7
MN/m^2 ($\leq$1000 psi at 500μm)). This indicated that the
microstructure consisted of grain or cell boundaries
of sufficient misorientation to influence yielding by
dislocation glide. The Petch plot slope value
(~5x10^{4}N/m$^{3/2}$, ~44 psi-in$^{1/2}$) suggested that the grain
boundary resistance to slip was low, as would be expected for low angle boundaries.

     Alloying KCl crystals increased their yield
strength with divalent cations being particularly effective (1). The strontium-doped KCl crystals grown
for the present study were particularly attractive in
that not only did they exhibit significant solid solution strengthening effects (7a), but exhibited low
optical absorption, which was a critical feature (9).

Forging such alloys resulted in increased yield
strengths, Fig. 3. Yield strengths as high as $\sim$ 53
MN/m^2 (7500) psi were achieved with $\sim$ 25 ppma Sr^{+2}
levels in 5μm grain size bodies.

It was expected that a family of lines on the
Petch plot should result in alloyed forgings with the
intercept value increasing with increased strontium
content. Instead, the strontium alloy ($\leq$25 ppma)
forgings exhibited two types of Petch behavior (7a),
either 1) higher intercept value (expected as a result
of solid solution hardening) with a slope comparable to
that in KC1 when forged at temperatures >250°C, or 2)
both higher intercept and slope values when forged at
temperatures <250°C. Post-forging annealing of set 2
showed that strengths could be reduced without grain
growth indicating recovery of residual strain hardening in the strontium-doped KC1 forgings (7a). Also,
if the yield stress for strain-hardened alloyed crystals were used as the Petch intercept, the slope of set 2
became comparable to that in KC1 forgings. Subsequent
forging studies of strontium-doped (up to $\sim$500 ppma)
KC1 have shown that increasing the forging temperature
(>200°C) caused a decrease in the yield strength and
an increase in ductility, with increased strontium content causing a slight increase in strength (11). At
about 250°C, the birefringence resulting from deformation bands also rapidly diminished, again consistent
with a loss in yield strength from recovery of strain
hardening.

As an added benefit to strengthening, strontium
diminished, and in some cases prevented, the loss of
strength by grain growth during post-forging annealing
at temperatures $\geq$350°C. On the other hand, "pure" KC1
forgings exhibited dramatic strength losses from grain
growth at annealing temperatures of $\geq$250°C where
strengths rapidly approach single crystal values.

The fracture energy of KC1 single crystals followed two limit boundaries with increasing resolved
shear stress from alloying or irradiation (19). The
upper limit observed showed the fracture energy (γ_c)
to be dependent upon the second power of the resolved
shear stress (τ_y) with a maximum γ_c averaging $\sim$4 J/m^2
at τ_y = 9 MN/m^2 (1300 psi). Above τ_y = 9 MN/m^2, the
upper limit of γ_c dropped rapidly. For the lower
boundary, the fracture energy was independent of resolved shear stress, γ_c = 0.25 to 0.3 J/m^2. Thus, the

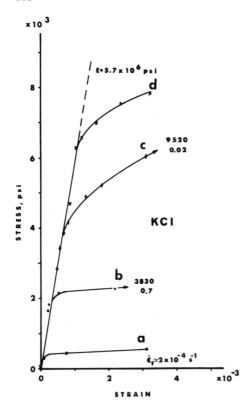

Figure 3. Room Temperature
Yield Points in KCl. Three
point bend test outer fiber
stress-strain curves for
single crystals (a - pure
and b - ~300 ppma Sr^{+2}) and
~7μm grain size press forged
bodies (c - pure and d - 20
ppma Sr^{+2}).

fracture energy of KCl single crystals at a fixed
resolved shear stress could vary anywhere between these
two limits at a displacement rate of 0.2 cm/min. The
fracture energy also was minimized by increasing the
displacement rate but could be enhanced by introducing
dislocations by prestraining. This behavior in single
crystals was shown to be a result of crack blunting by
dislocation glide from the crack tip which was required
to increase fracture energy and dislocation focusing at
the crack tip which sets the upper limit of fracture
energy ($\gamma_c \, \alpha \, \tau_y^2$).

        The fracture energy of pure and alloyed press-
forged and hot rolled KCl having yield stress $\geq$ 18
MN/m^2 ($\geq$2700 psi) was consistently high (1.5 to 4 J/m^2)
compared to single crystals (Table 1). In fact, the
KCl forgings with strontium levels from 0 to 25 ppma
were approximately tenfold tougher than their parent
single crystals. The consistently high fracture toughness in hot worked KCl was attributed to their

Table 1. FRACTURE ENERGY OF KCl*

Single Crystal			Hot Worked		
Sample	τ_y MN/m²	Range of γ_c J/m²	Sample	σ_y MN/m²	γ_c J/m²
			Press Forged		
Pure KCl	1.3	0.27	KCl	~18	2 to 3
KCl ≈25 ppma Sr^{+2}	~3.5	0.3 to 0.5	KCl + ≈ 25 ppma SR^{+2}	~30	3 to 4
			Hot Rolled		
–	–		KCl + 0.01 w/o KBr**	~19	1.5 to 2.5
–	–		+ 0.1 w/o KBr**	~23	3 to 4

* Displacement rate, 0.2 cm/min.

** Samples courtesy of Dr. B. G. Koepke, Honeywell Corporate Research Center, Bloomington, Minn.

substantial dislocation substructure, in the form of boundaries and individual dislocations, which readily blunted cracks.

SUMMARY

The microstructure of forged pure KCl at height reduction of ≦60% (≦ -0.9 plastic strain) is a result of simultaneous {110}<1$\bar{1}$0> slip and polygonization. A "grain" structure is initially formed at strains > -0.3 and is refined to ≦10μm with increasing strain with only a minor influence of forging temperature in the range of 125° to 200°C. At strains > -0.9, inhomogeneous recrystallization is initiated and grains are nucleated which grow rapidly into the surrounding strained, fine-grained matrix. KCl forgings exhibit strong textures, but as was discussed, show improved strength and fracture toughness. This indicates that even though polygonized structures would contain low angle boundaries, there is sufficient boundary mis-orientation to affect mechanical properties. The strontium alloys, although requiring higher forging temperatures, have very stable fine-grained microstructures, as well as improved strengths, due to solid solution hardening, which can be augmented by strain hardening.

ACKNOWLEDGEMENTS

The research was supported by the Advanced Research Projects Agency of the Department of Defense and was monitored by Dr. C. M. Stickley under ARPA Order No. 2031. The contributions of Dr. M. Krulfeld and Mr. W. L. Newell of NRL in portions of the experimental work are acknowledged, as are discussions with Drs. B. G. Koepke of Honeywell Corporation Research Center and H. K. Bowen of M.I.T.

REFERENCES

1. G. Y. Chin, L. G. VanUitert, M. L. Green, G. T. Zydzik and T. Y. Kometani, J. Am. Ceram. Soc. 56, 369 (1973).

2. R. D. Carnahan, T. L. Johnston, R. J. Stokes and C. H. Li, Trans. AIME 221, 46 (1961).

3. R. J. Stokes, Proc. Brit. Ceram. Soc. 6, 189 (1966).

4. V. Yu. Traskin, Z. N. Skvortsova, N. V. Pertsov and E. D. Shchukin, Soviet Phys.-Cryst. 15, 733 (1971).

5. P. F. Becher and R. W. Rice, J. Appl. Phys. 44, 2915 (1973).

6. R. H. Anderson, B. G. Koepke, E. Bernal G. and R. J. Stokes, J. Am. Ceram. Soc. 56, 287 (1973).

7a. P. F. Becher, S. W. Freiman, P. H. Klein, and R. W. Rice, pp. 579-600 in Proc. Third Conf. on High Power Infrared Laser Window Materials, Vol. II, C. A. Pitha, A. Armington and H. Posen (eds.), AFCRL-TR-74-0085, (1974).

7b. P. F. Becher and R. W. Rice, pp. 449-461 in Proc. Second Conf. on High Power Infrared Laser Window Materials, Vol. II, C. A. Pitha (ed.), AFCRL-TR-73-0372 (1973).

8. B. G. Koepke, R. H. Anderson and E. Bernal G., pp. 601-614 in Proc. Third Conf. on High Power Infrared Laser Window Materials, Vol. II, C. A. Pitha, A. Armington and H. Posen (eds.), AFCRL-TR-74-0085 (1974).

9. J. W. Davisson, M. Hass, P. H. Klein and M. Krulfeld, pp. 31-42 in Proc. Third Conf. on High Power Infrared Laser Window Materials, Vol. I, C. A. Pitha, A. Armington and H. Posen (eds.) AFCRL-TR-74-0085 (1974).

10. E. Bernal G., B. G. Koepke, R. H. Anderson and R. J. Stokes, Preparation and Characterization of Polycrystalline Halides for Use in High Power Laser Windows, Quart. Tech. Report No. 3, Corporate Research Center, Honeywell, Inc., December, 1972.

11. P. F. Becher, R. W. Rice and M. Krulfeld, High Energy Laser Windows, Semi-Annual Report No. 4, U. S. Naval Research Laboratory, July, 1974.

12. S. W. Freiman, D. R. Mulville and P. W. Mast, J. Mater. Sci. $\underline{8}$, 1527 (1973).

13. B. G. Koepke, R. H. Anderson, E. Bernal G. and R. J. Stokes, J. Appl. Phys. $\underline{45}$, 969 (1974).

14. H. Mecking and F. Kirch, pp. 257-288 in Recrystallization of Metallic Materials, F. Haessner (ed.), Dr. Riederer-Verlag GmbH, Stuttgart (1971).

15. H. M. Miekk-Oja and V. K. Lindroos, Surf. Sci. $\underline{31}$, 442 (1972).

16. E. Bernal G., B. G. Koepke, R. J. Stokes and R. H. Anderson, pp. 413-447 in Proc. Second Conf. on High Power Infrared Laser Window Materials, Vol. II, C. A. Pitha (ed.), AFCRL-TR-73-0372 (1973).

17. J. Comer, H. Posen and A. Kulin, pp. 683-692 in
     Proc. Third Conf. on High Power Infrared Laser
     Window Materials, Vol. II, C. A. Pitha, A. Arming-
     ton and H. Posen (eds.), AFCRL-TR-74-0085 (1974).

18. H. K. Bowen, private communication (1974).

19. S. W. Freiman, P. F. Becher and P. H. Klein, High
     Energy Laser Windows, Semi-Annual Report No. 4,
     U. S. Naval Research Laboratory, July, 1974.

PRESS FORGING OF SINGLE CRYSTAL CALCIUM FLUORIDE

Roger R. Turk

Hughes Research Laboratories
3011 Malibu Canyon Road
Malibu, California 90265

ABSTRACT

Single crystals of high-purity calcium fluoride have been deformed uniaxially in an attempt to improve strength and resistance to cleavage, without impairing infrared transmission. Order of magnitude increases in strength, such as those found in forged KCl, have not been attained, but fine-grained polycrystalling material has been produced which is resistant to crystalline cleavage. Deformation rates of 10^{-2} min^{-1}, reductions of 10 to 73% in height, and deformation temperatures of 550 to 1000°C have been used. Flexural strengths over 13,000 psi and grain sizes down to 5 μm have been obtained. Reduction of residual stress through heat treatment has been studied, and resultant techniques applied before, during, and after deformation. No increase in infrared absorption has been noted at the CO laser wavelength of 5.3 μm.

1. INTRODUCTION

This paper describes part of an overall program of development of alkaline-earth fluorides for use as laser window materials for the 2 to 6 μm range. These wavelengths include HF, DF, and CO chemical lasers. Window requirements are low optical absorption high strength, and scalability to large sizes. Our approach to these goals is twofold: growth of crystals by a limited reactive atmosphere process (RAP) to ensure a low level of impurity and thus low absorption, followed by deformation of the crystals to obtain a fine-grained, polycrystalline material of high strength.

2. BACKGROUND

Deformation plays a multiple role in attaining our objectives. Initially, it polygonizes the crystal boule sufficiently to remove the tendency of the CaF_2 crystal toward catastrophic cleavage on the (111) plane, interposing grain boundaries and varying grain orientations. In addition, strengthening results from grain refinement, as newly formed grains deforming by slip must adjust to neighboring grains to maintain coherency, with smaller grains giving higher resistance to flow. This is stated in the Hall-Petch relationship, $\sigma_{ys} = \sigma_o + kd^{-1/2}$, which relates increasing yield stress to decreasing grain diameter, showing the grain size dependence of lengths of dislocation pile-ups formed within the grains.[1] Our strength measurements do not confirm or deny this relationship, because no yield point was reached. Finally, deformation offers a convenient method of forming a large window blank from a crystal boule of high-purity, RAP-grown material with low optical absorption.

Of the alkaline-earth fluorides, we are studying CaF_2 most intensively as a promising candidate for a high strength window. Unlike most ionic crystals having the rock salt structure, CaF_2 does not cleave on the (100) plane, and slip along the (110) plane in the [110] direction. Instead, it cleaves on the (111) plane and slips along the (100) plane in a [110] direction.[2] It does, however, add the (110) [110] slip system at temperatures greater than $200°C$.[3] According to Evans, Roy, and Pratt,[3] polycrystalline plasticity occurs above 320 C, with slip on (100) planes (three independent systems) and on (110) planes (two more independent systems), giving a total of five possible independent slip systems. This ensures ductile behavior of all crystal orientations. Rate of work-hardening, according to Phillips,[4] decreases rapidly with temperature increase, but is still high near the melting point. Therefore, it would be expected that increased strength should result from deformation at all temperatures below the melting point.

With this background in mind, our press-forging work is tailored toward rapid, easy polygonization and forming at high temperatures, followed by lower temperature strengthening and grain refinement (in later work). This approach has the advantage of requiring no specific initial orientation of the crystal, or even single crystallinity, since polycrystallinity and directional flow patterns are established before much work-hardening occurs. Scale-up is simplified also, with the abandonment of the need for seeded, oriented, single-crystal boules with no grain boundaries in the larger sizes.

3. CRYSTAL PREPARATION

Crystal ingots of CaF_2 to be used for deformation strengthening must meet two requirements: they must be of sufficient purity to ensure low optical absorption, and they must be unstrained enough to be removed from the furnace intact. Since growth rate influences both impurity rejection and strain, these requirements are both satisfied by careful control of this variable.

Calcium-fluoride ingots of 3 cm diameter were grown by a limited reactive atmosphere process (RAP), developed at Hughes by Dr. Ricardo Pastor. The reactive atmosphere is HF (99.9% with a water content of a few parts per million) in a helium carrier. This produces some degree of cation and anion purification while excluding water and oxygen from the process.

Strains, inherent in directionally grown Bridgman crystals are removed using a vacuum anneal. The main feature of this anneal is slow, gradient-free cooling from a high temperature. This is achieved by enclosing ingots in an evacuable box constructed of 1/2 in. thick tungsten slabs within the hot zone of the vacuum furnace, and cooling from the anneal temperature at a rate dictated by the size of the ingot (Table I). Ingot strain is revealed by strong birefringence patterns when viewed between crossed polarizers (Fig. 1). Proper annealing causes all birefringence to disappear. We examine all crystal boules for "veils" and clarity as a measure of thoroughness of RAP purification, and for bubbles and grain boundaries as a result of high growth rate and spurious crystal nucleation. We use visual examination, employing crossed polarizers and high intensity conical illumination.

Forging lengths of 2.5 to 3.0 cm are cut from 3 cm diameter ingots with a reciprocating wire saw equipped with diamond-coated wire (Figs. 2 and 3). All surfaces are then ground and polished, and imperfections removed prior to forging (Fig. 4).

TABLE I

Ingot Cooling Rates During Vacuum Annealing from 1000 °C

Diameter, cm	Cooling Rate, °C/hour
3.0	15
4.3	10
5.5	3 (estimated)

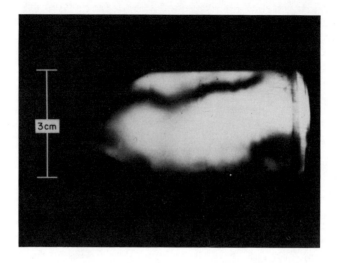

Fig. 1. Birefringence in strained as-grown
crystal of CaF$_2$.

Fig. 2. Reciprocating wire saw,
diamond-coated wire.

Fig. 3. CaF$_2$ crystal cut for forging.

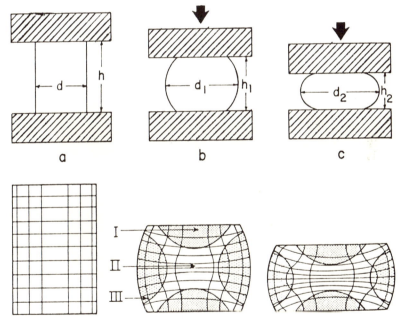

Fig. 4. Flow in flat-die forging
(from Forging Handbook).

4.   FORGING APPARATUS

We forged crystal boules in a die made of high strength, fine-grained graphite. The die consists of a cylindrical graphite shell as a ram guide and two flat cylindrical rams. This type of deformation, flat-die forging, exhibits nonhomogenous flow of material for non-lubricated surfaces, as shown in Fig. 4. We use a graphite foil interface lubricant to reduce this effect. Note that maximum material flow occurs in the center of the pressed volume, minimum at the faces, and an intermediate amount around the circumference, which develops circumferential tensile stresses.[5]

Resistance heaters around the die shell and within each ram heat the die and specimen (Figs. 5 and 6). Thermocouples record temperatures within each ram near its face and within the die cavity at the specimen surface. Pyrolytic graphite and insulating discs of ZrO_2 in the pressing column minimize heat loss. Radiation baffles consisting of concentric shells of gold-coated sand quartz contain the high temperatures. An argon atmosphere contained in a stainless steel dry box protects sample, press, and tooling from oxidation at high temperature. We use a manual 20-ton press for forging and monitor platen movement with a dial gauge. All deformation is conducted at a rate of 10^{-2} min^{-1}, and we lower temperatures with a motor-driven Variac.

5.   FORGING

We forged more than 62 specimens during this program at temperatures ranging from 550 to 1000°C. Reductions ranged from 10 to 73%, but the dual hot-cold work process was used for only the last twelve runs. A typical forged blank is shown before and after removal of the graphite lubricant in Figs. 7 and 8. This crystal was forged 47% at 900°C. Just as typical are Figs. 9 and 10, showing minor and severe cracking, respectively, in blanks forged 25% at 750°C. Significantly, almost all cracking in forged crystals occurred after removal from the forging die. This implies a residual strain problem caused by a combination of circumferential tensile forging strain and temperature gradients during cooling. We believe that this is aggravated by lower temperatures (which emphasize effects of poor crystal orientation) and by smaller deformations (which give lower strength and less polygonization to block cleavage). A greater percentage of uncracked pieces has resulted from material given greater deformations at higher temperatures. We feel that the elliptical shape of most forgings is a result of initial crystal orientation, a phenomenon we have encountered during forging of KC1 crystals. This should be eliminated if uniform, high temperatures are used which are

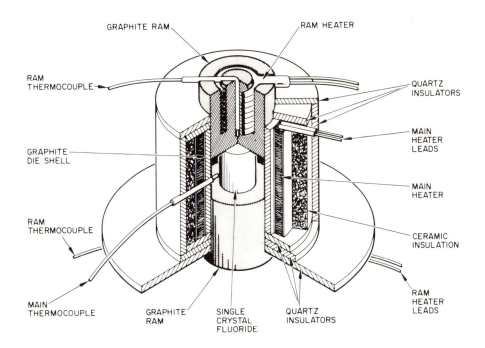

Fig. 5. Press forging apparatus.

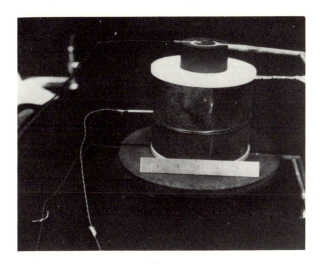

Fig. 6. Press forging apparatus.

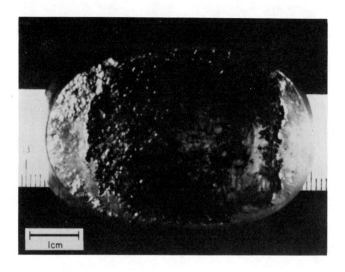

Fig. 7.   Forged CaF$_2$, reduced 47%
at 900°C, graphite lubricant still
visible.

Fig. 8.   Forged piece from Fig. 7,
polished.

Fig. 9. Forged window, slight crack.

Fig. 10. Forged window, large crack.

definitely above the recrystallization temperature. Greater reductions and lower temperatures generally give finer grain size after forging (see Table II). Variances found in Table II are due to extreme nonuniformity of grain size found in flat-forged specimens.

6. RECRYSTALLIZATION STUDIES

Strain in the forged blank cannot be tolerated in a finished window. It is removable by annealing, but this process usually removes all additional strength gained through deformation and results in a coarse-grained structure. It is our belief that a quantitative knowledge of the recrystallization characteristics of a forged piece should permit stress-relieving at a temperature low enough to retain either a worked or fine-grained structure, along with the accompanying high strength and cleavage resistance.

A typical curve of hardness (strength versus temperature) is shown in Fig. 11. It can be seen that a stress-relief temperature region exists for polycrystalline materials which does not require nucleation and growth of new, small grains. If further reduction in stress is necessary, it can be attained with minimal growth in grain size along with minimal loss in hardness (yield strength). These are the temperatures which must be determined if optimum heat treating cycles are to be developed.

TABLE II

Grain Size of Forged Calcium Fluoride

Date	Forging		Grain Size, μm
	Temperature, °C	Reduction, %	
6/22/73	1000	73	26 to 90
6/28/73	800	45	44 to 85
7/10/73	800	43	25 to 40
7/13/73	800	55	14 to 45
7/12/73	800	38	11 to 25
7/24/73	650	42	4 to 7

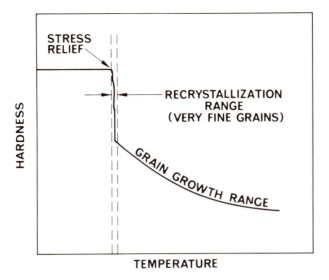

Fig. 11. Plot of hardness versus
temperature showing drop.

Studies of recrystallization temperatures as a function of
degree of deformation and deformation temperature are under way.
Our original plan was to conduct a series of forging runs at three
temperatures below recrystallization and at three reductions. A
determination of the recrystallization curves for each of these
nine runs would give information which, when cross-plotted, will
give accurate recrystallization data for all intermediate forging
conditions. From this data, stress-relief temperatures may be
chosen which are optimum for removal of strain and retention of
strength.

In accordance with the above plan, nine deformation runs were
made (Table III) at temperatures of 700, 625, and 550°C, with
"cold*" deformations of approximately 3, 6, and 10%.† These

---

*Below recrystallization.

†In view of subsequent studies, we believe that these deformations
are not large enough to provide definite hardness changes at
recrystallization. Reductions on the order of 50% are probably
necessary.

TABLE III

Forging Runs for Recrystallization Studies

| Temperature, C | | Reductions, % | | Average Hardness |
Hot	Cold	Hot	Cold	Knoop, 200 g
880	700	60	9.0	182
900	700	60	7.5	187
900	700	60	3.8	197
900	625	60	6.0	180
900	625	60	6.0	176
900	625	60	2.8	180
900	550	60	9.4	162
890	550	60	4.5	179
900	550	60	2.8	165

followed preliminary "hot**" deformations of 60% in each run to reduce effects of initial crystal orientation and polygonize crystal structure before starting cold work.

Samples were cut, polished and marked after forging (Fig. 12), then tested for as-forged hardness, using a Leitz Miniload Hardness Tester.

After initial hardness testing, individual marked pieces were cut and then heat-treated at different temperatures to find the drop in hardness signifying recrystallization. Fig. 13 shows plots of temperature versus hardness corresponding to 700° forgings at 3.8, 7.5, and 9% deformations. Recrystallization appears to occur at approximately 870, 850, and 807°C, respectively. We attribute the initial high hardness of the sample reduced 3.8% to surface finish effects, and assume recrystallization occurs when hardness drops from near 180 to 170 Knoop. Scatter between individual hardness readings has necessitated careful surface finishing of original

---

** Above recrystallization-900°C was used, since recrystallization occurs ~850°C.

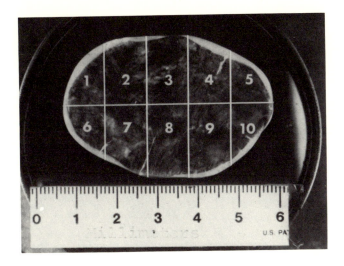

Fig. 12. Forged sample marked for
recrystallization specimens.

forgings to avoid introduction of spurious hardnesses and re
recrystallization effects. We now believe greater cold deformations are both feasible and necessary, probably in the range of
50%. These should result in higher strength and hardness, with
marked lowering of recrystallization temperature and more uniform
working. Anisotropy effects, as described by O'Neill, Redfern,
and Brookes,[6] should be minimized by a more uniformly worked
structure.

7. STRENGTH STUDIES

We have cut and polished bars for testing in four-point
flexure to determine modulus of rupture. Table IV gives values
obtained for various forging conditions. Testing was done on an
Instron Universal Testing Machine, at a rate of 0.002 in./min,
using a self-aligning fixture of tungsten carbide (Fig. 14). This
nonyielding fixture (modulus ~10^8 psi) should permit accurate
measurements of Young's modulus with very minor corrections for
yielding of apparatus.

The spread in strength values is partly traceable to visible
imperfections within bars (bubbles, etc.). However, these values,
coupled with the lack of any noticeable yield point, indicate
extreme sensitivity to surface finish (rupture is a surface

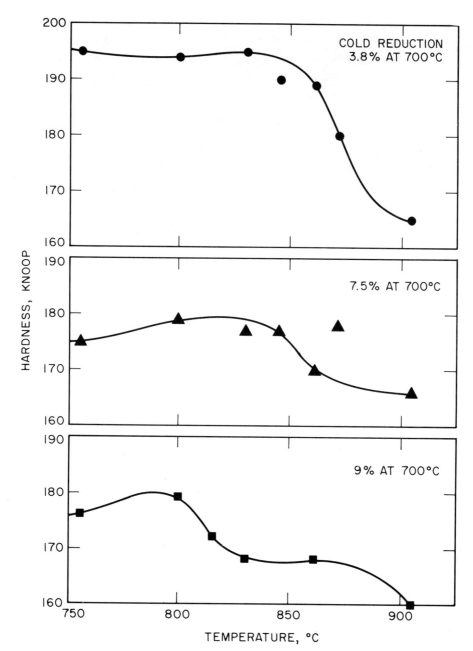

Fig. 13. Recrystallization plots of
hardness versus temperature for
specimens forged 3.8, 7.5 and 9.0%
at 700°C.

TABLE IV

Calcium Fluoride Modulus of Rupture versus
Forging Conditions

(Listed in Order of Percent Reduction)

Temperature, °C	Reduction, %	Modulus of Rupture, psi
780	68	12,100
1000	66	9,700
900-550	60-9.4	6,000
880-700	60-9.0	10,200
900-700	60-7.5	7,600
890-550	60-4.5	8,600
900-700	60-3.8	12,100
800	55	6,800
850-550	50-45	9,700
800	50	13,900
800	43	6,200
780	26	9,700
800	25	13,000
755	25	9,600
755	25	7,300
850	22	12,600
850	19	10,800

These values are in the same range as those measured for
single crystal CaF_2, which implies improvement only in
cleavage resistance.

Fig. 14. Four-point flexure test
apparatus of tungsten carbide.

phenomenon, yield is determined by bulk) and/or a ductile-brittle
transition temperature just above the test temperature. This last
possibility should be studied through warm tests, although Evans,
Roy, and Pratt,[3] using compressive test data, claim the ductile-
brittle transition occurs at 300°C, below which no yield point is
found for polycrystalline materials. Note again that the Hall-Petch
relationship, $\sigma_{ys} = \sigma_0 + kd^{-1/2}$, relates to yield strength, and that
all listed values are for rupture with no yield, so grain sizes
would not be expected to correlate with strength.

8. ABSORPTION

Tests of absorption of radiation were made at 5.3 μm using a
calorimetric technique with a CO laser. All tests have shown no
increase in absorption of forged material over that of the original
single crystal. On the contrary, slightly lower absorption values
have been found for forged windows. We believe this may be attributable to a surface layer less deeply disturbed by polishing
because of the polycrystalline, harder nature of the forged
material.

9. DISCUSSION AND SUMMARY

In forging of calcium fluoride crystals, a common problem in all stages of window preparation is that of strain which leads to cracking. This strain can result from growth rate or growth impurities, forging deformation, nonuniform or rapid cooling rates, or even severe surface deformation during polishing. Calcium fluoride has an unfortunate combination of high thermal coefficient of expansion (23×10^{-6} per °C) and high modulus of elasticity, or stiffness (21×10^6 psi). These properties, along with its propensity for (111) cleavage, require that it be made strong, polycrystalline and strain-free before it can be used as a laser window. Our approach to this problem has been to relieve strain by anneals where necessary to facilitate fabrication, and to determine optimum final stress-relief temperatures which can be applied with minimal loss of properties gained through deformation.

10. ACKNOWLEDGMENTS

This work has been carried out in connection with AFML Contract F33615-73-C-5075, with Dr. G. Edward Kuhl as technical monitor. The author also wishes to acknowledge contributions made to the program by Mr. Ronald Scholl for his recrystallization work; Mr. Robert Joyce, who performed all forging and preparatory work; Mr. Arthur Timper for hardness tests and etching; Mr. Matthew Himber for polishing of all flexure samples and windows; and Dr. Harvey Winston, for his helpful suggestions during the program. Special thanks are due to Dr. Ricardo Pastor, who developed the Reactive Atmosphere Process used to produce high-purity crystals, and to Mr. Kaneto Arita and Mr. Morton Robinson for growing all crystals used for this program.

11. BIBLIOGRAPHY

1. R. W. Cahn, Physical Metallurgy (Interscience Publ. Div., John Wiley & Sons, Inc., New York, 1965).

2. E. Schmid and W. Boas, Plasticity of Crystals (F. A. Hughes, London, 1950).

3. A.G. Evans, C. Roy, and P. L. Pratt, "The Role of Grain Boundaries in the Plastic Deformation of Calcium Fluoride," Proceedings of British Ceramic Society, No. 6, 173-188 (1966).

4. W. L. Phillips, Jr., "Deformation and Fracture Processes in Calcium Fluoride Single Crystals," Journal of the American Ceramic Society 44, 499 (1961).

5.   J. E. Jensen, <u>Forging Industry Handbook</u> (1966).

6.   J. B. O'Neill, B. A. W. Redfern, and C. A. Brookes, "Anisotropy in the Hardness and Friction of Calcium Fluoride Crystals," Journal of Materials Science $\underline{8}$, 47-58 (1973).

GRAIN BOUNDARIES AND GRAIN BOUNDARY MOBILITY IN HOT-FORGED

ALKALI HALIDES

M. F. Yan, R. M. Cannon, H. K. Bowen and R. L. Coble

Massachusetts Institute of Technology

Cambridge, Massachusetts

INTRODUCTION

Recent interest in alkali halide crystals as infrared windows for high power lasers has led to the development of high temperature deformation processing of halide single crystals to improve their mechanical properties. From these studies, Pitha et al (1974), yield strengths and fracture strengths have been improved; but a complete understanding on the relationship between hot-forging parameters and the resultant microstructures has not been suggested nor have quantitative relationships for grain boundary mobility been reported. The substructure development in hot-worked KCl crystals depends directly on the stress and strain and only indirectly on the temperature, crystallographic orientation and dopant concentration. The observed boundary mobility depends on the temperature, driving forces, (grain boundary area, residual strain, etc.) and impurity drag forces exerted on moving boundaries. The purpose of this paper is to elucidate the substructure formation and subsequent motion of boundaries in hot-forged KCl.

HOT-FORGING STUDIES

Potassium chloride crystals of various dopant levels and impurity levels were used. High purity single crystals were purchased from Optovac [1]. Strontium-doped and bromine-doped crystals were obtained from Raytheon[2] and AFCRL[3], respectively.

---

[1] Brookfield, Mass.
[2] D. Readey, Waltham, Mass.
[3] H. Posen, Bedford, Mass.

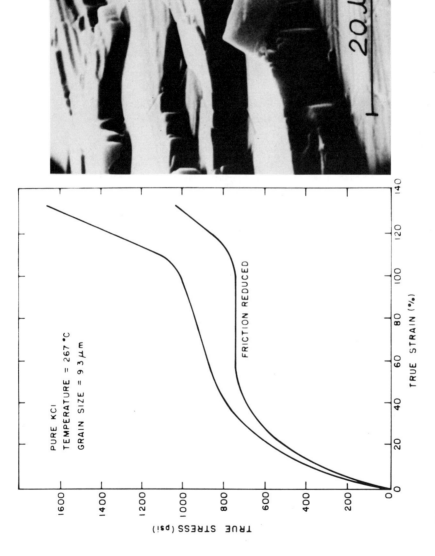

Figure 2 Fractured surface of hot forged KCl showing cleavage plane and subgrain boundaries. Note small misorientation between cleavage plane and adjacent subgrains.

Figure 1 Stress-strain relation during hot forging of "pure" KCl. The increase in stress after the steady state region results from die wall constraint.

Calcium and strontium-doped KCl crystals were also grown in our
laboratory from Fisher reagent grade powders by the Czochralski
method while under a nitrogen atmosphere. The Sr-doped crystal
from Raytheon was grown from a melt concentration of 200 molar ppm;
while the Sr- and Ca-doped crystals grown in our laboratory had
melt concentrations of 470 and 670 molar ppm, respectively. The
crystal-melt distribution coefficients for Sr in KCl were reported
by Chin et al (1973), Miles et al (1973) and Ikeya and Itoh (1968)
and vary from 0.14 to 0.25. We have assumed 0.20 as the distribution coefficient to calculate the impurity concentration in
crystals. Electrical conductivity measurements on the Raytheon
crystal gave an approximate agreement with this calculation.

The details of the forging procedures have been described
elsewhere, Bowen et al (1973). Most crystals described here were
forged in the <100> direction to a final true strain of 100-140%
at a strain rate of 0.05-0.2 min^{-1}. Typical runs were carried
out at 200-300°C but some were conducted over the range of 180-700°C
to observe changes in the substructure. Samples were hot-ejected
from the die and quenched in the ambient atmosphere. Polishing
and controlled etching by alcohol diluted water revealed grain
boundaries and substructures. Grain sizes were reported as the
average linear intercept.

The starting crystals typically had diameter to height ratios,
D/H, of 1 to 2, so that by the end of the forging D/H was typically 10 or more. This causes a significant hydrostatic constraint to
develop because of friction between the punch faces and sample.
Flow stresses were corrected for this effect using the approximate
relation from Hill (1950)

$$\sigma_y = \frac{P}{(1 + \frac{\mu}{3} \frac{D}{H})} \tag{1}$$

where σ_y is the flow stress, and P is the average forging pressure.
The coefficient of friction μ was taken as 0.12 for the Grafoil[4]
liners. A typical stress-strain curve during forging, Figure 1,
shows that after about 40% strain the flow stress approached a
constant value.

A typical substructure is shown in Figure 2 which shows the
fracture surface of a pure KCl sample with a grain size $\sim 10\mu$m.
The tendency of KCl to fracture on (100) faces and the similarity
in the cleavage steps between this and single crystal fracture
surfaces indicates a very small angular misorientation exists
between adjacent grains. In tact pole figures of forgings conducted
in the <100>, <110>, and <111> directions of pure KCl crystals show
a high degree of texture (Figure 3).

The subgrains represent a stable deformation structure; only
infrequently, did recrystallization, in the form of secondary

---

[4] Union Carbide pyrolytic graphite paper 0.01 inch thick.

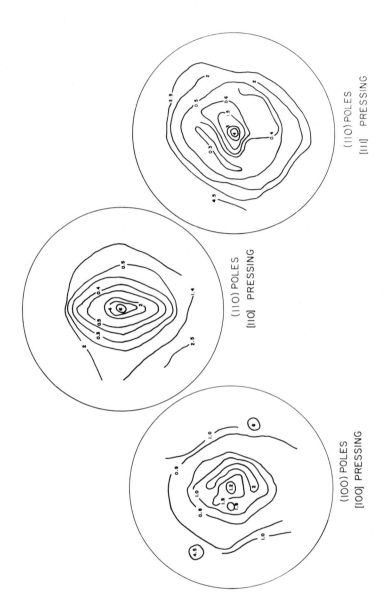

Figure 3   Pole figures for hot worked KCl

grain growth, occur during forging or subsequent rapid cooling. However, secondary grain growth does occur during annealing, even at room temperature in pure materials. Although the size of the subgrains changes with changes of deformation stress, we found that the low angle boundaries were virtually immobile during subsequent annealing.

For the entire range of stress and temperature investigated the subgrains within each sample were observed to be reasonably uniform in size, and equiaxed although infrequently they tended to be slightly elongated. In some studies of single crystals of NaCl, Poirier (1972), and metals, Clauer et al (1970), nearly square cells with a bimodal size distribution have been observed. That this was not observed in the present study may be in part due to the high D/H ratios of our samples combined with the total imposed strains which require multiple slip instead of single or duplex slip.

The greatest variability observed in the microstructures is an indication of a wide range, perhaps even a bimodal distribution, in the misorientation angle of the subgrain boundaries. The difference in etching intensity (Figure 4) is attributed to the existance of higher angle boundaries surrounding patches of subgrains with smaller misorientation angles between them. Streb and Reppich (1973) reported a more extreme variability in LiF creep specimens. They found a cell structure within subgrains in which they were able to resolve the individual dislocation etch pits in the very low angle cell walls. We have not been able to similarly resolve any of the boundaries found in forged KCl into etch pits.

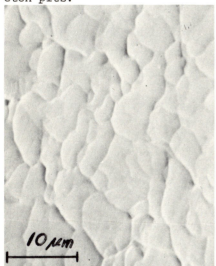 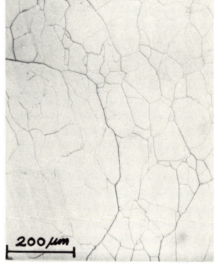

Figure 4: Microstructures showing apparent variation in subgrain boundary misorientation angle in Ca-doped crystal forged at 3640 psi and 242°C, (left), and undoped KCl forged at 390 psi and 635°C, (right).

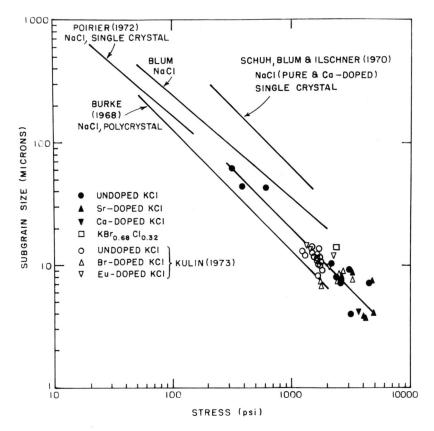

Figure 5   Subgrain size versus forging stress for forged
KCl compared with subgrain size curves from NaCl creep
studies, data of Blum reported by Streb and Reppich (1973).

Subgrain sizes, d, measured from both pure and doped forgings
are plotted against the steady state flow stress in Figure 5.
Similar data from forgings of several types of KCl crystals by
Kulin et al (1974) are also included. All of the results appear
to fit a single curve with

$$d \propto \sigma^{-1} \qquad\qquad\qquad (2)$$

as has typically been found for other materials, Bird et al (1969).
Some of the scatter in the data is due to difficulty in uniformly
etching all of the low angle boundaries. The results of several
creep studies on NaCl are also shown. The results of the present
study would agree very well with those of Burke (1968) for polycrystalline NaCl if the stresses were nomalized by the shear modulus
which is higher for NaCl than for KCl.

Within the scatter in the data, no additional effects of temperature or doping are apparent so that the effects of purity and
temperature on subgrain size originate from their effect of the
flow stress. These data include samples forged over a wide range
of temperature to vary the flow stress. Further, they include
materials with significant concentrations of monovalent dopants and
with divalent and trivalent dopants which have been shown to be
solid solution strengtheners, Chin et al (1973), Harrison et al
(1974).   A few data points from samples of <110> or <111>
orientations agreed with the results from <100> crystals.   For LiF
Streb and Reppich (1973) observed a shift of the d - σ curves from
changes in strain rate, but the present results are for a narrow
range of strain rates; no influence was observed.

Creep studies in both metals and ceramics have shown that
subgrains typically form during the transient stage, Sherby and
Burke (1967). This structure frequently takes as much as 40% strain
to fully develop, but then remains stable, Ilschner (1973). This is
consistent with the strains required to reach the steady state
forging stress. However, for metals at lower temperatures or
higher strain rates which are beyond the steady state creep range
similar reciprocal relations between the dislocation cell size
and the flow stress are still frequently observed especially in
cyclic deformation studies. However, the cell walls become diffuse
dislocation tangles rather than tightly knit subboundaries. This
behavior indicates that the cell formation per se is not a result
of climb recovery processes, but that the cell size is related to
the internal stress field of the dislocations. Climb recovery does
however, affect the size by limiting continued hardening at high
temperatures.

Both Burke (1969) and Blum (1973) have shown for NaCl that
at low stresses the steady state flow stresses fit a simple power
law stress dependence with strain rate, $\dot{\varepsilon} \propto \sigma^{n}$; but that at higher
stresses this relation breaks down as n increases with stress, and
the data are better represented by an exponential relation, $\dot{\varepsilon} \propto e^{a\sigma}$.

Blum and Burke estimated the transition stress to be about 300 and 1200 psi, respectively, It is likely to be lower for KCl, because of its low shear modulus. The controlling deformation mechanism in this high stress range is not well understood. Initially it may include modified climb mechanisms, Weertman (1968), but at higher stresses it may not even be climb limited, Blum (1973).

Our forging data span the stress range from power law creep to stresses approaching the athermal flow stress. Further, the data include samples with a significant variation in the contribution of solid solution strengthening. Surprisingly, they all appear to fall on a single line in the σ-d plot.   At high stresses and lower temperatures others have indicated that forged KCl tends to develop a banded structure rather than equiaxed subgrains, Becher et al (1974).   Behavior in metals, Garofalo (1965), suggests this is more likely due to increasingly difficult cross-slip on secondary systems than to the fact that the temperature is too low for polygonization.

The low temperature strengthening which results from these subgrains in forged KCl is consistent with experience with metals. However, Abson and Jonas (1970) have recently suggested that the relatively perfect sub-boundaries formed at higher temperatures may be less effective barriers to slip than the tangled cell walls which form at lower temperatures and higher stresses in metals. This may partially explain the fact that the room temperature strain to fracture in hot-forged KCl is larger than would be expected for a fine grain size polycrystal with only two active slip systems, since large pile up stresses do not develop, Bowen et al (1973).

GRAIN BOUNDARY MOBILITY STUDIES

Grain boundary migration mobilities of "pure" and doped KCl were measured by the secondary grain growth process. The hot-forged samples were annealed within 0.5°C of the desired temperature and under a dried nitrogen atmosphere. The annealed samples showed several big (mostly isolated) grains grown among the small grains as shown in Figure 6.   The grain boundary energy and misorientation angle of the subgrains was evaluated indirectly. Measurements of Class and Machlin (1966) showed that the grain boundary energy for twist boundaries >10° was about 110 erg/cm^2. The growing grains must have high angle boundaries as suggested by the etching patterns. Therefore, we assumed that the grain boundary energy of the growing grain was 110 erg/cm^2. The dihedral angle between the growing grain boundary and the surrounding subgrain boundary was about 160°. Therefore, the grain boundary energy of the subgrains is about 1/3 that of the growing grain, i.e. about 37 erg/cm^2. This value of the grain boundary energy corresponds to 2-3° misoriented boundaries, as extrapolated from the data of

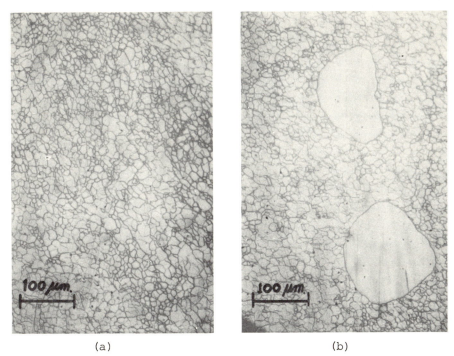

(a)   (b)

Figure 6: Microstructures of hot-forged KCl (a) 470 molar ppm
$SrCl_2$ (in melt) before anneal and (b) after 1 hr anneal at 448°C.

Class and Machlin, (1966). The grain to grain misorientation predicted here is also consistent with that determined from Laue patterns, Bowen et al (1973).
     In computing the grain growth velocity, the area of the largest
grain was chosen because of geometrical and grain growth incubation
time considerations. The grain growth velocity, V, is given by

$$V = \frac{(A/\pi)^{1/2} - r}{t} \tag{3}$$

where A is the area of largest grain found;
     r, the neighboring small grain radius; and
     t, the annealing time.

     The driving force, ΔF, for the motion of high angle boundaries
is due to the surface energy of the neighboring subgrain boundaries,

$$\Delta F = S_V \gamma_{GB} = \frac{2\gamma_{GB}}{L} \tag{4}$$

where S_V, the surface area per unit volume is related to

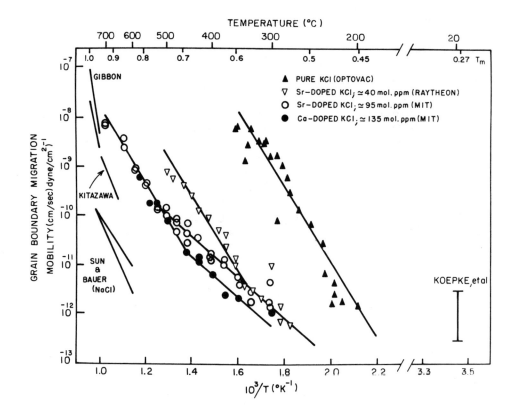

Figure 7   Grain boundary mobility of hot forged KCl.   Sources of
single crystal before forging are in parentheses.   Typical
driving force is about 10^5 dyne/cm^2 for mobility data in forged
samples.   Data of Gibbon (1968), Kitazawa (1974)   are   for   KCl
bicrystals; Koepke (1974), for hot forged KCl; Sun   and Bauer
(1970a) for NaCl bicrystals.

L, the average length of grain intercepts, Underwood (1970), and where γ_{GB} was estimated above as 37 erg/cm^2.

Each value of grain boundary mobility data plotted in Figure 7 is computed from

$$M \equiv \frac{V}{\Delta F} = (\frac{(A/\pi)^{1/2} - r}{t}) \ (\frac{2\gamma_{GB}}{L})^{-1} \tag{5}$$

Several features of the grain boundary mobility data are noted: (1) The apparent activation energy for boundary migration in undoped KCl at the temperature range investigated is about 35.4 kcal/mole. The boundaries in Ca- and Sr-doped KCl also have the same mobility activation energy above some transition temperature. (2) The 45 molar ppm Sr-doped and 95 molar Sr-doped KCl have the same boundary mobility at low temperatures. The 135 molar ppm Ca-doped samples have a lower mobility. However, all doped samples have the same mobility activation energy of 18.6 kcal/mole. (3) The present mobility data and their extrapolations are higher than those observed by Gibbon (1968) and Kitazawa (1974) for KCl and Sun and Bauer (1970 a) for NaCl. (4) However, mobilities extrapolated from our data for doped and undoped samples are well below the room temperature mobility reported by Koepke et al (1974). It is well known from experiment and theory, Lücke and Detert (1957), Cahn (1962) and Lücke and Stüwe (1962), that the interactions of an impurity atmosphere with a migrating boundary exert a drag on that boundary and strongly affect the apparent boundary mobility. In the low velocity limit, the boundary velocity, V, is linearly related to the driving force, ΔF, and the concentration of segregated solute remains essentially the same as at static equilibration. For this case, the velocity and mobility are,

$$V = \frac{\Delta F}{RT} \ \frac{\Omega}{a} \ \frac{D}{C_{GB}}$$

$$M = \frac{V}{\Delta F} = \frac{1}{RT} \ \frac{\Omega}{a} \ \frac{D}{C_{GB}} \tag{6}$$

where D is the diffusion coefficient of the impurity
  C_{GB}, is the impurity concentration about the boundary
  Ω, molar volume of the bulk material; and
  a, the boundary width.

Equation (6) can be modified to include the spatial variation of the diffusion coefficient and impurity concentration:

$$M = \frac{\Omega}{RT} \left(2 \int_0^\delta \frac{C_{GB}^{excess}}{D} \, dx \right)^{-1}$$ (7)

where C_{GB}^{excess} is the impurity concentration in excess of the bulk concentration; and

δ, the characteristic distance beyond which $C_{GB}^{excess} = 0$.

The diffusion coefficient D of cation divalant impurities in an ionic crystal with the NaCl structure was calculated by Lidiard (1955)

$$D = D_o p$$ (8)

where $D_o = 1/3 \, b^2 \omega$;

p is the degree of impurity-vacancy association;

ω is the frequency of exchange between impurity atoms and the associated vacancy; and b is the atomic jump distance.

The two limiting cases for the degree of association are:

$$p = KC_V \quad \text{for} \quad KC_V \ll 1$$ (9)

$$p = 1 - (KC_V)^{-1/2} \quad \text{for} \quad KC_V \gg 1$$ (10)

where C_V is the cation vacancy concentration due to either Schottky defects or divalent cation impurities; and K is the equilibrium constant for impurity-vacancy association.

When the space charge layers about the grain boundaries are considered, the excess impurity concentration C_{GB}^{excess} is evaluated by the same treatment of Kliewer and Koehler (1965) for free surfaces:

$$C_{GB}^{excess} = C_\infty [\exp(\frac{-(e\phi(x) - e\phi(\infty))}{kT}) -1]$$ (11)

where C_∞ is the bulk concentration of impurity;

$(\phi(x) - \phi(\infty))$ is the potential difference between the point location x and the interior of the sample;

e is the charge of an electron; and

$e\phi(o) \equiv 0$.

However, the surface potential does not affect the distribution of the uncharged vacancy-impurity complexes. Therefore, the degree of impurity vacancy association in Eqn.(8) is modified to p',

$$p' = \frac{p}{p + (1 - p) \exp(\frac{-(e\phi - e\phi_\infty)}{kT})} \tag{12}$$

With Eqns. (11), (12) and the modified Eqn. (8), the integral in Eqn. (7) can be evaluated by the technique suggested in Lehovec (1953),

$$\int_0^\delta \frac{C_{GB}^{excess}}{D} dx = \frac{4\delta C_\infty}{2D_o p} [(\exp(\frac{e\phi_\infty}{2kT}) - 1) p$$

$$+ \frac{(1 - p)}{3} (\exp(\frac{3e\phi_\infty}{2kT}) - 1)] \tag{13}$$

Several important limiting cases of Eqn. (13) will be considered:

1) for $T > T_{tr} > T_o$

where T_{tr} is the intrinsic-extrinsic transition temperature and T_o is the isoelectric temperature at which $e\phi_\infty = 0$.

The degree of association becomes

$$p \simeq KC_V = KK_s^{1/2}$$

where K_s is the Schottky constant.

For $e\phi_\infty < 0$,

$$\int_0^\delta \frac{C_{GB}^{excess}}{D} dx = \frac{2\delta C_\infty}{3D_o KK_s^{1/2}}$$

and thus the grain boundary mobility is

$$M \simeq \frac{\Omega}{RT} \frac{3}{4\delta C_\infty} D_o KK_s^{1/2} \tag{14}$$

According to Fuller et al (1968), the Schottky defect formation enthalpy is 57.4 kcal/mole and the impurity (Sr^+) cation vacancy complex binding enthalpy is 13.1 kcal/mole. de Souza (1969) found that the Sr^+ diffusion in KCl has an activation energy of 17.75 kcal/mole. Therefore, we estimate that the activation energies of D_o and $D_o K$ are 30.9 and 17.75 kcal/mole, respectively.

The δ^{-1} factor in Eqn. (14) generally increases the mobility activation energy. From Kliewer and Koehler (1965) for NaCl in the intrinsic temperature range, δ^{-1} has activation energies of 11.7, 7.3 and 5.6 kcal/mole for 1 ppm, 10 ppm, and 100 ppm doped samples, respectively. Consequently, the mobility energies in this temperature range would be 57.7, 53.3 and 51.6 kcal/mole for 1 ppm 10 ppm and 100 ppm impurity sample respectively.

2) For $T_{tr} > T > T_o$ the association is

$p \simeq KC\infty$. Thus for

$e\phi\infty < 0$,

$$M \simeq \frac{\Omega}{RT} \frac{3}{4\delta} D_o K \qquad\qquad (15)$$

The mobility is independent of impurity concentration and has an activation energy of 17.7 kcal/mole. The contribution of δ^{-1} to mobility activation energy is significant in this and other lower temperature ranges discussed.

3) For $T_{tr} > T_o > T$ and $KC\infty << 1$, the degree of association is

$p \simeq KC\infty$. Thus for

$e\phi\infty > 0$,

$$M = \frac{\Omega}{RT} \frac{3D_o K}{4\delta} \exp\left(-\frac{3}{2}\frac{e\phi\infty}{kT}\right) \qquad\qquad (16)$$

The temperature dependence of $e\phi\infty$ in Kliewer and Koehler can be approximated by

$$e\phi\infty = A\left(1 - \frac{T}{T_o}\right) \qquad\qquad (17)$$

where T_o is the isoelectric temperature and A ranges from 18.0 to 16.1 kcal/mole for NaCl with 1 ppm to 100 ppm impurity. The mobility, then, has an activation energy of about 41.5 kcal/mole.

4) For $T_{tr} > T_o > T$ and $KC\infty >> 1$, the association is approximately

$p \simeq 1 - (KC\infty)^{-1/2}$ and when

$e\phi\infty > 0$

$$M \simeq \frac{\Omega}{RT} \frac{D_o}{4\delta C\infty} \exp\left(\frac{-e\phi\infty}{2kT}\right) \qquad\qquad (18)$$

The mobility in this low temperature range is inversely proportional to $C\infty$ and has an activation energy of 39.1 kcal/mole.

Before we compare our data with the above theory, we should note two major limitations. Firstly, for the space change theories developed by Lehovec (1953) and Kliewer and Koehler (1965), the properties of surfaces other than being perfect sinks and sources of

vacancies were not considered. Lifshitz et al (1965, 1967) were
the first to suggest such a deficiency. They proposed a general
solution but not in a workable form to this problem. Poeppel and
Blakely (1969) and Blakely and Danyluk (1973) made certain simpli-
fying assumptions about the surafce structures and were able to
obtain a solution of the space charge potential with the surface
site density and surface binding energy as parameters. However,
with little knowledge about grain boundary structures of ionic
materials, we did not apply their theory to the boundary mobility
problem.

     Other than the conceptual difficulties discussed in the pre-
ceeding paragraph, we also note that Eqn.(11) does not apply below
100°C for tight binding (13.8 kcal/mole) between impurity and
vacancy. In addition, the excessive concentration of free impurity concentration about the surface demands modifications
described in Kliewer and Koehler (1965). Kliewer (1965) solved
this problem by modifying the charge term in Poisson's equation.
This will change the differential equation on which the integral
in Eqn.(13) is evaluated. For the purity of our sample, such a
modification is only required at room temperature or below.

     The second important limitation in our derivation is related
to the assumption of the low velocity limit. According to Cahn
(1962) the impurity drag reaches a maximum when the boundary
velocity, $V = \beta^{-1}$ where the parameter β^{-1} is essentially equal to
the impurity atom drift velocity across the grain boundary. At
higher velocities the boundary breakes away from the impurity
atmosphere and approaches its intrinsic mobility. Gordon and Vandermeer (1966) calculated β^{-1} for several assumed cases for metal. An
approximate calculation for β^{-1} using the space charge attraction
(for Sr^+ diffusion coefficient, D^{Sr}) gives

$$\beta^{-1} \approx \begin{cases} \dfrac{\sqrt{3}\, D_o p}{\delta}\ \exp(\dfrac{-3e\phi\infty}{4kT}) \approx 10^2\ \dfrac{D^{Sr}}{\delta} & T > T_o \\[2em] \dfrac{\sqrt{3}\, D_o p}{\delta} \approx \dfrac{D^{Sr}}{\delta} & T < T_o \end{cases} \qquad (19)$$

Although the Cahn theory does not adequately describe the present case
where the attraction energy is modified by the impurity distribution,
we assume β^{-1} as a guide to the expected break away velocity.

     For a pure material or for an impure material above break
away the space charge cloud consists only of vacancies. Assuming
reasonable values of the grain boundary diffusivity, D_b, of 10^{-5}
cm^2/s, we can calculate the intrinsic boundary mobility from a
relaxation similar to that of Turnbull (1951)

$$M = \frac{\Omega D_b}{aRT} \qquad (20)$$

For most achievable driving forces this intrinsic mobility will be too slow to cause the boundary to break away from the vacancy cloud except at low temperatures; further the drag from the vacancy cloud will be insignificant compared to the intrinsic boundary resistance.

Having presented the extension of the impurity drag theory in the low velocity limit to include the space charge phenomena, we analyze the theory in the context of the data in Figure 6.

(1) It correctly includes the dependence of mobility on impurity concentration in the high and low temperature ranges.

(2) A significant increase in activation energy is predicted only in the intrinsic temperature range for doped or undoped KCl. Our data show an increase of activation energy from 18.6 kcal/mole to 35.4 kcal/mole at temperatures as low as 350°C. However, Kitazawa (1974) reported 57 kcal/mole for undoped KCl between 650°C and 725°C consistent with Eqn.(14).

(3) The activation energy 17.7 kcal/mole, predicted from Eqn. (15) agrees fairly well with our low temperature data, 18.6 kcal/ mole.

(4) When $e\phi_\infty = 0$ at the isoelectric temperature, an infinite boundary mobility is predicted in Eqn.(13). This unrealistic prediction can be removed when the strain field interactions in addition to the electrostatic interactions between grain boundary and impurity atoms are included in our treatment.

(5) When the observed boundary velocity data are compared with β^{-1} as defined in Eqn.(19) we find that all our data in undoped KCl and most of the high temperature data for doped KCl are above the break away velocity. This is also confirmed by irregular and jerky grain boundaries observed in these specimens. In view of the relatively high activation energy, the boundaries are probably in the low to high velocity transition region rather than in the boundary diffusion controlled high velocity limit. However, similar comparison suggests the low temperature data in doped KCl are in the low velocity limit.

Müller (1935) studied the recrystallization rate of NaCl crystals which had been deformed under varying stresses 650-4000 gm/mm^2 (920-5700 psi) at 400°C. He observed a change in the activation energies similar to our data. The activation energy in the low temperature range was 17±3 kcal/mole and without any systematic variation with the deformation stresses. However, the activation energy at high temperatures increased from 30 to 60 kcal/mole as the deformation stress decreased from 4000 gm/mm^2 to 800 gm/mm^2. Müller's data confirm our predictions in the low temperature range and indicate the strong driving force dependence of the mobility activation energies. For some of his samples produced at low deformation stresses, the low velocity - low driving force requirements may have been fulfilled and the activation energy fits our prediction. However, the samples formed at high deformation stresses were probably in the same low to high velocity transition region as in our samples, and indeed these two sets of

TABLE 1

GRAIN BOUNDARY MOBILITY DATA: EXPERIMENTAL AND THEORY

Sample KCl	Temperature Range °C	Experimental		Theory*
		Q kcal/mole	$M_o(\frac{cm}{sec})(\frac{dynes}{cm^2})^{-1}$	Q kcal/mole $M_o(\frac{cm}{sec})(\frac{dynes}{cm^2})^{-1}$
undoped (Kitazawa)	650-725	57	4.3×10^3	58 5.8×10^5
Sr-doped (470 ppm in melt)	250-500	18.6	1.7×10^{-5}	17.7 3.3×10^{-5}
Sr-doped (200 ppm in melt)	250-350	18.6	1.7×10^{-5}	17.7 3.7×10^{-5}
Sr-doped (470 ppm in melt)	500-700	35.4	1.0	Break away
Sr-doped (200 ppm in melt)	350-500	35.4	18	
undoped	200-350	35.4	2.8×10^4	

* The predictions are based on:

$D_o K = 2.5 \times 10^{-2} \exp(\frac{-17.7}{RT})$

de Souza (1969) and Lidiard (1955)

$K_s^{1/2} = 45.6 \exp(\frac{-28.6}{RT})$

Fuller at al (1968)

$\delta \overset{\sim}{=} 6.9 \times 10^{-10} \exp(\frac{11.7}{RT})$ cm

$C_\infty = 10^{-6}$; T > 650°C

$\delta \overset{\sim}{=} 4 \times 10^{-7}$ cm

$C_\infty = 5 \times 10^{-5}$; 500 > T > 200°C

Kleiwer and Koehler (1965)

data have similar mobility activation energies; but they also deviate significantly from our prediction for low velocity behavior.

The better fit of Kitazawa'a data to the low velocity limit may be justified since he used the Sun and Bauer (1970b) method in which the capillary force is about 10^4 dynes/cm^2. Therefore, the boundary velocity is only about 10^{-3} β^{-1}.

The high boundary velocity of Koepke et al (1974) is probably due to boundary break away. The break away may be initiated by the residual stress stored in the hot-forged specimens. It is likely that in the initial migration stage, the boundary is in the unstable region of the velocity driving force curve postulated in Cahn (1962) and Lücke and Stüwe (1962). The boundary may be stabilized after the residual stress is consumed or a significant amount of impurity is accumulated by the migrating boundary.

(6)   Table 1 lists experimental and predicted activation energies, Q, and the pre-exponential terms M_o of the mobility data. The weak T^{-1} dependence of the predicted M_o is approximated by the average of a given temperature range. The pre-exponential term for Kitazawa's data gives fair agreement with the predicted value, and our data from doped specimens at low temperatures give a more satisfying agreement.

SUMMARY

The boundaries which form in hot-forged KCl are related to the steady state creep substructure and the grain size is inversely proportional to the steady state forging stress. The subgrains are stable against secondary recrystallization; however, occasional high angle boundaries serve as nuclei for rapid boundary migration and this secondary recrystallization can occur in undoped specimens at room temperature.

The mobility of the boundaries as a function of dopant concentration and temperature are consistent with grain boundary drag theories when those theories which were derived essentially for metals are modified for the ionic lattice. Our data and the data of Kitazawa were shown to have the qualitative behavior predicted by the equations and in addition, to show good quantitative agreement for the low velocity region based on data (diffusion coefficients, etc.) taken from the literature.

The final implication from the data is that a stable fine grained structure can be obtained by forging Sr-or Ca-doped KCl in the range of 300-400°C. The resulting ∿10μm subgrain size has low secondary recrystallization rates (low grain boundary mobility) and the boundaries are in the non-break away regime. Thus, forging followed by adequate cooling time to room temperature to eliminate thermal strains results in the stable, fine grained structure at room temperature.

ACKNOWLEDGEMENTS

   We express thanks for financial support by the Advanced
Research Projects Agency, the contract monitored by Air Force
Cambridge Research Labs (No. Fl9628-72-C-0304). We wish also to
thank Dr. K. Kitazawa of the University of Tokyo for his unpublished data.

REFERENCES

Abson, D. J. and Jonas, J. J., (1970), "The Hall-Petch Relation
      and High Temperature Subgrains," Met. Sci. J., 4, 24.
Becker, P. F., Freiman, S. W., Klein, P. H. and Rice, R. W., (1974),
      "Strengthening Behavior in Polycrystalline KCl" in the reference
      of Pitha et al (1974), p. 579.
Bird, J. E., Mukherjee, A. K. and Dorn, J. E. (1969), "Correlation
      Between High Temperature Creep Behavior and Structure," in
      "Qualitative Relation Between Properties and Microstructure,"
      ed. D. G. Brandon and A. Rosen, Israel University Press, p. 255.
Bowen, H. K., Singh, R. N., Posen, H., Armington, A. and Kulin, S.
      A., (1973), "Fabrication and Properties of Polycrystalline
      Alkali Halides," Mat. Res. Bull., 8, 1389.
Blakely, J. M. and Danyluk, S. A., (1973), "Space Charge Regions at
      Silver Halide Surfaces: Effect of Divalent Impurities and
      Halogen Pressures," Surface Sci., 40, 37.
Blum, W., (1973), "Activation Analysis of the Steady State Deforma-
      tion of Single and Polycrystalline Sodium Chloride," Phil.
      Mag., 28, 245.
Burke, P. M. (1968), "High Temperature Creep of Polycrystalline
      Sodium Chloride," Ph.D. Thesis, Stanford University.
Cahn, J. W., (1962), "The Impurity - Drag Effect in Grain Boundary
      Motion," Acta. Met., 10, 789.
Chin, G. Y., VanUitert, L. G., Green, M. L., Zydzik, G. L. and
      Kometani, T. Y., (1973), "Strengthening of Alkali Halides
      by Divalent-Ion Additions," J. Am. Ceram. Soc. 56, 369.
Class, W. H. and Machlin, E. S., (1966), "Crack Propagation Method
      for Measuring Grain Boundary Energies in Brittle Materials,"
      J. Am. Ceram. Soc., 49, 306.
Clauer, A. H., Wilcox, B. A. and Hirth, J. P., (1970), "Dislocation
      Substructure Induced by Creep in Molybdenum Single Crystals,"
      Acta. Met., 18, 381.
Fuller, R. G., Marquardt, C. L., Reilly, M. H. and Wells, J. C. Jr.,
      (1968), "Ionic Transport in Potassium Chloride," Phys. Rev.,
      176, 1036.
Garofalo, F., (1965), "Fundamentals of Creep and Creep-Rupture in
      Metals," McMillan, New York.
Gibbon, C. F., (1968), "Technique for Measuring Grain Boundary
      Mobility and its Application to Potassium Chloride," J. Am.

Ceram. Soc., 51, 273.

Gordon, P. and Vandermeer, R. A., (1966), "Grain Boundary Migra-
     tion," in "Recrystallization, Grain Growth amd Textures,"
     A.S.M., Metals Park, Ohio, p. 205.

Harrison. W. B., Hendrickson, G. O. and Starling, J. E., (1974),
     " Mechanical and Optical Properties of Recrystallized Alkali
     Halide Alloys," in the reference of Pitha et al (1974), p.615.

Hill, R., (1950), "Plasticity," Oxford University Press, London.

Ikeya, M. and Itoh, N., (1968), "Distribution Coefficients of
     Various Impurities in Alakli Halides," Jap. J. Appl. Phys.,
     7, 837.

Ilschner, B., (1973), "Hochtemperatur - Plastizität," Springer-
     Verlag, Berlin.

Kitazawa, K., (1974), Private Communication.

Kliewer, K. L. and Keohler, J. S., (1965), "Space Charge in Ionic
     Crystals, I: General Approach with Application to NaCl."
     Phys. Rev., 140, A, 1226

Kliewer, K. L., (1965), "Space Charge in Ionic Crystals, II:
     The Electron Affinity and Impurity Accumulation," Phys. Rev.
     140, A, 1241.

Koepke, B. G., Anderson, R. H., Bernal, G. E. and Stokes, R. J.,
     (1974), "Room Temperature Grain Growth in Potassium Chloride,"
     J. Appl. Phys., 45, 969.

Kulin, S. A., Neshe, P. P. and Kreder, K., (1974), "Development of
     Polycrystalline Alkali Halides by Strain Recrystallization for
     Use as High Energy Infrared Laser Windows," Technical Report,
     AFML-TR-74-17., ManLabs, Inc., Cambridge, Mass.

Lehovec, K., (1953), "Space Charge Layer and Distribution of Lattice
     Defects at the Surface of Ionic Crystals," J. Chem. Phys.,
     21, 1123.

Lidiard, A. B., (1955), "Impurity Diffusion in Crystals (Mainly
     Ionic Crystals with the Sodium Chloride Structure)," Phil.
     Mag., 46, 1218.

Lifshitz, L. M. and Geguzin, Ya. E., (1965), "Surface Phenomena
     in Ionic Crystals," Sov. Phy.-Sol. St., 1, 44.

Lifshitz, L. M. , Kossevich, A. M. and Geguzin, Ya. E., (1967),
     "Surface Phenomena and Diffusion in Ionic Crystals," J.
     Phys. Chem. Solids, 28, 783.

Lücke, K. and Detert, K., (1957), "A Quantitative Theory of Grain-
     Boundary Motion and Recrystallization in Metals in the Presence
     of Impurities," Acta Met., 5, 628.

Lücke, K. and Stüwe, H. P., (1962), "On the Theory of Grain
     Boundary Motion," in "Recovery and Recrystallization of
     Metals," ed. L. Himmel, AIME Symposium, Feb. 1962.

Miles, P. A., Readey, D. W. and Newberg, R. T., (1973), "Research
     on Halide Superalloy Windows," AFCRL-TR-73-0758, Raytheon
     Research Division, Waltham, Mass.

Müller, H. G., (1935), "The Nature of Recryatallization Processes:
     Experiments on Pressed Rocksalt," Zeitsch. f. Physik., 96, 279.

Pitha, C. A., Armington, A. and Posen, H., (1974), "Third Con-
    ference on High Power Infrared Laser Window Materials,"
    AFCRL Special Report No. 174.

Poeppel, R. B. and Blakely, J. M., (1969), "Origin of Equilibrium
    Space Charge Potentials in Ionic Crystals," Surface Sci., $\underline{15}$,
    507.

Poirier, J. P., (1972), "High Temperature Creep of Single Crystalline
    Sodium Chloride II: Investigation of the Creep Structure,"
    Phil. Mag., $\underline{26}$, 713.

de Souza, M. F., (1969), "Effective Volumes of Vacancies of Divalent
    Ions and Their Diffusion Coefficients in Alakli Halides,"
    Phys. Rev., $\underline{188}$, 1367.

Schuh, F., Blum, W. and Ilschner, B., (1970), "Steady State Creep
    Rate, Impurities and Diffusion in the Rock-Salt Structure,"
    Proc. of Brit, Ceram. Soc, No. 15, 143.

Sherby, O. D. and Burke, P. M., (1967), "Mechanical Behavior of
    Crystalline Solids at Elevated Temperature," Prog. in Mat.
    Sci., $\underline{13}$, 325.

Streb, G. and Reppich, B., (1973), "Steady State Deformation and
    Dislocation Structure of Pure and Mg-Doped LiF Single Crystals,"
    Phys. Stat. Sol. (2), 16, 493.

Sun, R. C. and Bauer, C. L., (1970a), "Tilt Boundary Migration in
    NaCl Bicrystals," Acta Met., $\underline{18}$, 639.

Sun, R. C. and Bauer, C. L., (1970b), "Measurement of Grain
    Boundary Mobilities Through Magnification of Capillary
    Forces," Acta Met. $\underline{18}$, 635.

Turnbull, D., (1951), "Theory of Grain Boundary Migration Rates,"
    Trans. AIME, $\underline{191}$, 661.

Underwood, E. E., (1970), "Quantitative Stereology," Addison-
    Wesley Publishing Co., Reading, Mass.

Weertman, J., (1968), "Dislocation Climb and Theory of Steady
    State Creep," Trans. A.S.M., $\underline{61}$, 681.

CONTRIBUTORS

Co-Chairmen

R. C. Bradt, Associate Professor, Department of Material
    Sciences, Ceramic Science Section, The Pennsylvania State
    University, University Park, Pennsylvania

R. E. Tressler, Assistant Professor, Department of Material
    Sciences, Ceramic Science Section, The Pennsylvania State
    University, University Park, Pennsylvania

Conference Session Chairmen

J. C. Hurt, Army Research Office, Durham, North Carolina

J. T. A. Roberts, Argonne National Laboratory, Argonne,
    Illinois

R. M. Spriggs, Lehigh University, Bethlehem, Pennsylvania

P. K. Talty, Aerospace Research Laboratories, Wright-Patterson
    Air Force Base, Ohio

B. A. Wilcox, Battelle Laboratories, Columbus, Ohio

Arrangements

Patricia Ewing, Coordinator, J. Orvis Keller Conference
    Center, The Pennsylvania State University, University
    Park, Pennsylvania

Authors

C. N. Ahlquist, University of Colorado, Boulder, Colorado

C. A. Andersson, Westinghouse Research Laboratory, Pittsburgh, Pennsylvania

R. H. Anderson, Honeywell Research Center, Bloomington, Minnesota

P. Beardmore, Ford Scientific Laboratory, Dearborn, Michigan

P. F. Becher, U. S. Naval Research Laboratory, Washington, D. C.

W. R. Bitler, Pennsylvania State University, University Park, Pennsylvania

H. K. Bowen, Massachusetts Institute of Technology, Cambridge, Massachusetts

E. Butler, Pilkington Bros. Ltd., Lancashire, England

R. M. Cannon, Massachusetts Institute of Technology, Cambridge, Massachusetts

A. C. D. Chaklader, University of British Columbia, Vancouver, Canada

G. Y. Chin, Bell Laboratories, Murray Hill, New Jersey

R. L. Coble, Massachusetts Institute of Technology, Cambridge, Massachusetts

A. G. Evans, National Bureau of Standards, Washington, D. C.

S. W. Freiman, U. S. Naval Research Laboratory, Washington, D. C.

J. E. Funk, Alfred University, Alfred, New York

R. S. Gordon, University of Utah, Salt Lake City, Utah

R. K. Govila, Oakland University, Oakland, Michigan

A. H. Heuer, Case-Western Reserve University, Cleveland, Ohio

B. J. Hockey, National Bureau of Standards, Washington, D. C.

M. H. Hodge, Thomson-Ramo-Woolridge, Philadelphia, Pennsylvania

J. H. Hoke, The Pennsylvania State University, University Park, Pennsylvania

L. Hwang, Case-Western Reserve University, Cleveland, Ohio

C. A. Johnson, General Electric Research Center, Schenectady, New York

H. F. Kay, University of Bristol, Bristol, England

K. R. Kinsman, Ford Scientific Laboratory, Dearborn, Michigan

P. H. Klein, U. S. Naval Research Laboratory, Washington, D. C.

H. Knoch, University of Erlangen-Nurnberg, Erlangen, West Germany

B. G. Koepke, Honeywell Research Center, Bloomington, Minnesota

R. Kossowsky, Westinghouse Research Laboratory, Pittsburgh, Pennsylvania

T. G. Langdon, University of Southern California, Los Angeles, California

F. F. Lange, Westinghouse Research Laboratory, Pittsburgh Pennsylvania

P. A. Lessing, University of Utah, Salt Lake City, Utah

E. M. Lenoe, Army Mechanics and Materials Research Center, Watertown, Massachusetts

D. J. Michael, PPG Industries, Pittsburgh, Pennsylvania

T. E. Mitchell, Case-Western Reserve University, Cleveland, Ohio

J. Nemeth, Champion Sparkplug, Fling, Michigan

M. R. Notis, Lehigh University, Bethlehem, Pennsylvania

B. J. Pletka, Case-Western Reserve University, Cleveland, Ohio

G. D. Quinn, Army Mechanics and Materials Research Center, Watertown, Massachusetts

B. Reppich, University of Erlangen-Nurnberg, Erlangen, West Germany

R. W. Rice, U. S. Naval Research Laboratory, Washington, D. C.

J. T. A. Roberts, Argonne National Laboratory, Argonne, Illinois

W. D. Scott, University of Washington, Seattle, Washington

M. S. Seltzer, Battelle Laboratories, Columbus, Ohio

R. N. Singh, Argonne National Laboratory, Argonne, Illinois

J. R. Smyth, Iowa State University, Ames, Iowa

A. A. Solomon, Argonne National Laboratory, Argonne, Illinois

R. J. Stokes, Honeywell Research Center, Bloomington, Minnesota

P. K. Talty, Air Force Materials Laboratory, Wright Patterson
    Air Force Base, Ohio

J. R. Tinklepaugh, Alfred University, Alfred, New York

L. J. Trostel, Jr., The Norton Company, Worcester, Massachusetts

R. R. Turk, Hughes Research, Malibu, California

D. P. Wei, Westinghouse Research Laboratory, Pittsburgh,
    Pennsylvania

M. F. Yan, Massachusetts Institute of Technology, Cambridge,
    Massachusetts

SUBJECT INDEX